Technical Principles
of
Forest Management

Feng Zhongke

China Forestry Publishing House

图书在版编目(CIP)数据

森林经理学技术原理 = Technical Principles of Forest Management:英文 / 冯仲科著. —北京:中国林业出版社,2017.12

ISBN 978-7-5038-8913-4

Ⅰ.①森… Ⅱ.①冯… Ⅲ.①森林经理—英文 Ⅳ.①S757

中国版本图书馆 CIP 数据核字(2017)第 022010 号

COPYRIGHT© Feng Zhongke and China Forestry Publishing House.

All RIGHT RESERVED No part of this work covered by the copyright hereon may be reproduced or used in any form or by any means—graphic, electronic, or mechanical, including but not limited to photocopying, recording, taping, Web distribution, information networks, or information storage and retrieval systems—without the written permission of the publisher.

National Archives of Publications and Culture Cataloging in Publishing Data:
Technical Principles of Forest Management, First Edition
ISBN:978-7-5038-8913-4

First Printed in the P. R. China in Jan, 2018 by China Forestry Publishing House
No. 7 Liuhai Hutong, Xicheng District, Beijing (100009)
Http://www.forestry.gov.cn/lycb.html
Email:jiaocaipublic@163.com

PREFACE

I am Feng Zhongke, I want to tell a story that how I become a scholar of a forest management from mine surveying student. As far as I remembered, I lived in a flat plateau of the Northwest Loess Plateau in China. This county is about 2000 meters away from the town, and there is an ancient secret country named Eld Mixu, which is 7000 meters south of that town and about 15000 meters west of county, having a very good virgin forest until now. When I was 12 years old first time I entered into this piece of primeval forest, since then my view with the forest is to establish a difficult fate.

In 1983, I graduated from the Western Mining Institute of mine surveying. Department of geology laid the academic foundation of my life to in surveying & mapping, geographic information, resource management. After that, I worked as a teacher in Beijing Coal Industry University, Shougang Institute of Technology. In this period of teaching, scientific research and engineering, I worked a lot of mine surveying and mapping, ecological restoration, metallurgical construction, development of surveying instruments the teaching, scientific research and engineering work.

In 1997, I entered the Beijing Forestry University for my doctorate in forest management, since then, I entered in a magical forest management world. At first, I was engaged in forest investigation, the instrument, technology principles and methods, especially the 3S(GNSS, GIS, RS) application of forest survey, and soon to precision forestry subject research breakthrough. Then the development in mobile phone and CCD, EDM, GNSS plate, GRY (gyro) combined with one tree measuring flat instrument realize the automatic observation at the same time on the basis of Internet plus technology, the research of forest planning and design is promoted to a new level. The real problem to study this key issue is the last two or three years that is how to make the rotation of forest plantation shortest and make farmers annual income more and also how to select cutting of annual growth cutting innatural and overmature forest so that the stand density is optimal and harvest and growth are equal to the amount of wood material. Whether it is in theoretical complication, model calculation, instrument design, practice, I have reached the goal to harvest more wood, because I know China and the world we need to harvest more timber by selective cutting. I'm just like a student learning, thinking, performing experiment, development, validation, and today I have become a forest management scholar. I would like to discuss this issue with everyone in the world who is interested. This is the English version of *Technical Principles of Forest Management* is the result of 20 years of my efforts engaged in forestry research. Thanks to my student Abdul Mannan from Pakistan for the proofreading of the book in English and also thanks to my Master's and PhD students from 2012 -2017 years to participate in relevant laboratory activities.

Feng Zhongke

Jan-01, 2018

Beijing Forestry University

CONTENTS

PREFACE

INTRODUCTION ··· 1
    0.1 What is forest management? ·· 1
    0.2 Forest management theory and method ································· 2
    0.3 The standard model of single tree normalization ···················· 10
    0.4 Pattern research of large area grid trapezoidal continuous ecological environment model ··· 12
    0.5 Forest management equipment technology and system platform ·········· 17

CHAPTER 1 The Mathematical Foundation of Forest Management ············ 30
    1.1 Linearization of the non-linear equations ······························ 30
    1.2 Multivariate robust estimation ·· 34
    1.3 Genetic algorithm ·· 37
    1.4 Artificial neural network ·· 40
    1.5 Internet of Things ·· 42
    1.6 Cloud Computing ·· 44

CHAPTER 2 Technical Foundation of Forest Management (Focus on General Principle) ··· 49
    2.1 Principle of electronic theodolite / total station /CCD superstation ········ 49
    2.2 MINI total station / superstation instrument ·························· 60
    2.3 GNSS principle ··· 90
    2.4 3D laser scanning measurement principle ····························· 110
    2.5 Space remote sensing platform ·· 116
    2.6 Photography measurement principle ····································· 123
    2.7 Digital camera calibration ·· 139

CHAPTER 3 Forest Manager's Basic Numerical Tables and Model ... 149
   3.1 The summary of forest numerical table ... 149
   3.2 Tree volume built research model ... 152
   3.3 Bamboo model to study ... 168
   3.4 Shrubs model to study ... 172
   3.5 Chinese dominant tree growth model ... 173

CHAPTER 4 Forest Division ... 180
   4.1 Forests division ... 180
   4.2 Forestland planning principle ... 190
   4.3 Planning principles of forest species ... 201
   4.4 Sub-compartment afforestation design ... 209
   4.5 Small-scale precision business planning and design ... 211

CHAPTER 5 Technical Principles of Forest Inventory ... 215
   5.1 Forest survey techniques overview ... 215
   5.2 The second class forest management survey ... 222
   5.3 Forest manager's three kinds of investigation ... 244
   5.4 Forest parameters inversion through remote sensing image pair ... 252
   5.5 Forest parameter inversion by UAV image pair ... 261
   5.6 Ground photography tree measurement technology ... 263
   5.7 Modernization of ground forest observation ... 265
   5.8 Mobile/tablet/GNSS forest survey counted measuring system ... 271
   5.9 Forest map drawing and GIS spatial analysis ... 272

CHAPTER 6 System Design of Forest Precision Manangement and Manangement Platform ... 277
   6.1 Forest precise management system design ... 277
   6.2 Evaluation of forest ... 279
   6.3 The forest management precision sub-compartment cluster analysis ... 293
   6.4 The cluster analysis of forest in small class based on forest type ... 300
   6.5 Carbon sink analysis for Chinese forest vegetation ... 310

CHAPTER 7 The Precise Wisdom of the Chinese Forest Management ... 322
   7.1 Forest manager technical problems and further perfect way ... 323
   7.2 System innovation project of the forest survey ... 331
   7.3 To the wisdom of forest management ... 332
   7.4 Prospect: forest management technology development ... 337

INTRODUCTION

0.1 What is forest management?

Forest management is a cross-disciplinary application of forestry, economic management, geometrics, computer Information, systems engineering and other disciplines. The research objective of forest management is to conserve forest resources and ecological environment (including vegetation, climate, topography, soil, water, etc). forest management is the combination of technology and application of forest geometry, physics and ecology under static and dynamic conditions. This includes forest resources and ecological environment investigation, planning, design, evaluation, prediction and decision-making by the tool of geomatics, forest mensuration, and computer.

Generally speaking, the main contents of forest management are in the following areas:

(1) Forest geomatics;
(2) Forest mensuration;
(3) Forest administration;
(4) Forest systematic.

Forest administration is the core content of forest management. Forestry is a classic as well as a modern science. The classical part of the administration is that the forest appeared on earth much earlier than humans and the modern part is the today's forestry science that has been closely linked with the modern biotechnology, information technology, and other disciplines. According to the current China's national classification forestry discipline is divided into the following categories.

Forestry is divided into two levels i. e. level 1 and level 2.

Level 1 includes forestry, forestry engineering and economic management of agriculture and forestry. Forestry has further seven branches in level 2 i. e. forest tree genetics and breeding, forest silviculture, forest protection, forest management, wild fauna and flora protection, ornamental botany and horticulture, soil water conservation and desertification control. Forestry engineering has three branches i. e. forest engineering, xylology and technology and forest chemical engineering. Forestry economics and management is the second-level discipline belonged to economics and management of agriculture and forestry (Table 0-1).

What we mentioned above almost cover the basic academic and technical fields of the forestry industry. These eleven disciplines are not difficult to understand, now the main question is what is the forest management?

Generally speaking, the forest management is to investigate the quantity, quality, distribution,

Table 0-1 The division of forestry

1st LEVEL	Forestry	Forestry engineering	Agricultural economic management & forestry
2nd LEVEL	i. Forest tree genetics and breeding	i. Xylology and technology	i. Economic management of forestry
	ii. Forest silviculture	ii. Forest engineering	
	iii. Forest protection	iii. Forestchemical engineering	
	iv. Forest management		
	v. Wild flora and fauna protection		
	vi. Ornamental botany and horticulture		
	vii. Soil water conservationand desertification control		

dynamic change, environmental effects, spatial planning and design of forest.

In recent years, with the relatively healthy development in China's forest management, the majority of researchers expand the theories, techniques, and application research for national forest resource management. Thus the progress of China's forest management technology and also making the systematic contribution on forest reserves, biomass, carbon sequestration, solid anti-sand, absorption of CO_2, the output of O_2, dealing with PM2.5, etc.

At present the main research direction of forest management discipline focuses on the following aspects:

(1) Forest resources and ecological environment survey & monitoring;

(2) Forestry planning and design;

(3) Government forest resource management and forest farm operation;

(4) Global forest resources and climate change.

0.2 Forest management theory and method

From the perspective of science and technology, forest management from the simple to the complex can be summarized as in Figure 0-1.

Figure 0-1 Forest awareness system

From the perspective of forest management, the main characteristics that should be kept in mind are as follows is a diversity of forest functions i.e. long-term forest-system cycle; the dynamics of forest resources, forests aging uncertainties, value addition functions and functions of forests particularly its spatial effects in time and space. It is difficult to express the forest quantitatively and scientifically. Forest ecosystems are not simple but dependent on atmospheric climate, topography, soil & water resources, vegetation and human role is also one of the key factors. To understand the complex ecological system it can be summarized in Figure 0-2.

Figure 0-2 Forest ecosystems

Since entering the new century the biggest change in the understanding of forest is the change from pure timber production to the comprehensive system of forest ecological system of the earth as a whole. When we look at the problem from the aspect of earth observation from the simple to the complex can be summarized as follows:

(1) Single tree before was examined by the only volume, DBH i. e. 1.3m, and tree height H but now research of measurement has expanded i. e. model and methods for shrub, grass, branches, leaves, roots and other organ's volume, biomass, structure, leaf area index and CVI etc are also examined these days.

(2) Forest at different levels requires field observation or angle gauge technology to determine the average diameter and its distribution, density n I and its distribution, stock biomass m, moisture w, CVI and other parameters. Now extend to the system determination of various vegetation groups related characteristic parameters and model estimates.

(3) In the past forest are more concerned about quantity, quality and distribution of forest vegetation and measuring methods while now we are more concerned about the vegetation and its ecological environment elements (such as atmospheric climate, topography, soil, hydrology and water resources, economic etc) relationships and associations between.

(4) In past forest measuring methods were manual by using simple tape, calipers, altimeters and optical-mechanical instrument for technical workers that required hard labor survey work to obtain data especially in the large field area. Now general 3S technology (containing general of remote sensing RS and digital aviation photography measurement DAP; geographic information system GIS, expert system ES, decision support system DSS, real networking, cloud computing, satellite navigation GNSS (containing GPS and Beidou), the superstation instrument and 3D laser scan ground fine measuring system), change the whole load in an efficient way to new level.

(5) Innovation in instruments and tools is also one of the key improvements of ecological systems for earth observation, surveying, and mapping. Beijing Forestry University with the 3S research center and development team are using this equipment for more than last 10 years for research and development of environmental monitoring equipment and software platform and now has achieved encouraging results.

0.2.1 Forest management theory and model

We develop many theories of forest management and many of these theories are very realistic to solve a series of production problems. In fact, scientific system of forest management has lacked the scientific theoretical system. What we call theory is a pattern or model, but this does not affect the progress of the development of forestry and forest science because today conditions seem not mature for building a universality of a strong theoretical model of forest management. However the issue-specific forest management models and methods to get satisfactory results will help.

Pattern original meant imitation perfect specimen is now understood as a kind of standard form. If a proven effective model is widely accepted, it becomes a pattern. Although the definitions of the model are varied but its core meaning is still the same that model is the abstract of real things and phenomena or simplified in the real world. Forest management theory research is the relationship between forest-environment-economy-people. It is impossible to express the forest world freely and as like the artists, but to establish model (concept model, physical model, a mathematical model) simulations form a useful pattern and promote the results.

Forest management scholars on variety of issues through scientific research and production practice, setting up a series of theories and models, including:

1) The sustainable use of forest

This theory started back in more than 2000 years ago in China during the warring States period, mentioned in the book of Mencius. After the industrial revolution in Europe and particular in Germany, scholars systematically have studied the sustainable use of the forest for more than 200 years and continue to give new content to it. Today's popular theory for sustainable forest management is the upgraded version of the sustainable use of the forest. Today's sustainable use of forests could be understood as forest used for continuous output of wood and other forest products, to play an important role in ecological, economic and social benefits, also to increase the forest productive use of forests and improve the ecological environment.

2) Natural forest—continuous forest management

In 1863, professor Geyer from the University of Munich published the book of *Forest Utilization* and he also published the book of *Forest Silviculture* in 1880. Professor Geyer in these two books proposed the theory that forest productivity should be utilized as much as possible in forest management and second aim is to protect and maintain the forest, to get more of harvest and finally to return to the thought prototype of nature.

Professor Geyer also opposes the idea of the artificial pure forest of the same age, which dominates the management mode and he encourages the idea of the artificial mixed forest of different ages, which use the natural growth of forest and gradually cutting patterns. Thinking of forests as complex organism's theory and practice, constant renewal in forest management guidelines, selection cutting is a good forest management. In 1922, German scholar published the book named as continuous forest, describing the formation of a natural forest continuous management system and the sustainable use of forests. If we look to nature and constant renewal of forestry management seems to have no scientific definition but does have real ideas and methods that still occupy a certain position of forest management in continuous forest management ideology. Corresponding to both forest management practices include zoning rotation method, volume framework, generalized normal forest method etc.

3) Ecosystem management

Last century during 70's and 80's the key points of ecosystem management theory includes two fundamental branches, one is forest ecology and other is landscape ecology. Absorbing the reasonable part of forest going concern theory to achieved an economic value, ecological value and social value of the forest. The improved ecological environment promotes forest ecological system health. There is nothing new neither in Ecosystem management nor for the corresponding technical system and mode of operation.

4) Forest classification management

The forest is divided into five large categories based on the role they play and its production namely timber, conservational forest, economical forest, fuel wood forest and special-purpose forests. It is further divided into two functional directions namely ecological forest and commercial forest. The so-called forest classification management depends on upon the forest's natural environment, the socio-economic conditions the structural characteristics of the forest is divided into several different types, according to their respective objectives and take the appropriate business model, implementation of management by objective.

5) The sustainable management of forests

Sustainable forest management is the fashionable academic terms of forest management that have been commonly used in the last 20 years. The main concept of sustainable forest management is the sustainable management of forest resources. The forest lands should be a safeguard in such a manner that the next generation will be able to enjoy the social, economic, ecological, cultural, and spiritual needs.

The forest provides different kind of products and services such as timber, animal feed, pharmaceutical, fuel, shelter, employment, open space, wildlife habitats, landscape diversity, carbon storage and the nature reserves etc. Now we have to adopt appropriate measures to protect forests against harmful effects of pollution such as from atmospheric pollution and fires, pests and diseases to keep the multiple values of forests. But there is no recognition of its matching technology system. By comprehensive study and analysis, we can draw the following changes:

(1) Time limitation. Thoughts, theories, and models of forest management are progressing and upgrading constantly and to take account of the benefits of ecology, economy, society and ascend to the safety of the ecological system.

(2) Field extension. In the past, we study only the small scale problems in forest tracts, pure forest, forest clear-felling of the same age but today based on the regional, national and global climate change on the scale of research questions.

(3) Technical progress. In the past, the angle gauge, compass, and clinometers were the main tools for forest management but today 3S technology, cloud computing, and the internet make a global forest village botanical garden.

(4) Promotion of awareness. Single, simple and qualitative understanding of forest management now changed to today's diverse and complex, quantitative understanding of forest management.

0.2.2 Forest woodland planning

Regulations promulgated by the former ministry of forestry the classification of land use and land according to the following Figure 0-3 and Table 0-2 classification.

Suppose a total study area land S classified by remote sensing systems all over. Suppose the area S_1 is woodland, S_2 wasteland, S_3 agricultural, S_4 difficult to use land, S_5 soil, and S_6 inland water respectively. Which can be written in the formula $S_1+S_2+S_3+S_4+S_5+S_6=S$. As land use and forest planning problem, we need to look at the demographic, social, economic, environmental and other factors on the ecological environment security.

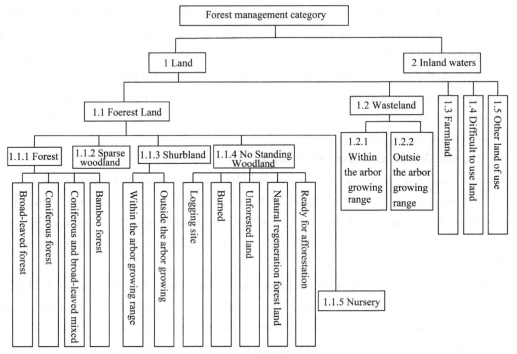

Figure 0-3 Classification

Table 0-2 Technical criteria for land types

No.	Land type	Technical standards
1	Land surface	Perennial exposed surface of the land and the beach
2	Forests	Tree species composition having canopy density ≥ 0.2 of or crown width > 10m
3	Coniferous forest	Coniferous accumulation proportion $\geq 65\%$
4	Broad-leaved forest	Accumulation proportion of broad-leaved $\geq 65\%$
5	Mixed coniferous forest	The accumulation of coniferous, broad-leaved trees proportion of $<65\%$
6	Bamboo forest	Bamboo forest made of bamboo excluding $DBH < 2$ cm
7	Open woodland	In a wood land, the canopy density is 0.10 to 0.19
8	Tree growth within the shrubland	Site conditions suitable for the growth of certain tree species of shrub
9	Tree growth outside the bush	Site conditions are not suitable for tree growth of shrubland
10	Deforested land	Deforested land is the land that after the cutting of trees can not meet the standard of opening blanks and not more than five years
11	Not forested	Land which is not used for afforestation
12	Natural regeneration of forest land	Natural regeneration grade $\geq$, but did not meet the forest standards of forest land
13	Ready to plantation	Land that has been prepared but not reforested at the time of the survey

(continue)

No.	Land type	Technical standards
14	Nursery ground	Fixed land for a tree nursery.
15	Wasteland	The land on which survey is not yet done, the surface of soil, can grow vegetation land
16	Arbor within the scope of growth	Site conditions suitable for the growth of certain tree species of plots
17	Arbor growth beyond the wastelands	Site conditions are not suitable for the growth of tree species
18	Agricultural land	A landused for crop cultivation and grazing
19	Hard to use	The land that is difficult to use under current condition sincludes tidal flat, saline land, marshes, bare land, rocky land, desert, desert, Gobi, tundra, etc
20	Other land	Including towns, settlements, industrial and mining land, traffic land, and land not included in the land
21	Inland waters	Inland natural and artificial waters, waters normally maintained throughout the year, including rivers, lakes, reservoirs, ponds and other water surface, forest streams do not fall into this category

Optimization of ecological environment of the forest vegetation in the most basic configuration requiredthe following aspects like quantity and quality, area, volume, biomass, cover, carbon sinks and so on. Research and planning on this issue have two points: one is the evaluation issues, the number of existing forest cover, quality, optimizing distribution, spatial and temporal effects; the second is to make more efforts in a barren land. So we have multi-targeted planning.

The main objectives of multi-targeted planning are:

(1) Ecological environment security in the region;

(2) The best environmental conditions of the region;

(3) Ecological, social and economic benefits to forest vegetation of the region.

The main constraints are:

(1) The necessary farmland for grain, meat product guarantee of the region;

(2) Town, residential areas and mining and traffic land which satisfy the need of the local people's industry, life, and social development;

(3) To meet the need of a tree, shrub, grass, greening system suitability and layout planning in local ecological security and ecological optimization.

0.2.3 Forest planning

Basic principal requirements and constraints of forest planning are:

(1) Ecological requirements;

(2) The environmental protection requirements;

(3) Climate and weather constraint;

(4) Topography constraint;

(5) Soil and geology constraints;

(6) Constraints to economic and social developmentneed.

In China, forest species are classified according to the Figure 0-4:

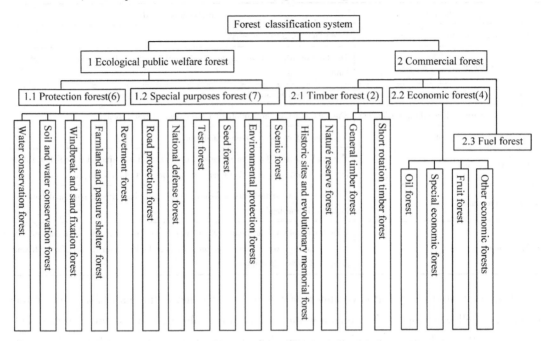

Figure 0-4　Forest classification map

Thus the regional demand for quantity, quality, ecological environment protection, economic development and satisfy the people's life can be achieved by the above classification in China. It is worth to point that these two groups of five type's forest actually have common benefits of economy, ecology, and society.

0.2.4　Forest management practices

1) The normal forest management methods

The normal forest is an idealistic, linear, limited, sustainable, pure, even-aged and thinning business model. So far, there is no forest stand achieve the meaningful normal forest operation, but it doubtless the ideal technology and method system of forest management theory and pattern. Also laid the great mathematics, ecology, and engineering is the foundation for forest sustainable operation.

A case study of age class, normal forest model can be expressed as Figure 0-5.

It is easy to imaginenormal forest but it is difficult to form an operational system because for a small area and divided area into n equal parts, and every part is planted each year, next part next year and so on. We are checking method based on the analysis of advantages and disadvantages of the establishment of the forest sub-compartment precision management methodologies as a realistic method for peer exchange reference and practical application.

Fazhenglin growth can be expressed as Figure 0-6:

2) Examination of management methods

Forest management under the guidance of models and patterns forming a series of practical

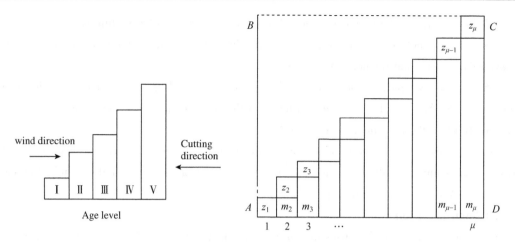

Figure 0-5 Fazhenglin diameters, volume

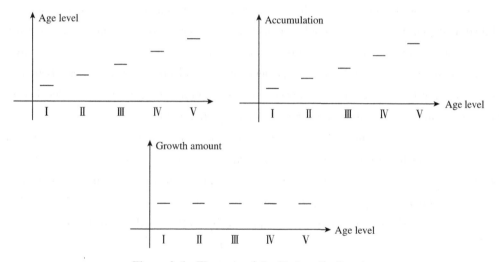

Figure 0-6 Elements of the Fazhenglin forest

forest management practices. One of the best methods was proposed by France A. Gurnard in the 19th century. After that, a Switzerland's forester named H. Biolley the experiments performed by him still has a profound influence on the inspection law.

The normal forest is suitable for the pure age and clear cutting jobs. But in many cases the real forest was uneven-aged, mixed, and require scientific selective cutting, the inspection method is the most effective ways and means to solve the problem.

The core idea of the inspection method and technical performance are:

(1) Forest zoning for small area $12 \sim 15$ hm^2;

(2) Measured diameter of all the trees, estimated volume, the general manager of $5 \sim 7$ years, not more than 10 years;

(3) Set initial forest stock as M_1, M_2 at the end, after a selective felling volume during the years of operation c, annual growth of $Z = (M_2 - M_1 + C)/a$. Z is a period of growth on the next scheduled selection cutting;

(4) The growth of the previous period Z a is the next pre-selected cutting volume;

(5) Selective cutting can be adjusted to form the ideal structure of the uneven-aged forest. Between the density and diameter $N = ae - bd$;

(6) H. Biolley believes that small mixed forest of spruce and fir (20~30 cm), middle-diameter timber (35~50 cm), large-diameter timber (>55 cm) can maintain the highest forest stand productivity when the accumulation ratio of 2 : 3 : 5.

Beijing forestry University Surveying & Mapping Establishment of 3S Centre electronic stereo angle measurement in small classes to count tree diameter at breast height and tree height of small precision work method has a broad prospect.

3) Small precision work

Class precision job method uses measuring tree gun (containing electronic stereo angle rule) for basic field survey, sub-compartment as survey unit in each class periodically (generally 5 years). Observation includes angle rule point count of DBH and the tree high information. The combined analytical wood data is based on density n, diameter d, average high h and material product model and growth volume model. Sub-compartment accurate working believe that selective cutting is a necessary means for forest management, through selective cutting we can both regulate density and harvest timber, as well as subcompartments optimization, can be controlled because of density and diameter that makes larger growth than selective cutting.

The basic idea of the small class precision operation method is to use the small class for the business unit, through the establishment of small class fixed point (x_0, y_0, z_0) and sizing modeling tracking observation points to electronic stereo angle gauge (measuring gun) observation count wood i basic information $(x_i, y_i, z_i, D_i, H_i)$, through the built-in (average tree height) $N \smallsetminus \overline{D} \smallsetminus \overline{D}_i \smallsetminus V \smallsetminus W \smallsetminus C$ and so on. The dynamic model is established by multi-period observation:

$$\begin{cases} \overline{H}(t+1) = f_H[H(t)] \\ N(t+1) = f_N[N(t)] \\ D(t+1) = f_D[D(t)] \\ V(t+1) = f_V[V(t)] \end{cases}$$

By building a dynamic model based on material requirements such as market, economy, the lowest rate of scrap wood and other constraints to ensure the optimization of forest structure, (density, diameter distribution, and mingling, angular scales) are most reasonable (Figure 0-7). The best method should have clear thinking, strategy improvement, advanced instrument, timely observation, design accuracy, accurate operation, continuous breeding, ecological and reasonable.

0.3 The standard model of single tree normalization

M. Kunze stem curve is generally expressed as:

$$y^2 = px^r \tag{0-1}$$

And its volume formula is,

INTRODUCTION

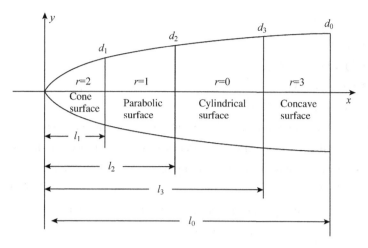

Figure 0-7 Tree trunk structure diagram

$$v = \int_0^{l_0} \pi y^2 dx = \frac{1}{r+1} g_0 l_0 \tag{0-2}$$

Test
$$g_0 = \frac{1}{4}\pi d_0^2$$

It is contemplated that a tree height l_0 can be divided into N equal parts of the tree, where $l_i = i\frac{l_0}{N}$ ($i = 1, 2, \cdots, N$), height, so $H_0 = \left(\frac{1}{N}, \frac{2}{N}, \cdots, 1\right)$, while their relative tree height, its diameter is d_i, and the relative diameters are $\left(\frac{d_1}{d_0}, \frac{d_2}{d_0}, \cdots, \frac{d_0}{d_0} = 1\right)$, successively. In this way, any tree can be regarded as a normalized single-tree standard model with relative height $H = \frac{i}{N}(\leq 1)$, and relative diameter $\frac{d_i}{d_0} \leq 1$. We know that a tree in different places have different dry index r, as shown in Figure 0-7, the $r = 0, 1, 2, 4$ of the volume can be expressed as:

$$V = g_0 l_0 - \frac{1}{4}g_1 l_2 - \frac{1}{6}g_3(5l_3 - 2l_3) \tag{0-3}$$

If one tree is considered to be divided into N equal parts and we know that $r=0$ is part of, $r=1$ is part of $m=1$, besides $r=2$ is part of m_2, where $r=3$ is the part of m_3 while the volume formula of individual tree:

$$V = m_0 g_0 l_0 - \frac{1}{2}m_1 g_1 l_1 - \frac{1}{6}g_3(5m_2 l_3 - 2m_3 l_2) \tag{0-4}$$

Regarded it as the normalized individual tree, and its volume formula is:

$$V = \frac{m_0^2 g_0 + \frac{m_1^2}{2}g_1 - \frac{1}{2}g_3 m_2 m_3}{m_0 + m_1 + m_1 + m_2 + m_3} \tag{0-5}$$

How do we confirm $m_i = ?$ ($i = 0, 1, 2, 3$), it is easy to prove from $r = 2\ln(y_1/y_2)/\ln(x_1/x_2)$, let observation x_i, y_i, errors of δ_x, δ_y.

To make

$r=0$, that is $y_1 = y_2$

$r=1$, that is $(y_1/y_2)^2 = x_1/x_2$

$r=2$, that is $y_1/y_2 = x_1/x_2$

$r=3$, that is $(y_1/y_2)^2 = (x_1/x_2)^3$

It can be easily proved that applying the law of error propagation to know $m_i = ?$ ($i = 0, 1, 2, 3$) and then a tree will be divided into different surfaces sections. Can we rapidly estimate the volume of a tree? Here is a way to measure the volume by using a 2-m steel ruler.

After choosing the target tree, we use the steel ruler to measure the circumference of ground with trunk l_0 (in m) and circumference l_1 (in m) at $h = 1.51$ m, calculating the range $\Delta l = l_0 - l_1$ (in cm). From M. Kunze curve equation, we can get:

The volume and tree height H for the conifers are:

$$\begin{cases} V_c = 4 \dfrac{l_0^5}{\Delta l} \\ H_c = h \dfrac{l_0}{l_0 - l_2} \end{cases} \tag{0-6}$$

For the broad-leaved tree,

$$\begin{cases} V_b = 6 \dfrac{l_0^3}{\Delta l} \\ H_b = h \dfrac{l_0^2}{l_0^2 - l_2^2} \end{cases} \tag{0-7}$$

It can be used with the phone calculator quickly estimate standing volume.

0.4 Pattern research of large area grid trapezoidal continuous ecological environment model

China has a vast territory, complex terrain, varied natural conditions, rich in forest resources and uneven distribution of vegetation. Most of the area is mountainous in the west and mainly plains and hills in east. Characterized by the mountains and high altitude in the western side while plains and low altitude in the eastern side. From north to the south it comes across five temperature zones such as the boreal, temperate, warm temperate, subtropical and tropical. So China's forest resources include almost all types of forests that are present in the world. China lies in the eastern part of the Eurasian continent. China is affected by the subtropical monsoon climate and various types of soil. Meanwhile, with the development of spatial information technology the hidden knowledge of a large amount of data resources is far from being fully excavated and utilized. People just use the little knowledge of data, leading to large-scale data waste. In order to make better use of data knowledge, data mining technologies have emerged which are the means of extracting useful information from the massive database. Data mining can be used to understand the spatial data and the relationship between spatial and non-spatial. Spatial data is based on the reorganization and spatial query optimization.

To manifest all kinds of conditional factors associated with the forest growth in different regions by the database, the researchers introduced a trapezoidal grid method. The scale we use for national mapping 1 : 10000, while the scale used worldwide, for drawing maps of different countries 1 : 100000. In Beijing, the scale of 1 : 10000 is divided into trapezoidal lattices with equal latitude and longitude $(2n-1)$ in the top and bottom of the trapezoid. The data in each grid is consistent, and the data mining technique is used to obtain the main factors related to forest growth and calculate the area of existing forest land in China. At the same time, the correlation model between vegetation cover NDVI and key influencing factors was established. According to the physiological conditions and related models of different tree species, the suitable area was selected for planting, which is called 'suitable tree' in forest science and 'appropriate to fit the tree', a combination of forms to achieve our 'precision forestry' objectives(Figure 0-8).

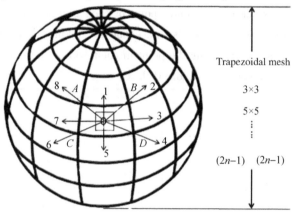

Figure 0-8 The grid diagram

By dividing the total territory of China into thousands of average, continuous, bounded trapezoidal grid by 1 : 10000 scale, we can count the soil conditions (soil type, parent material, pH, NPK and other organic matter content, etc.), climate condition (annual maximum and minimum temperature, average temperature, rainfall), topography (elevation, slope, aspect), vegetation (vegetation type, coverage) and other attribute data.

Use of the remote sensing technology to input all kinds of data systematically, establish forest resource database discovery and utilize the potential knowledge in the database. By data mining we can extract these data with the use of certain association rules and apply for the selection of China's inland forest species planting area and regional deployment to meet the principle 'suitable tree for a suitable place'. This will provide adequate theories and practical basis for selection and planting of trees in future.

0.4.1 Rectangular grid method

For local statistics and interpolation analysis to one variable, we need to use the grid method because the size and density of the grid seriously affect the accuracy of the research data and model. There are 3 methods mainly used: regular grid method (GRID), irregular triangular grid method (TIN) and mixing method (GRID-TIN) in the configuration of the terrain model. The digital terrain model (DTM) is one of the fundamental methods to describe ground features, which is the digital representation of surface morphology and feature morphological attribute information.

The proposed model was used to design a digital high way by Miller in 1956. It has been widely used in various types of circuit selection design and engineering aspects of the terrain data area, volume, slope calculations and it is also widely used in military intelligence navigation, disaster prediction etc. To research the spatial interaction between forest vegetation and environment of the central capital region, we take the geographical location center of Beijing as the center of the research range.

In order to analyze spatial correlation of each factor in different grid scale, we divide the area in accordance with the rules of a rectangular grid in 100 km range into 1 km, 3 km, 5 km, 7 km, 10 km and 20 km of different grids, taking the center point of each grid as sample point to facilitate the study of sample point's interpolation analysis. Based on the 1 km, 3 km, 5 km, 7 km, 10 km and 20 km grid size grid, we randomly sample the point in each grid on the scale by the grid method of 3×3, 5×5, 7×7 and 9×9, as shown in Figure 0-9.

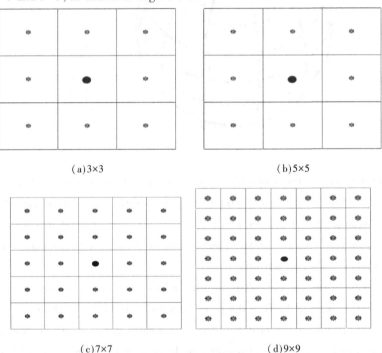

(a)3×3　　　　　　　　　　　　(b)5×5

(c)7×7　　　　　　　　　　　　(d)9×9

Figure 0-9　Rectangular grid method

0.4.2　The law of gravity extension method

Newton's Gravitational Law is one of the greatest achievements of science in the 17th century that explained the basic laws of fundamental interactions namely two objects are attracted to each other in nature and gravity is directly proportional to the product of two object's mass and inversely proportional to the square of the distance between two objects. The basic expression is as follows:

$$F = G\frac{m_1 m_2}{r^2} \tag{0-8}$$

$$\vec{F} = \frac{Gmm}{\vec{r}} \tag{0-9}$$

In formula, F—The interaction force between two objects, namely Gravity;

G—Gravitational constant;

m_1—The mass of the object 1;

m_2—The mass of the object 2;

r—The distance between two objects;

$\vec{F}$—the force of radial vector;

$\vec{r}$—The distance of radial vector.

This research is based on the Law of Gravity, based on the principle of interaction forces exist in the sample points of each grid in different grid scale, within each square grid. It is assumed that in each square grid if the indexes are evenly distributed, the interaction between each grid and the associated grid can be achieved by studying the spatial correlation between the grid sample points in each grid. Specific formula is as follows:

$$C_i \times B_i \times Y_i = \sum_{j=1}^{n} \frac{C_j \times B_j \times x_j}{D^r} \tag{0-10}$$

In formula, C_i—Vegetation coverage of point I;

B_i—Biomass of point I;

Y_i—Some environmental indicators of point I;

C_j—Vegetation coverage of point j;

B_j—Biomass of point j;

x_j—Some environmental indicators of point i;

D^r—The distance from j to i.

The formula indicates a law of the spatial correlation of centers of sample points in each grid with its surrounding sample points changing the distance between sample points. The types of indicators of the sample points are equal within each grid, so the sample points are grid center. The research is about spatial relationship between each grid center point and the center and surrounding center point. Formula will be revised as follows:

$$C_i \times B_i \times Y_i = \sum_{j=1}^{n} \frac{\overline{C}_j \times \overline{B}_j \times \overline{x}_j}{\overline{D}^r} \tag{0-11}$$

Among them, $\overline{C}_j = \frac{1}{n}\sum_{j=1}^{n} C_j, \overline{B}_j = \frac{1}{n}\sum_{j=1}^{n} B_j, \overline{x}_j = \frac{1}{n}\sum_{j=1}^{n} x_j, \overline{D}_j = \frac{1}{n}\sum_{j=1}^{n} D_j.$

In the case of vegetation coverage C_i, C_j of sample points in each grid, biomass B_i, B_j, some environmental pollution indicators Y_i, x_j, and the average distance between center sample points and surrounding center point $\overline{D}$, we can calculate the distance exponent r, judging the spatial correlation between environmental pollution index and vegetation coverage and biomass. The mutual relationship will transit from the single factor of spatial correlation to the distribution of vegetation and environmental effects.

6 grid scales as 1km, 3km, 5km, 7km, 10km, and 20km are taken to the research scale, 4 palaces as 3 × 3(9 palace), 5 × 5(25 palace), 7 × 7(49 palace) and 9 × 9(81 palace) are taken to research dimensions, so to the different grid scales and palaces, each indicator of environmental pollution has 24 results of r. Finding the changes with distance, spatial relationship between environmental pollution and vegetation is the focus, we will specifically analyze the data and algorithms in the following.

0.4.3 Moran's index method

Moran's index method is a way to study relativity issues of certain variables in different spatial locations, which is auto-correlation of different sample points of the same variable with sample points space, a specific formula is as(0-12).

$$I = \frac{n}{\sum_{i=1}^{n}\sum_{j=1}^{n} W_{i,j}} \cdot \frac{\sum_{i=1}^{n}\sum_{j=1}^{n} W_{i,j}(x_i - \bar{x})(x_j - \bar{x})}{\sum_{i=1}^{n}(x_i - \bar{x})^2} \quad (\forall j \neq i) \quad (0\text{-}12)$$

In formula, x_i—The value of a variable in the location I;

$\bar{x}$—The average value of a variable x;

$W_{i,j}$—The weight of space.

In normal circumstances, the Moran's index I is normalized to the value Z, which is used to determine the positive & negative and significance of space autocorrelation of the variable. When $Z>0$ is significant, indicating the variable among sample points exist positive spatial autocorrelation, otherwise negative, when $Z = 0$, variable sample points don't exist spatial autocorrelations, but a random distribution.

We take 3×3, 5×5, 7×7 and 9×9 palace scale, a sample point within the range of 100 km of any scale in a rectangular grid as the research center, 8grid, 24 grid, 48 grid, and 80 grid around of the sample point as the study area. The spatial correlation between vegetation factors and environmental factors was studied. It was found that with the increase of the distance between the sample center point and the surrounding sample, the influence degree of the vegetation factor to the environment was smaller, and the influence range of the vegetation to the inhalable particulate matter in the air was rectangular grid side 10km. The range of the grid is calculated to be 41.87 km.

The Moran's I method is used to analyze the spatial autocorrelation degree of each factor, that is to analyze and study the correlation between each sample point and the point in different grid scale. With the increase of the distance between the sample points, the Moran's I index decreases gradually, which shows that the spatial correlation and auto-correlation decrease with the increase of the relative radius.

Based on the virtual radius of 100km in the central capital region, the sampling points were randomly sampled with different rectangular grids and different grid scales. The vegetation factors (vegetation coverage and biomass) and environmental factors (respirable particulate matter) the results showed that the influence of vegetation on the environmental effect was consistent with the spatial correlation of the above factors. All of which had the spatial correlation decreasing with the increase of distance. At last, the maximum impact range D_{pm} is 41.87 km between vegetation factors and environmental factors interaction, after the spatial autocorrelation of all kinds of factors in the research area and spatial correlation analysis of vegetation and environmental factors, the conclusion is that various spatial correlation factors decrease with increasing distance.

China's territory is divided in accordance with the latitude and longitude of full and seamless coverage by Rectangular Grid Method, after analyzing the soil conditions (soil nitrogen, phosphorus, and organic matter content), climatic conditions (average annual rainfall, maximum temperature, and minimum temperature) and vegetation coverage conditions of each grid. It is clearly indicated that the current China's vegetation coverage per capita has not yet reached the international advanced level. In order to better the foundation of forestry, the database of forest vegetation has been established, which contains two factors soil conditions and climatic conditions. That mainly affects forest coverage, and it is also the theoretical foundation, basis for future afforestation and fundamental guarantee to fulfill the principle 'suitable tree for suitable place'.

0.5 Forest management equipment technology and system platform

Forest measurement is the basic mean of forest information. Forest measuring equipment for obtaining forest information and data precision plays a decisive role in forest management. In order to realize the digitization of forest survey precisely both inside and outside theindustry, the classical forest survey methods need innovative ideas. For this already invent a series of new measuring equipment including multi-functional electronic measuring gun, three-dimensional angle gauge measuring and the tree superstation meter, etc, boosting our country forestry informatization, to provide a guarantee for our country to realize precision and digital forestry.

The information with each passing day today, every industry is trying to chase the trend of the times and under the forestry information into the overall strategic concept, the construction of modern forestry management platform is particularly important. With the diversity of the function of forest resources the long-term growth cycle and the state of the dynamic and the uncertainty of mature forest, wide distribution of mergence and space structural characteristics.

Forest measuring technology should be a leap-forward development, i.e. without optical micrometer era, directly by the mechanical cursor measurement tools in the modern electronic grating and computer technology, information technology, space technology, video technology, the combination of graphics technology integration and age. The forest measuring, surveying and mapping Science and technology and other multi-disciplinary integrated high technology is a perfect improvement in forest classical measurement techniques. At the same time through the integration of the system, unit innovation and practical research to build a perfect forest survey inside and outside the industry, integration information management platform.

0.5.1 Forest survey inside and outside integration platform introduction

The key laboratory of precision forestry in Beijing Forestry University and the team of Prof. Feng zhongke with the hard work and achievements of many years of research have made development in six fields. The first one is namely software and five kinds of hardware products refer to:

(1) Electronic measuring tree gun (MPTS-2);

(2) Electricity by tree meter (FET-2);

(3) Total station measuring tree (FTS-2);

(4) Mobile acquisition recorder;

(5) UAV quarter 3D camera system.

First, one software platform is inside and outside integration platform (FSIM. Shop V2.0). At the same time, the preservation and improvement of the classical forest survey methods, innovative ideas can improve the classic form of forestry. In the depth of the automatic data acquisition, infor-

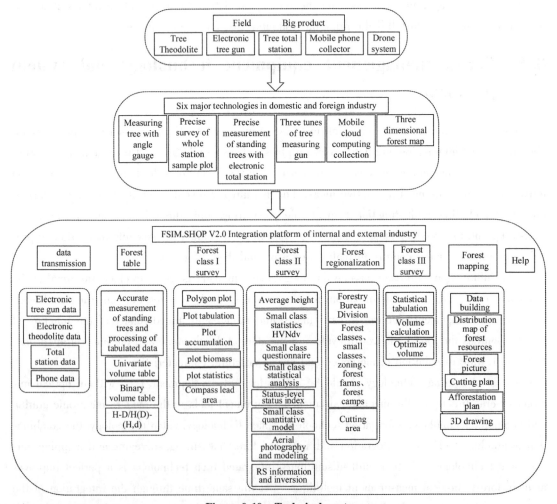

Figure 0-10 Technical route

mation and knowledge mining the platform is the foundation of forestry information and is the embodiment of forestry information technology. IT industry vision is not limited to forestry but also take into account other industry applications in the land survey, construction of water conservancy project, power transmission line project, rural land survey, demarcation of mines resources, civil engineering, the full application of the precision measurement, digital signal processing technology(Figure 0-10).

0.5.2 Electronic measurement instrument design tree

Angle gauge and traditional instruments such as forest compass are widely used in forestry for measuring forest stand as a basic measuring tool. But traditional instruments in the field investigation inthe process of actual operation, there are still many problems like the traditional measuring tools are without data storage and digital display function can only rely on paper-based manual records. Data recording and reading sometimes also becomes a big error recording is not easy to preserve, data need to manually input in the industry. Because of the influence of the terrain, understory shrubs measurement difficulties the precision is low. Traditional instruments operation cycle is long, low efficiency, poor accuracy and not much effective for the monitoring of forest resources and ecological environment demand. At the same time, due to the forest survey work classification, the different resource survey needs different accuracy. Usually, high precision products having precision $\pm 0.5''+0.5$ mm$+0.5 \cdot 10^{-6}D$ d above, precision products having to measure accuracy higher than $\pm 2'' + 2$ mm $+ 1 \cdot 10^{-6}D$ that of low precision of product are better than $\pm 1° + 1$ cm/100 m. Internationally, the output value of the three kinds of precision products accounts for about 1/3.

China is the world's superpower of surveying and mapping. The demand for surveying and mapping products is huge especially in low precision of surveying and mapping instruments in forestry application has a large amount, wide face, do not need more training can be used etc. Therefore, there is urgent need of development of such products to meet the demand of industry application.

Handheld total station have core components like EDM + MEMS (T, delta) + SD card+ CPU + LCD (LED) + USB, electronic three-dimensional Angle gauge and electronic compass (Figure 0-11). The handheld total station can measure average height, arbitrary height, diameter, density,

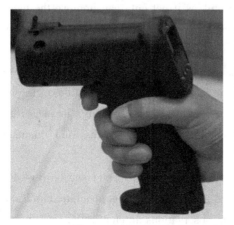 

Figure 0-11 Handheld total station tree gun (test) diagram

basic measuring magnetic (distance, azimuth angle and tilt angle), small class demarcation (compass lead/area measurement), crown density and other functions that can meet the requirement of common in forestry investigation(Table 0-3). This reduced the workload in the field is and due to its own data store communication function the traditional paper-based manual record of work can be avoided. This machine is easy to carry and lighter at the same time can reduce the burden of the field and has general applicability to the investigation of forestry in China.

Table 0-3 Handheld total station (tree gun) measurement specification parameter table

Features	Parts	Parameters
Distance measurement	Tree	0.5~80m
	Reflector plate	0.5~100m
	Smallest unit of display	1mm
	Laser level	II Level
	Measurement error	2mm
Tilt angle	Range	−75°~75°
	Accuracy	0.3°
Azimuth	Accuracy	0.1°
Display screen	Type	LCD
Input	Type	button
Communication	USB	
On-Board battery	Power supply	Yes
	Continuous measuring Range	Lithium, 5 hours
Operating environment	Ambient Temp	−20~50℃

Stock volume is one of the important basic information of forest investigation. The accuracy in determining the traditional method of the stock volume data is not high, due to limiting conditions, operating more cumbersome. After every ten years conduct once division survey and every time cutting down nearly 300 thousand trees to revise the volume table and by volume formulas we determine volume, the accuracy is relatively high, but the damage to forest resources in our country is too big. So it does not meet our purpose of protecting the ecological environment. The application of electronic theodolite for this test tree is very useful, as it not only can improve the measurement speed, accuracy and efficiency but also the most important thing is we don't have to cut down a tree or climb trees.

Theodolite while measuring tree usually choose high precision and is mainly composed of angle gauge, and proprietary software. With automatic measurement without cut down tree height, diameter, crown, crown volume, crown volume, the surface area of functionality. Its measuring principle is pillar method to calculate volume, triangle method for calculating the trunk diameter(Figure 0-12).

As compared to the electronic theodolite, the total station tree meter can be used for precise distance measurement and has many characteristics that are automatic, accurate, rapid and nondestructive. Total station can be used to measure the observations automatically, coordinates of each tree diameter(x_i, y_i), breast height D_i , tree height H_i . The units used in the total station are centimeter (cm) and meter (m) (Figure 0-13).

 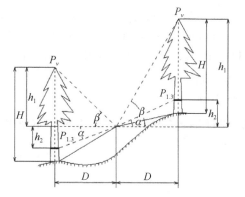

Figure 0-12 Measurable electronic theodolite measuring tree

Figure 0-13 Total station measuring tree diagram

In UAV 3D photography a GPS navigation system, automatic pose velocity measurement, remote numerical control and monitoring of unmanned aerial vehicle low timing camera system are used. The system uses unmanned aerial vehicle flight for flight platform, with high resolution digital remote sensing equipment for airborne sensors in order to get a low high-resolution remote sensing data for the application goal, mainly used for rapid geographic data acquisition and processing.

The system uses 3D digital camera, GPS, pose speed automatic test equipment, digital radio accessquartet required image data, station coordinates, photography attitude. The use of related equipment and procedures realize image correction parameter initial standardization. The use of digital photogrammetry includes software and hardware for image correction of splicing. To create images based on orthogonal projection image and the ground model of forestry surveying and mapping to provide the most simple and most reliable and intuitive application data.

Unmanned aerial vehicle photography system consists of a hardware system and software system. Hardware system mainly includes three parts the unmanned aerial vehicle, the airborne system and monitoring system. The software system includes five aspects route design, flight control, remote monitoring, data pretreatment and flight inspection. By the combination of hardware and software, we can effectively achieve the quick route design, aerial coverage checking, real-time data transmission, data pretreatment, and such related content and also effectively solve the programmable in parallel flight and the difficulties of the SPC attitude stability.

0.5.3 Internal and external integration platform

In the aspect of the automatic data acquisition, efficient informationprocessing, and knowledge depth mining, this platform is the foundation of forestry information and the embodiment of forestry information technology core, meanwhile, it is a practice and a useful attempt in the implementation of the national strategy of informatization.

1) The non-destructive precision of stumpage and modeling

The main function of non-destructive precision of stumpage and modeling software is to obtain non-destructive measurement data of stumpage by using total station and electronic theodolite, to calculate the volume of wood, tree crown volume and other basic information by using algorithm and also to build the three-dimensional tree model to achieve one way volume table and binary volume table. The software achieves the algorithm of non-destructive precision of the basic information of stumpage, measurement data encapsulation, security, and simplifies computational complexity (Figure 0-14).

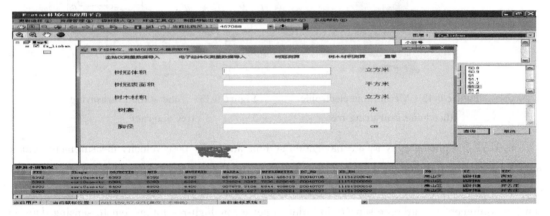

Figure 0-14　Software interface

The non-destructive precision of standing timber refers to the measurement of angles of the visible characteristics of the trunk and characteristics of the crown by using the total station, electronic theodolite, tree-measuring gun and other instruments. It can calculate tree factors through zenith distance and horizontally contained angle to obtain the data of tree crown volume, superficial area, tree volume, tree height, and *DBH*. The terminal data input of total station and electronic theodolite enables to achieve data management, data analysis of non-destructive precision of standing timber and the establishment of single tree volume model (one-way volume table and binary volume table).

2) The main function

The main function of first, second, third forest resource inventory and graphics software is to achieve the measures of basic sample information in the first forest resource survey by tachometer, tree measuring gun and 3D camera pro to achieve database creation, statistical analysis of the survey data based on sub-compartment. In the second forest resource survey to achieve the tree measurement, forestry logging design and plantation layout in third forest inventory by the tachometer, electronic theodolite, tree measuring gun and to achieve the forest resources distribution, special map and stock map based on ArcGIS engine (Figure 0-15).

The forest resource management system is the secondary development project based on ArcGIS Engine, which is mainly applied in the first, second and third investigation in forest resources survey. In the forest resource management, it is supposed to be able to process two-speed data of sub-compartment hierarchical rendering, Class Break Renderer, sub-compartment statistic management (the management of attribute data and spatial data), SuperMap Ajax, forest plot (stock map, sub-

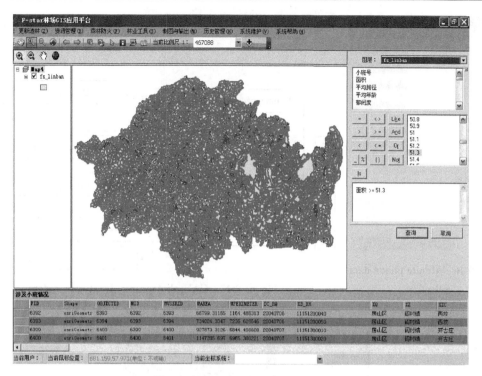

Figure 0-15 Application platform

compartment distribution maps, species distribution maps and afforestation design), thematic mapping and polygon sample plot method.

3) Stump volume calculation

The main function of the root volume calculation software is to use the local data to establish the direct relationship between the root diameter and DBH of the felled trees. The software is a part of the forest resource survey by inputting the measured ground diameter data and obtaining the calculation software for the tree volume, tree factors such as the tree height and diameter at breast height. Inversion of stand density, average stand height, stand volume, stand biomass, diameter order distribution using polygon sampling method.

4) Cloud Computing of tree measurement

Cloud Computing is also known as 'on-demand computing', is a kind of internet-based computing where shared resources, data, and information are provided to computers and other devices on-demand. It is mainly based on the increase of internet related services use and delivery mode to provide dynamic, easily expandable and often virtual resources via the internet.

Mobile cloud computing platform can be connected with the backstage database of various web portal on the basis of cloud computing platform like a cloud computing platform. Mobile cloud computing platform enables data to be available on the Internet and be transmitted point to point meanwhile, it enables people to download and upload data as they want at any time and in any place only if they have a mobile phone capable of surfing on the internet, as shown from Figure 0-16 to Figure 0-18.

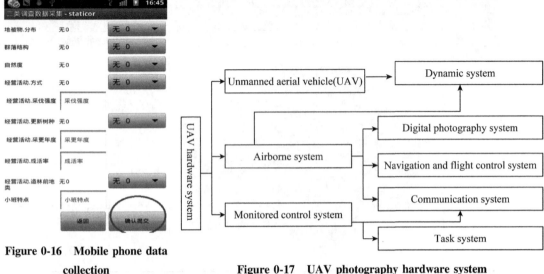

Figure 0-16 Mobile phone data collection

Figure 0-17 UAV photography hardware system

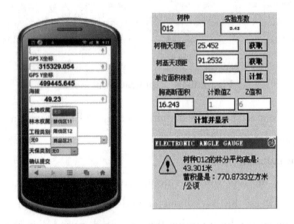

Figure 0-18 FMCC. V1. 0

Measurable cloud computing android version (FMCC. V1. 0) is based on android mobile phone platform used in the field of data collection. The main function is to make a sample from the area, obtain felled tree measurement data from the field by the mobile phone incoming server and data on the server side to build statistical analysis. A large amount of data parallel distribution is on the server side. Specific to complete the following tasks i. e. survey field data acquisition; data wireless transmission; data warehousing management; field data real-time processing (single timber volume calculation, sample data analysis).

5) Digital up shot forest measurement

The major function of Digital up shot forest measurement software (DCPFM. V1. 0) is to calculate the volume of stumpage, crown, and diameter at breast high or height of optional position by photo arithmetic of common digital camera, 3D camera. It can also be used to calculate the average height, diameter class distribution, amount of growing stock, and biomass obtains of basic forest's forest stand with a message in small class(Figure 0-19).

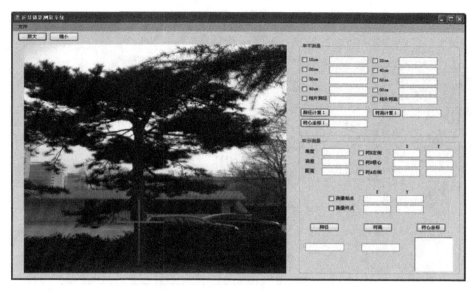

Figure 0-19 Digitalup Shot Forest Measurement (DCPFM. V1.0)

This system adopts common digital camera, to assistspecially made cross ruler to collect digital photos of an individual tree or sample plot and through software to enable individual tree measurement, sample plot measurement, the measurement of height and timber volume when it is covered by a crown. It has following major functions: the function of individual tree measurement, the height, and diameter at an optional position, pith coordinate, and volume of timber.

Sample plot measure measurement function: the height of any position of the trees and the diameter of any position of the trees; the coordinates of the heart of any tree; area of sample plot; stock volume; height and volume of a tree when it is covered by a crown.

6) Forest fire prevention system

Based on ArcGIS Engine forest fire prevention system is developed which include the basic function of mapping, in accordance with the Octree theory on the basis of the forest fire spreading model proposed by Wang Zhengfei and Mao Xianmin. A forest fire spreading model which points out the impact factors of forest fire spreading including slope, slope aspect, wind speed, wind direction, surface fuels, and air humidity. With the help of spatial analysis function of ArcGIS, forest fire prevention system can calculate the best path of forest fire suppression, analyze past data on forest fires and relevant impact factors, and thus build fire spot forecast model by identifying the geographical attributes and fire danger ratings of those fire spots(Figure 0-20).

An intelligent led forest fire prevention system with the combination of designing fire observation station, detecting fire spot, ranking and forecasting fire danger, modeling fire spreading as well as evaluating post-fire loss and forest fire management information with the action of tracking, dispatching, and commanding.

7) Wildlife protection system

Map relative operation can achieve all basic functions, like amplification, hawk-eye, property inspector, editorial elements; achieving real-time upload, edit, delete of thewild Amur tiger and the

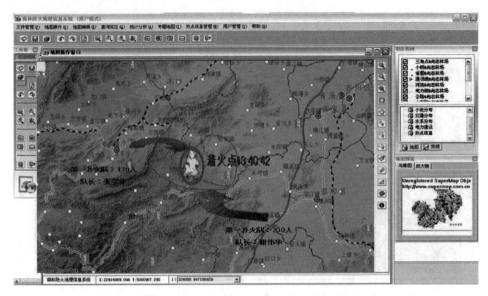

Figure 0-20   Forest fire prevention information system

data can connect with the database to build the space data of SDE. SDE space data which is built based on the survey data of Manchurian tiger can conduct the space analysis, buffer analysis, and overlay analysis. Also, space analysis can be used to plan the habitat of the wild Manchurian tiger. Graphs and histogram achieved by statistics function according to the wild survey data of Manchurian tiger and real-time count of the information of river distance. Altitude and distance towards settlement place of each survey spot can be used to analyze the regularities of distribution of Manchurian tiger. Type thematic map can be made according to the distributed information of each area (Figure 0-21).

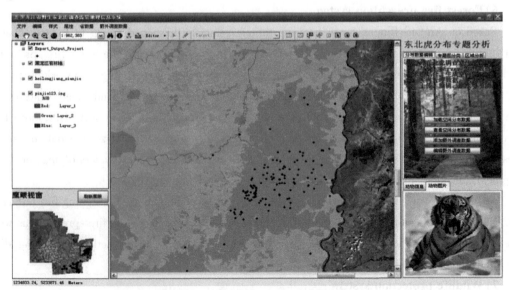

Figure 0-21   Wildlife protection system

8) County environmental system

This program includes OA office system and geographic information system. OA office system mainly achieves the needs of daily work i. e. the storage of many tables and data, inquiry and statistic analysis. Geographic information system achieves the basic geographic information mapping service based on the ArcGIS Server, space inquiry, and property inspector, the connection between map and table, buffer analysis of emergency, storage, inquiry and statistic analysis of environmental data and basic data thematic map (Figure 0-22).

Figure 0-22 County environmental system

9) The aerial remote sensing mapping system

Forestry maps are not only the main findings of forest resource investigation but also the fundamental materials used in forest resources managements and operation in forestry. Traditional handmade forestry maps involve relatively long production span, poorquality, and poor recyclability. This kind of maps cannot meet the needs of modern forestry production. However, the Aerial Remote Sensing Mapping System, based on ArcGIS Engine has not only innovated the forestry mapping function but also explore new mapping techniques. ArcGIS Engine, with its powerful graphic editing and strong graphical visualization functions, has become a graphic plotting software for data collection, spatial database construction, and mapping. Therefore, this portion will be devoted to the integration of secondary development on the basis of the software ArcGIS Engine, that is the development of forestry mapping software for integration of drawing, filling, attribute input, automatic injection, measurement area, area adjustment, print out maps and other functions through the establishment of forestry graphic symbol library (Figure 0-23).

0.5.4 Costs Requirements from Different Forestry Businesses

We use modern technologies to study measurement techniques and develop new forest measurement devices in order to realize the automation and digitization of forest resources. We also establish a software platform of forest investigation technique and forest observation technique. This platform is characterized by its automation and accuracy in forest investigating. Through combining, integrating, upgrading and innovating the key technology and equipment, this platform didquite a good job in forest measurement (Table 0-4). Also, the platform offers strong technological support for forest resources investigation and management.

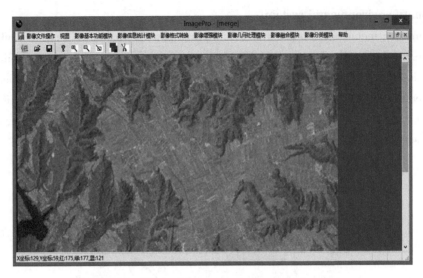

Figure 0-23　The aerial remote sensing mapping system

Table 0-4　Technology and equipment suitability comparison

Equipment technology	Business									
	No felling volume model		First forest resource inventory investigation		Second forest resource inventory investigation			Third forest resource inventory investigation		
Characteristics	Crown	Trunk	Sample plot observation	Angle observation	Small class precision measurement	Digital aerial photography	Aerospace Remote Sensing Inversion	Stem form	Logging survey	Forest selection design
3D scanner	A/A*	B/A*	B/A*	C/B*	B/A*	B/A*	B/A*	A/A*	B/A*	C/A*
Tree theodolite	B/A*	A/A*	B/D*	A/D*	B/D*	B/D*	B/D*	B/D*	B/D*	C/D*
Total station measuring instrument	B/A*	A/A*	A/A*	A/A*	B/A*	B/A*	B/A*	B/A*	B/A*	C/D*
Tree measuring gun	C/B*	B/C*	A/B*	A/B*	A/B*	A/B*	A/B*	A/B*	A/B*	C/A*
FSIM shopV2.0	A	A	A	A	A	A	A	A	A	A

Notes: A* suitable, B* basically suitable, C* not suitable, A* high, B* middle, C* low.

References

陈明艳,2001.森林资源规划设计调查信息管理软件[J].东北林业大学学报(04):67-68.

陈谋询,2001.我国21世纪森林经理发展趋势分析[J].华东森林经理(02):1-5.

达来,袁桂芬,1998.布鲁莱斯测高器野外测树误差的室内修正[J].内蒙古林业调查设计(S1):68-72.

邸冰,王静,2012.浅谈利用地理信息软件绘制森林资源规划设计调查林相图[J].防护林科技(04):2.

邸凯昌,2003.空间数据发掘与知识发现[M].武汉:武汉大学出版社.

冯仲科,聂伟,姚山,等,2008.造林决策信息系统框架设计[J].北京林业大学学报,30(S1):7-10.

冯仲科,余新晓,2000."3S"技术及其应用[M].北京:中国林业出版社.

韩琳,2012.浅谈数据挖掘与数据仓库[J].无线互联科技(03):70.

贺振平,2012.内蒙古数字林业建设对策研究[D].呼和浩特:内蒙古大学.
蒯汉军,2010.浅议森林资源二类调查技术[J].广东科技,19(24):34-36.
李明江,唐颖,周力军,2012.数据挖掘技术及应用[J].中国新通信,14(22):66-67,74.
李星敏,郑有飞,刘安麟,2002.遥感植被分类方法的概述及发展[J].陕西气象(03):20-23.
李兆莲,2013.浅析3S与PDA在资源调查中的应用[J].林业勘查设计(01):103-104.
刘文虹,2011.森林资源规划设计调查总结与反思[J].经济研究导刊(23):222-223.
刘旭升,张晓丽,2004.森林植被遥感分类研究进展与对策[J].林业资源管理(01):61-64.
马超飞,刘建强,2003.遥感图像多维量化关联规则挖掘[J].遥感技术与应用,18(04):243-247.
聂玉藻,马小军,冯仲科,等,2002.精准林业技术的设计与实践[J].北京林业大学学报(03):89-93.
任树军,曹俊茹,黄文俐,等,2004.遥感影像关联规则数据挖掘[J].山东理工大学学报(自然科学版),18(06):93-97.
孙国力,郑姝婷,高源,等,2011.森林经营方案编制和管理存在的问题及对策[C].辽宁省植物保护学会、辽宁省昆虫学会2011年学术交流研讨会(西安)论文集,88-89.
佟乃兴,赵晶明,张鹏,等,2013.浅谈细颗粒物(PM2.5)[J].化工管理(08):143.
王佳,臧淑英,2006.基于SuperMap Objects的森林资源管理系统设计与实现——以黑龙江省851农场林业局为例[J].哈尔滨师范大学自然科学学报(01):90-93.
王宇,2013.北京市近10年空气环境质量变化特征及影响因素分析[C].2013中国环境科学学会学术年会.云南昆明.
韦艳艳,张超群,2011."数据仓库与数据挖掘"课程教学实践与探索[J].高教论坛(01):94-96,99.
谢阳生,2010.大都市森林及绿地资源监测信息协同技术体系研究[D].北京:中国林业科学研究院.
杨洁,王国平,2005.计算空气污染指数的简捷方法[J].环境研究与监测,18(2):3.
于政中,李海文,赵世华,1993.研究我国林业科技论文探讨森林经理学科的发展[J].林业资源理(02):28-32.
张春梅,连凤宝,2001.大气污染浓度值与空气质量真实值偏离分析[J].太原科技(02):25-26.
赵宪文,1995.中国热带林遥感探索:分类方法与调查方案[J].林业科学研究(04):373-379.
周小成,庄海东,陈铭潮,等,2013.面向小班对象的森林资源变化遥感监测方法——以福建省厦门市为例[J].资源科学,35(08):1710-1718.
朱传凤,赵和平,1998.用空气污染指数评价城市空气质量[J].甘肃环境研究与监测,11(2):2.
曾伟生,周佑明,2003.森林资源一类和二类调查存在的主要问题与对策[J].中南林业调查规划(04):8-11.
Boyd D S, Foody G M, Ripple W J, 2002. Evaluation of approaches for forest cover estimation in the Pacific Northwest, USA, using remote sensing [J]. Applied Geography, 22(4): 375-392.
Franklin S E, Peddle D R, 1990. Classification of SPOT HRV imagery and texture features [J]. International Journal of Remote Sensing, 11(3): 551-556.
Peddle D R, Frankli S E, 1992. Multisource evidential classification of surface cover and frozen ground [J]. International Journal of Remote Sensing, 13(17): 3375-3380.
Srinivasan A, Richards J A, 1990. Knowledge-based techniques for multi-source classification [J]. International Journal of Remote Sensing, 11(3): 505-525.
Sulong I, Mohd-Lokman H, Mohd-Tarmizi K, et al., 2002. Mangrove mapping using landsat imagery and aerial photographs: Kemaman District, Terengganu, Malaysia [J]. Environment Development & Sustainability, 4(2): 135-152.
Walsh T, 2001. Remote sensing of forested wetlands: application of multitemporal and multispectral satellite imagery to determine plant community composition and structure in southeastern USA [J]. Plant Ecology, 157(2): 129-149.

CHAPTER 1　The Mathematical Foundation of Forest Management

1.1　Linearization of the non-linear equations

The term linear refers to the mathematical relationship between variables, it is the attribute of a straight line. In a mathematical sense, the solution of the equation satisfies the linear superposition principle. The linear superposition of any two solutions of the equation is still a solution of the equation. Linearity means the simplicity of the system but natural phenomena are complex and non-linear fundamentally. Fortunately, many phenomena in nature can be approximated as linear to a certain degree. Traditional physics and Natural Sciences can build a linear model for all kinds of phenomena and has achieved great success. But with the human beings depth study of a variety of complex phenomena in nature, more and more non-linear phenomena begin to enter the field of vision.

Non-linear is the mathematical relationship between variables which are not a straight line but in the form of a curve surface or any uncertain attribute. Non-linearity is one of the typical properties of the complexity of nature. Compared with the linear the non-linearity is more close to the nature of objective things and it is one of the important methods to quantify the complex knowledge. The difference between linear and non-linear is usually used to distinguish the dependence of x on the independent variable in the function of $y=f(x)$. A linear function is a function of the first degree; its image is a straight line. The other function is a non-linear function and the image is of different shapes except the straight line.

The linearization of a nonlinear equation is approximated in a certain condition or narrow the scope of work and the nonlinear differential equation is approximated as a linear differential equation. Using a mathematical method to deal with the variable of the nonlinear function expanded into a Taylor series, broken down into these variables in the vicinity of a small increment of the expression, and then spent more than a small increment of the item, we can obtain approximate linear functions. A linear equation can be written in practical applications. Linearization method is a common method in natural science research. The method is to deal with the advanced thermodynamic and kinetic equations and then to the linear equations by the linearization operation (such as the logarithm, the reciprocal, the perturbation).

The different non-linear function has different transformation methods. Commonly used transformation methods are direct substitution method, function transformation method, series expansion method, etc.

The direct substitution method is the original nonlinear variable in the model. Its general applicability in the model is a reciprocal (hyperbolic) model, polynomial model, logarithm model and S-curve model.

Function transform method refers to the function changes such as logarithmic and transposition of the original model of deformation to obtain a linear model and the general application of the model such as the power function model and exponential function model.

Series expansion method is the function of the Taylor series expansion which gets a linear approximation.

Reciprocal (hyperbolic) model

$$\frac{1}{Q} = \beta_0 = \beta_1 \frac{1}{p} + u$$

The formula can be replaced by $Y = \frac{1}{Q}$, $X = \frac{1}{p}$ and the original model can be transformed into

$$Y = \beta_0 + \beta_1 X + u$$

Polynomial model

$$Y = \beta_0 + \beta_1 t + \beta_2 t^2 + u$$

The formula can be replaced by $x_1 = t$, $x_2 = t^2$, and the original model can be transformed into

$$Y = \beta_0 + \beta_1 X_2 + \beta_2 X_2 + u$$

Logarithmic model

$$Y = \beta_0 + \beta \ln X + u$$

take $X_1 = \ln X$ into the original type of replacement, the original model can be transformed into

$$Y = \beta_0 + \beta_1 X_1 + u$$

Power function model

$$Q = AK^\alpha L^\beta e^\alpha$$

both sides of the equation were taken can be obtained:

$$\ln Q = \ln A + \alpha \ln K + \beta \ln K + u$$

Then the model can be directly replaced by the logarithm model.

Exponential function model

$$C = ab^Q e^u$$

both sides of the equation were taken, can be obtained:

$$\ln C = \ln a + Q \ln b + u$$

Then the model can be directly replaced by the logarithm model.

Series expansion method. The function model of the Taylor series expansion gives the two higher elements, approximate reaching one linear approximation formula. Concrete steps: First step: Make the initialization assignment by selected unknown parameters based on existing knowledge or condition. Then take the nonlinear function of the model expanded by Taylor Series in the vicinity of the initial value of this group and abandon the higher order terms of two or above. The second step is the standard linear transform good regression model using an ordinary least squares estimation of the unknown parameter. A new set of sample observations is calculated from the given sample observations and initial values and a new set of least squares estimator of the linear regression model can be obtained by using this new set of samples. The third step is the nonlinear function in this group of new parameter estimation near the Taylor series expansion, after linearization to get a new standard linear regression model. The new standard linear regression model is applied to the ordinary least square method and a new set of least squares estimator is obtained. Repeat the process until the parameters estimated value to satisfy the increment was less than ε (ε is given an arbitrarily small positive number) the final estimator is the solution.

Taylor series expansion:

$$f(x) = f(x_0) + f'(x_0)(x - x_0) + \frac{1}{2!}f''(x_0)(x - x_0)^2 + \cdots + \frac{1}{n!}f^{(n)}(x_0)(x - x_0)^n + R_n(x)$$

Among them:

$$R_n(x) = \frac{f(n+1)(\xi)}{(n+1)!}(x - x_0)^{n+1}$$

known as the Lagrange remainder. Giving up the higher order terms of the then

$$f(x) = f(x_0) + f'(x_0)(x - x_0)$$

It is expressed as a linear functional form $Z = f(X_1, X_2, X_3, \cdots, X_i)$, Approximate value $X_i^0 = (X_1^0, X_2^0, X_3^0, \cdots, X_i^0)$

Then

$$Z = f(X_1^0, X_2^0, X_3^0, \cdots, X_i^0) + \sum_{i=1}^{n} \frac{\partial f}{\partial X_i^0}(X_i - X_i^0)$$

$$Z = \sum_{i=1}^{n} \frac{\partial f}{\partial X_i^0} X_i + f(X_1^0, X_2^0, X_3^0, \cdots, X_i^0) - \sum_{i=1}^{n} \frac{\partial f}{\partial X_i^0} X_i^0$$

For the convenience of this section, $X_i - X_i^0$ will be expressed as $\mathrm{d}X_i$, and $X_i = X_i^0 + \mathrm{d}X_i$

The different nonlinear functions in the linearized method are not identical, this section the binary quadratic equation as an example, we take the general differential expansion for solving the linearized dominated by solving N element, N times equation Jacobian matrix method and the general differential method as an illustration.

Example of two elements two equation:

$$Ax^2 + By^2 + Cxy + Dx + Ey + F = 0$$

Linearization method NO 1: The equation is assumed to be $F(x, y) = 0$

$$F(x, y) = \begin{bmatrix} f_1(x, y) \\ f_2(x, y) \end{bmatrix} = \begin{bmatrix} A_1 x^2 + B_1 y^2 + C_1 xy + D_1 x + E_1 y + F_1 \\ A_2 x^2 + B_2 y^2 + C_2 xy + D_2 x + E_2 y + F_2 \end{bmatrix}$$

$F(x, y)$ of the Jacobi matrix

$$F'(x, y) = \begin{bmatrix} \dfrac{\partial f_1(x, y)}{\partial x} & \dfrac{\partial f_1(x, y)}{\partial y} \\ \dfrac{\partial f_2(x, y)}{\partial x} & \dfrac{\partial f_2(x, y)}{\partial y} \end{bmatrix}$$

Then

$$F'(x, y) = \begin{bmatrix} 2A_1x + C_1y + D_1 & 2B_1x + C_1y + E_1 \\ 2A_2x + C_2y + D_2 & 2B_2x + C_2y + E_2 \end{bmatrix}$$

Established the iterative formula is

$$(x, y)^{(k+1)} = (x, y)^{(k)} - [F'(x, y)^{(k)}]^{-1} F(x, y)^{(k)}$$

$$X_{+1} = X - F'(X)^{-1} F(X)$$

In the formula: $X_{+1} = \begin{bmatrix} x_{+1} \\ y+1 \end{bmatrix}, \begin{bmatrix} x \\ y \end{bmatrix}$,

$$F(X) = \begin{bmatrix} A_1x^2 + B_1y^2 + C_1xy + D_1x + E_1y + F_1 \\ A_2x^2 + B_2y^2 + C_2xy + D_2x + E_2y + F_2 \end{bmatrix}$$

$$F'(X) = \begin{bmatrix} 2A_1x + C_1y + D_1 & 2B_1y + C_1x + E_1 \\ 2A_2x + C_2y + D_1 & 2B_2y + C_2x + E_2 \end{bmatrix}$$

Then iterative formula $X_{+1} = X$, until $|X_{+1} - X| \leq \varepsilon$ is given an arbitrarily small positive this section is unified 0.0001. Here X_{+1} is a set of optimal solutions of nonlinear equations.

Linearization method no 2: $F(x, y)$ can be differential expansion

$$F = (2Ax+Cy+D)dx + (2By+Cx+E)dy + F(x_0) = 0$$

If $X = \begin{bmatrix} x \\ y \end{bmatrix}, M = \begin{bmatrix} dx \\ dy \end{bmatrix}, N = X = \begin{bmatrix} 2A_1x + C_1y + D_1 & 2B_1y + C_1x + E_1 \\ 2A_2x + C_2y + D_2 & 2B_2y + C_2x + E_2 \end{bmatrix}$

$$\begin{bmatrix} 2A_1x + C_1y + D_1 & 2B_1y + C_1x + E_1 \\ 2A_2x + C_2y + D_2 & 2B_2y + C_2x + E_2 \end{bmatrix} \begin{bmatrix} dx \\ dy \end{bmatrix}$$

$$= \begin{bmatrix} -F(x) \\ -F(x) \end{bmatrix} = \begin{bmatrix} -(A_1x^2 + B_1y^2 + C_1xy + D_1x + E_1y + F_1) \\ -(A_2x^2 + B_2y^2 + C_2xy + D_2x + E_2y + F_2) \end{bmatrix}$$

The iterative formula is established as: $N \cdot M = C$

In order to solve the ordinary least squares method, simplify

$$M = (N^T N)^{-1} N^T C$$

$$X_{+1} = X + M$$

Then the iteration formula, order $X_{+1} = X$, until $|X_{+1} - X| \leq \varepsilon$, at this time x_{+1} is a set of optimal solutions for nonlinear equations

$$F(x, y) = \begin{bmatrix} f_1(x, y) \\ f_2(x, y) \end{bmatrix} = \begin{bmatrix} x^2 + xy - 12 \\ y^2 + xy - 4 \end{bmatrix}$$

Assumed initial value $x = -5, y = -3$, after one and two methods, the number of iterations is 5 times. The calculation result is $x = -3, y = -1$. Assume initial value $x = 1, y = 4$. After one and two methods, the number of iterations is 4 times. The calculation result is $x = 3, y = 1$.

For more complex non-linear function models, such as

$$V = \frac{\dfrac{1}{b+1} H \dfrac{1}{4} \pi d^2}{\left[1 - \left(\dfrac{1.3}{H}\right)^b\right]^2}$$

In formula d means tree diameter of 1.3 meters high, the unit is cm; H means tree height, the unit is m; V means tree volume, the unit is m³; given a set of values calculated $d = 0.2$, $H = 16.5427$, $V = 0.2449099$, to determine the value of b. First order $A = \frac{1}{4}\pi d^2 H$, $B = \frac{13}{H}$, then the original transformation is

$$V = \frac{A\frac{1}{b+1}}{(1-B^b)^2}$$

And then the model was obtained by micro-differentiation

$$F = \left[\frac{A(1-B^{b^k}) + 2Ab^k B^{b^k}\ln B(b^k+1)}{(1-B^{b^k})(b^k+1)^2}\right]db + \frac{A\frac{1}{b^k+1}}{1-B^{b^k}} = 0$$

At this time $N = [A(1-B^{b^k}) + 2Ab^k B^{b^k}\ln B(b^k+1)(1-B^{b^k})(b^k+1)^2]$, $M = [db]$, $C = -\frac{A\frac{1}{b^k+1}}{(1-B^{b^k})^2}$, application of least square method $M = (N^T N)^{-1} N^T C$ calculation can be obtained as M. Therefore $b^{k+1} = b^k + M$.

When $db \leq \varepsilon$ 时,b^{k+1} is the solution calculated by programming b which is equals 1.2908.

1.2 Multivariate robust estimation

1.2.1 Overview

Robust estimation is also known as resistance estimation of the gross error interference for the least square difference. Its aim is to construct an Estimation method that shows it has strong ability of resist for gross error. Since 1953, G. E. P. BOX first proposed the concept of Robustness. Tukey, Huber, Hampel, Rousseeuw and others done a comprehensive research on robust parameters estimation and after many decades of a pioneering mathematical statistician and cultivated robust estimation has been developed to catch the attention of the multidisciplinary branches.

In forest management investigation often encounter similar chemical photo interpretation of small volume and ground measured volume of the statistical relationship between them. Commonly used statistical method is based on the classic least square method. When the survey data contain no gross error the least squares estimate can give good results, however, due to various reasons video observation results in more or less will contain a gross error than the use of robust estimation to estimate gross error data values.

1.2.2 Robust estimation principle

Robust estimation is a hypothetical model for practical problems and also believed that this model is not accurate but only an approximation of the actual problem theory model. It requires estimation method to solve the problem of this class should achieve the following goals:

(1) Assume that observation distribution model the valuation should be optimal or near optimal;

(2) When the theoretical distribution of the hypothetical distribution model and actual model has a smaller difference valuation is under the influence of small gross error;

(3) When the theoretical distribution of the hypothetical distribution model and actual model has a great deviation, valuations are not affected by the devastating.

In Robust estimation, the basic idea is that in the case of gross error, it is inevitable to select the appropriate estimation method, the parameters of valuation as far as possible to avoid the influence of gross error to get the best value of normal mode. Robust estimation principle is to make full use of the observed data (or samples) in the effective information, restrictions on the use of available information, eliminate harmful information.

Supposea forest survey regression problem as follows:
$$Z_i = a_0 + a_1 x_1 + a_2 x_2 + \cdots + a_m x_m \quad (i = 1, 2, 3, \cdots, n) \tag{1-1}$$
When there are n points, can be expressed as a matrix style
$$V = BX - Z \tag{1-2}$$
In the formula (1-2), $\boldsymbol{V} = (V_1, V_2, \cdots, V_N)^{\mathrm{T}}$
$$\boldsymbol{X} = (a_0, a_1, \cdots, a_m)^{\mathrm{T}}$$
$$\boldsymbol{Z} = (Z_0, Z_1, \cdots, Z_m)^{\mathrm{T}}$$
$$\boldsymbol{B} = \begin{bmatrix} 1 & x_{11} & x_{12} & \cdots & x_{1m} \\ 1 & x_{21} & x_{22} & \cdots & x_{2m} \\ \vdots & \vdots & \vdots & \ddots & \vdots \\ 1 & x_{n1} & x_{22} & \cdots & x_{nm} \end{bmatrix}$$

According to the principle of least squares, the $\boldsymbol{V}^{\mathrm{T}} V = \min$, make $\dfrac{2\boldsymbol{V}^{\mathrm{T}} \mathrm{d}V}{\mathrm{d}X}$, then
$$\boldsymbol{V}^{\mathrm{T}} B = 0 \text{ or } \boldsymbol{B}^{\mathrm{T}} V = 0$$
$$\boldsymbol{B}^{\mathrm{T}} (BX - Z) = 0$$
And finally
$$X = (\boldsymbol{B}^{\mathrm{T}} B)^{-1} \boldsymbol{B}^{\mathrm{T}} Z \tag{1-3}$$

Formula (1-3) for the least squares solution, the biggest drawback is a relatively poor estimation of gross error in the data point of observation errors that can produce a great influence on the results. So the selection is not sensitive to gross error Robust estimation to establish investigation statistics space model.

Fixed power, in order to avoid because the $v = 0$ and the calculation problem, preferable
$$P_i = \frac{1}{|V_i| + C} \tag{1-4}$$

Above all a normal least squares estimation steps are:

(1) List the error equation;

(2) Make $P_1 = P_2 = \cdots = P_N = 1$, composition equation $\boldsymbol{B}^{\mathrm{T}} B_x = \boldsymbol{B}^{\mathrm{T}} Z$;

(3) Calculation of X and correct V;

(4) Calculate the weight function $P_1, P_2, \cdots, P_n$;

(5) Equation method $\boldsymbol{B}^T PBX = \boldsymbol{B}^T PZ$; $(\boldsymbol{B}^T PB)^{-1}(\boldsymbol{B}^T PZ)$;

(6) To calculate X and V, then the weight function P;

(7) Ditto step iteration, until the difference between the two iterative weight function, is less than the limit, X finally obtained the results of Robust estimation for it.

Robust estimation algorithms have Huber estimation method, Denmark, Wen-Jiang zhou and De-ren li method, each method has its own Robust estimation function.

The unit weight means square value as follows:

$$\sigma_0 = \frac{V^T pv}{n-t} = \frac{\sum_{i=1}^{N} |p_i v_i|}{n-t} \tag{1-5}$$

The unknown association factor matrix for:

$$\sigma_{xy} = (\boldsymbol{B}^T PB)^{1-1} \tag{1-6}$$

With unknown function phi delta $\varphi = \boldsymbol{f}^T \delta x + f_0$, the coordinated factor matrix for unknowns:

$$\theta_{\varphi\varphi} = \boldsymbol{f}^T Q_{xx} \boldsymbol{f} \tag{1-7}$$

Mean square error of unknown parameter functions are as follows:

$$\delta_\varphi = \delta_0 p \sqrt{Q_{\varphi\varphi}} \tag{1-8}$$

1.2.3 Example

In GPS RTD positioning, the seven epochs of the x coordinate are 3, 8, 5, 6, 2, 4, 5, and the known mantissa is 5. Made for some random disturbance effects (such as radio communication, mountain, forest canopy) the last epoch observations mantissa jump from 5 to 10, 15, one can use the least square method and minimum by Robust estimation results are shown in Table 1-1.

Table 1-1 Least-square method with Robust estimation of GPS data comparison

The last mantissa	Least square			Robust estimation		
	X	σ_x	σ_X	X	σ_x	σ_X
5	4.7	±1.2	±0.5	5.0	±1.3	±0.7
10	5.4	±1.9	±0.7	5.0	±1.6	±0.9
15	6.1	±3.6	±1.4	5.0	±1.8	±1.0

Note: X is the estimated value; σ_x is the mean square deviation; σ_X is the mean square error.

Table 1-1 shows:

(1) When the observed value does not contain gross errors, the two estimation methods are close to the results of least square estimation and Robust estimation. Robust estimation is closer to the true value and the least square estimation accuracy is also high;

(2) When the observations with gross error using the least squares estimation result is poor and precision is low, but the on the other hand, Robust estimation results is very good and high precision.

1.3 Genetic algorithm

Genetic algorithm is a simulation Darwin the evolution natural selection and genetic mechanism of biological evolution process calculation model is a kind of searching optimal solution by simulating the natural evolution process, in the university of Michigan professor J. HOll and first proposed and published an influential book 'The Adaptation In Natural And Artificial Systems' in 1975, the GA name became familiar which is simply Genetic Algorithm (SGA).

1.3.1 Basic concept

Genetic algorithm (based algorithm) is a kind of reference to the evolution rule of biology (survival of the fittest and elimination of the weakest genetic mechanisms) evolved a random search method. Professor J. Holland of the United States first proposed in 1975, the main characteristic is a direct operation to a structural object with no the limitation of function continuity or function derivativeness; Have implicit parallelism and better global searching ability. Using probabilistic optimization method, it can automatically acquire and guide the search space of optimization, and adjust the search direction adaptively, which does not need to be determined. The properties of the genetic algorithm have been widely used in combinatorial optimization, machine learning, signal processing, adaptive control and artificial life, etc. It is the key in modern technology of intelligent computing.

For a function or a maximum optimization problem (a function of the minimum or similar), generally can be described as the following mathematical programming model:

$$\max f(x) \tag{1-9}$$

$$x \in R \tag{1-10}$$

$$R \subset U \tag{1-11}$$

Genetic algorithms as decision variables, x type 1-4 as the objective function type 1 to 5, 1-6 as constraint conditions, U is the basic space, R is a subset of U. X is called the feasible solution that satisfies the constraint conditions of solution set R all meet the constraint conditions of solution set known as the feasible solution set.

The genetic algorithm in the field of computer science and artificial intelligence is used to solve the optimization of a heuristic search algorithm and is a kind of evolutionary algorithms. This heuristic is often used to generate useful solutions to optimization and search problems. Evolutionary algorithm originally borrowed some phenomenon in evolutionary biology and developed these phenomena including heredity, mutation, natural selection and hybridization, etc. Genetic algorithm (GA) in the case of wrong selecting fitness function is likely to converge to a local optimum, and cannot achieve the global optimal (Figure 1-1).

Basic operation process of genetic algorithm is as follows:

(1) Initialization: Set the evolution algebra counter $t = 0$, t set the maximum evolution algebra, randomly generated M individuals as the initial group $P(0)$.

(2) Individual evaluation: Calculate the fitness of the individuals in the group $P(t)$.

(3) Choose operation: Choosing operation is to choose operator acting on the group. The purpose of choosing the optimum individual inheritance directly to the next generation or through matching crosses to produce new individual inheritance for future generations. The selected operation is established on the basis of evaluating the fitness of individuals in the group.

(4) Crossover operation: Crossover operation is the crossover operator acting on the group. Crossover refers to the two parent individuals to replace part of the structure reorganization and generating new individuals. Genetic algorithm (GA) plays a role as the core is the crossover operator.

(5) Operation: Operation is the variation to mutation operator ACTS on groups. Is it a string to a group of individuals of certain genes on value changes? Group $P(t)$ after selection, crossover and mutation operation for the next generation of group $P(t_1)$.

(6) Termination conditions: If $t = t$, are given in the evolutionary process of mutated individuals the optimal solution has the largest output termination of calculation.

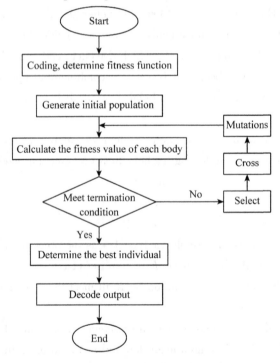

Figure 1-1 Genetic algorithm steps

1.3.2 Definition

The genetic algorithm starts with a population that represents a potential solution set and a population consists of a number of individuals that are encoded by a gene. Each individual is actually an entity with a characteristic chromosome. Chromosome as the main carrier of genetic material is a collection of multiple genes, their internal performance (genotype) is a combination of genes, and it determines the shape of the individual external performance. As the characteristics of the black hair are controlled by the chromosome in certain genes determine the characteristics. Therefore, in the beginning, it is needed to realize the mapping from the expression to the genotype, which is the coding work. Because of the fact that modeled gene coding is very complex, we tend to simplify such as binary code. After the generation of the initial population according to the principle ' survival of the fittest and elimination of the weakest' generational evolution produce a better approximate solution. In every generation, according to the problem domain of individual adaptation degree of individual choice and with the help of in nature genetics, genetic operators of a combination of crossover and mutation produce represents a new solution set of the population. This process will result in the population as a natural evolution of epigenetic population than the previous generation which is more adapted to the environment the last the best individual in the population after decoding can be used as an approximately optimal solution.

1.3.3 Characteristics

Genetic algorithm is a general algorithm to solve the problem of searching and can be used for a variety of common problems. The common features of search algorithm are:

(1) Firstly, formed a set of candidate solutions.

(2) The fitness of these candidate solutions is estimated according to some adaptive conditions.

(3) According to the fitness to retain some candidate solutions, to give up the other candidate solutions.

(4) Some operations are carried out to retain the candidate solutions, and the new candidate solutions are generated.

In genetic algorithm the above several characteristics together in a special way based on the parallel search of chromosome grouped with a nature of speculation selection operation, switching operation and mutation operation. This particular combination differentiates between genetic algorithm with other searching algorithm.

The characteristics of the genetic algorithm have the following several aspects:

(1) Genetic algorithm (GA) starting from the solution set of a string search, rather than start with a single solution. This is the genetic algorithm with the traditional optimization algorithm of great difference. The traditional optimization algorithm the optimal solution from a single initial value iteration; easily into the local optimal solution, starting from the string set search genetic algorithm, the coverage of big conducive to overall merit.

(2) The genetic algorithm to deal with more than one in the group at the same time namely to evaluate multiple solutions in the search space, reduce the risk of trapped in local optimal solution, and the algorithm itself is easy to realize parallelization.

(3) The genetic algorithm basically without the knowledge of the search space or other auxiliary information and only use to assess individual fitness function value on the basis of the genetic operation. The fitness function is not affected by not only continuously differentiable constraints, and its domain can be arbitrarily set. This characteristic makes the application of genetic algorithm

greatly expanded.

(4) The genetic algorithm is not using deterministic rules but the probability of transition rules to guide the search direction.

(5) Genetic algorithm has the self-organizing, self-adaptive and self-learning habits. A genetic algorithm using the evolutionary process of organizing the information obtained from a search, fitness of the individual has a higher probability of survival and obtains more genetic structure to adapt to the environment.

1.4 Artificial neural network

Artificial neural network referred to as the neural network (network neural abbreviated NN) is a kind of imitation of biological neural network structure; it is the function of the mathematical model or computational model. Neural networks are calculated by a large number of artificial neural networks. In most cases, the artificial neural network can change the internal structure based on the external information, which is a kind of adaptive system. The modern neural network is a nonlinear statistical data modeling tool, which is used to model the complex relationship between input and output or to explore the data of the model.

The neural network is a kind of operation model which consists of a large number of nodes (or 'neurons', or 'units') and the interaction between them. Each node represents a specific output function called the function activation. The connection between every two nodes represents a weighted value for the connection signal called the weight, which is equivalent to the memory of the artificial neural network. The output of the network is different from the connection mode the weight value and the incentive function of the network. The network itself is usually an approximation to the nature of some kind of algorithm or function or the expression of a logical strategy.

Its architecture is inspired by the functioning of the neural network. Artificial neural network is usually through a learning method based on mathematical statistics type (learning method) can be optimized. A practical application of artificial neural network through the application of mathematical statistics to make the decision problem of artificial sense, i.e. through statistics, the artificial neural network has the ability to make decisions as to the similar simple and easy judgment, the method of logic reasoning than formal calculus has more advantages (Figure 1-2).

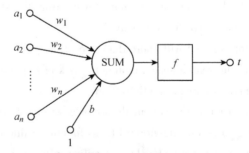

Figure 1-2 Neurons

1.4.1 Neurons

$a_1 \sim a_n$ is for each component of the input vector;

$w_1 \sim w_n$ is the weight of each synapse;

b is for bias;

f is a transfer function usually a nonlinear function. The followingdefaults to hardlim ( );

t is for neuron output.

Mathematical representation: $t = f(WA' + b)$

w as weight vector;

a as the input vector $A'A$ vector transpose;

b is for bias;

f as a transfer function visible, the function of a neuron is the inner product of the input vector and the weight vector is obtained, by a nonlinear transfer function obtained a scalar result.

The role of individual neurons to an n-dimensional vector space with ahyperplane splitting into two parts (called the decision boundary) given an input vector, neurons can judge this vector in ultra which side of the plane.

The hyperplane equation:

$$Wp + b = 0$$

In the above formula, w is weight vector, b is bias, p is vector on the hyperplane.

1.4.2 Basic feature

Artificial neural network is a nonlinear and adaptive information processing system which is composed of a large number of processing units. The artificial neural network put forward research results on the basis of the modern neuro science and tries to deal with the information of the processing of the neural network brain and the way of memorizing the information. Artificial neural network has four basic characteristics:

(1) The non-linear relation is a general characteristic of nature. The brain's intelligence is a nonlinear phenomenon. The artificial neuron is in two different states of activation or inhibition, which is a non-linear relation in mathematics. A network with a threshold value of neurons has better performance and can improve fault tolerance and storage capacity.

(2) A neural network is usually composed of a number of neurons. The overall behavior of a system depends not only on the characteristics of a single neuron but may also be determined by the interaction between the elements and the connection. The non-limitation of the brain is stimulated by a large number of connections between the units. Associative memory is a typical example of non-limitation.

(3) Artificial neural network qualitative with adaptive self-organization, self-learning ability. Artificial neural network not only can deal with the information that has a variety of changes but also with the processing of information. The non-linear dynamic system itself is constantly changing. The iterative process is often used to describe the evolution of dynamic systems.

(4) The evolution of a system with a non-convexity under certain conditions will depend on a particular state function i. e. the energy function which is corresponding to the extreme value of the system is relatively in a stable state. Non-convexity is the function of this kind of extreme value, so the system has more than one stable equilibrium state, which will lead to the diversity of the system evolution.

1.4.3 Characteristics and advantages

The characteristics and advantages of artificial neural networks are mainly manifested in three aspects:

First, it has self-learning function. For example, the realization of image recognition, only the first of many different image templates and corresponding to the results of the input of the artificial neural network. The network will learn through the function, slowly learn to identify similar images. Self-learning features are particularly important for prediction. Expected future artificial neural network computer will provide an economic forecast, market forecast, benefit forecast, its application prospect is very big.

Secondly is the associative memory function. This is the kind of association that can be realized by the feedback network of artificial neural network.

Third is with the ability to find the optimal solution. Looking for the optimization of a complex problem solution often requires a large amount of calculation. By using a designed to solve a problem feedback, the artificial neural network has computer high operation ability, may quickly find the optimal solution.

1.5 Internet of Things

The Internet of Things is an important part of the new generation of information technology. Its English name is 'The Internet of Things'. Thus, as the name suggests has two meanings:

First, the core of the Internet of Things and the foundation is still the Internet and is the extension and expansion of the network based on the internet.

Second, its user extends any item and carries on the information exchange and the communication. Internet of Things through intelligent sensing, recognition technology, and pervasive computing becomes the world's third wave of information industry development.

1.5.1 Definition

In 1999 initially the definition of the Internet of Things was the radio frequency identification (RFID), infrared sensors, global positioning system, laser scanner, gas sensors, etc. Information sensing equipment or anything connected to the Internet for information exchange and communication in order to realize the intelligent recognition, positioning, tracking and monitoring and management of a network.

A report released by the International Telecommunication Union (ITU) defined the Internet of

Things as follows: Internet of Things is based on two-dimensional code reading device, radio frequency identification (RFID) device, infrared sensors, global positioning system and laser scanner information sensing equipment. According to the agreement connected anything with the Internet of Things, exchange the information and communication in order to achieve intelligent identification, positioning, tracking, monitoring and management of a network.

1.5.2 Key Technology

There are three key technologies in the Internet of Things:

(1) Sensor technology. Which is also the key technology of computer application? As we all know so far most of the computer processing is by digital signals. So the computer has been required to convert analog signals into digital signals so that the computer data can be processed.

(2) RFID tagsensor technology. RFID technology is a combination of wireless radio frequency technology and embedded technology as one of the integrated technology. RFID in automatic identification, goods logistics management has a broad application prospects.

(3) Embedded system technology. It is a complex technology, which combines computer hardware and software, sensor technology, integrated circuit technology and electronic application technology. After many decades of evolution, intelligent terminal products characterized by embedded system can be seen everywhere. An embedded system is changing people's lives, promoting the development of industrial production and national defense industry.

1.5.3 Forestry application

1) The application of the Internet of Things also helpful in the management of old and famous trees

Rare and valuable trees are precious resources of society but with the rapid expansion of the scale of the city and with growing environment the ancient and famous trees have been destroyed in varying degrees. The traditional protection and conservation model of old and famous trees have been unable to meet the needs of modern urban development and planning. The need is for the more accurate tree information real-time access program. With the introduction of the Internet of Things the old trees for information implants. Through the management of management information systems, we can find the problem in a timely manner at the same time through the analysis of some growth data, for pest control, anti-theft management of old trees. Compared with the usual artificial maintenance, there is a precise record of the maintenance of old trees, such as irrigation, fertilization and so on.

2) The application of the Internet of Things in the management of wood tracking

Countries for the management of forest trees tend to more stringent and more accurate in direction. However, the forestry management department of our country for timber management is currently using the visual identification of trees mainly through the identification code. Counting the trees manually have a lot of errors, it is difficult to trace and unable to maintain a complete record of the entire timber supply chain. That also brings some available room for illegal logging. The introduction of the Internet of Things technology brings the development in the wood tracking management system, according to the RFID identification technology to carry out the source of wood and related aspects of the tracking management.

By using the Internet of Things a large number of sensors capture seedling growth process, temperature, humidity, light, soil and other data required for the analysis of seedling growth, take the linkage control. Through the intervention of the light, temperature, humidity, artificial regulation of seedling growth environment will affect the efficient and practical production.

3) The application of Internet of Things in forest fire prevention

The traditional forest fire monitoring system mainly uses the front camera acquisition system for forest fire information and the video capture module uninterrupted accept camera system for video data and stored data in the server. The video decoding module uses the video information collected by the video capture module and converts the video information to a predetermined format through the video decoding algorithm for the identification of the fire image. The short comings of this monitoring method are obvious, that take up a lot of bandwidth in the transmission of information. Also have low efficiency of fire image recognition, video decoding, poor resolution of fog, steam and other interference, the level of automation and intelligence of the forest fire warning low, can afford large area applications.

The Internet of Things mainly helps functionally in prevention, monitoring, and forest firefighting. The Internet of Things can construct the temporary, unexpected basic information gathering environment of the emergency response system. This established the early warning system for the information perception ability of the complex environment and the unexpected events through the wireless sensor. The use of Internet of Things technology can be used in the monitoring area of smoke, high temperature and so on. Sensors will be used to spread the information through the wireless network and feedback to the monitoring center. The monitoring center according to the information received will determine whether there is a fire and through a variety of ways to inform the monitoring personnel. At the same time, the network with GPS and GPRS communication module of the multi-mode mobile information acquisition terminal provide the whole network node localization and firefighting personnel real-time positioning tracking. Combined with GIS, fused the site dynamic information and emergency response database with all kinds of model base information.

1.6　Cloud Computing

Cloud Computing (Computing Cloud) is based on the internet related services, i. e. the use and delivery mode that usually involving the Internet to provide dynamic, easy to expand and often virtual resources. Cloud is a metaphor for the internet. In the past, the map is often used to represent the telecommunication network cloud and later also used to represent the underlying infrastructure of the internet and the abstract. Narrow cloud computing IT infrastructure refers to the delivery and usage patterns through the network to obtain the required resources according to the needs. Generalized cloud computing service refers to the delivery and usage patterns to through the network to obtain the required service according to the needs easy to expand. This service can be IT, software, and internet-related but also includes other services. It means that computing power can also be used as a commodity through the internet.

1.6.1 Profile

1) Concept

Cloud Computing is a pattern of pay per use. This pattern provides available, convenient on-demand network access. Enter the configurable computing resources shared pool (resources included network, servers, storage, applications, service) these resources can be quickly provided for just put a few management work and service providers with little interaction.

Cloud Computing is often confused with grid computing, utility computing, and autonomic computing. Grid Computing is a distributed computing composed of a group of loosely coupled computers and super virtual computer commonly used to perform some of the major tasks. Utility computing is a package of IT resources and billing methods, such as in accordance with the calculation, storage costs, such as the traditional power and other public facilities.

Autonomic computing is a computer system with self-management function. As a matter of fact, many Cloud Computing deployments depend on the computer cluster (with the grid composition, the architecture, the purpose and the working ways), but also absorb the characteristics of autonomous computing and utility computing.

2) Background

Cloud Computing includes a series of dynamic update and virtualization of resources. These all resources provide Cloud Computing users to share and conveniently through internet access and users do not need to master the Cloud Computing technology but only require individuals or groups to lease Cloud Computing resources. Cloud Computing is another great game changer after the big change of the client server in the 1980s. Cloud Computing is as early as in the 1960s, McCarthy raises the computing power just like water and electricity utilities to users concept, which became the origin of the idea of Cloud Computing. At the beginning of the 21st-century, virtualization technology, SOA, SaaS application support, Cloud Computing as a new resource use and delivery mode is gradually recognized by the academia and the industry. China Cloud Development and Innovation Industry Alliance evaluation of Cloud Computing now become as the innovation of business model in the information age.

Following the personal computer revolution, the internet revolution, cloud computing is seen as the third wave of IT and is an important part of China's strategic emerging industries. It will bring about the fundamental changes in the way of life, production and business model, Cloud Computing will become the focus of attention of the whole society.

Cloud Computing is a traditional computing technology, such as Distributed Computing, Parallel Computing, Utility Computing, Network Storage Technologies, Virtualization, Load Balance other traditional computer and network technology development integration of the product.

1.6.2 Cloud Computing

By making computing distributed in a large number of distributed computers rather than the local computer or remote server, the operation of the enterprise data center will be more similar to the internet. This allows companies to switch to the needs of the application, access to the computer and storage system based on demand like from the old single generator mode turned to the power plant centralized power supply mode. It means that computing power can also be used as a commodity to flow, like gas, water, and electricity but it is more convenient and low cost. The biggest difference is that it is transmitted through the internet.

Cloud Computing services on the internet and the nature of the cloud in water cycle has a certain similarity. Therefore, the 'cloud' is a very appropriate analogy. According to the definition of the National Institute of Standards and Technology Cloud Computing services should have the following characteristics:

1) Dynamic allocation of resources

Partition or release of different physical and virtual resources is according to consumer dynamic demand. When there is an increase in demand, the availability of resources can be increased by matching to achieve the rapid flexibility of resources available. Cloud Computing provides customers with the infinite capability to achieve the scalability of the IT resource utilization.

2) Demand services self-service

Cloud Computing is to provide customers with self-service resources, users do not need to interact with the provider can automatically get the ability to self-service computing resources. At the same time, the cloud system provides customers with a certain application services directory; customers can choose to use self-service options to meet their own needs of the service project and content.

3) Using network as the center

Cloud Computing components and the overall structure of the framework connected by the network and exist in the network, while providing services to the user through the network and customers can use different terminal equipment, through the application of the standard network access so that the cloud computing services everywhere.

4) Services can be quantified

In the process of providing cloud services different types of services for customers through the measurement of the method to automatically control and optimize the allocation of resources. The use of resources can be monitored and controlled, is a kind of pay for the service mode.

5) Pool and transparency of resources

For providers of cloud services a variety of underlying resources (Computational, storage, network, resource, logic, etc) some heterogeneity is shielded i.e. the boundary is broken, all resources can be unified and becomes the so-called 'resource pool' for users provide on-demand service to the user, these resources are transparent, infinite, users do not need to understand the internal structure, only care about their own needs are to be met.

1.6.3 Cloud Computing forestry applications

The following analysis and discussion include several typical applications of intelligent forestry Cloud Computing platform:

1) Massive data storage

Forest resource data storage is one of the most important and the most complicated part forest resources management, due to a large amount of data, higher degree of data dispersion characteristic making forest resources data storage is a key process in the forestry informatization. Forest resource data including the following monitoring data of forest resources like wetland and biodiversity data, desertification and sandy land monitoring data, data of forest fire prevention, comprehensive afforestation data, wild animal epidemic disease data, forestry harmful biological data. Forestry industry and economic operation number according to the enormous amount of information and its data type also includes attributes, documents, multimedia, statistical data and other types. This makes a great challenge to construct the data center of forest resources for the efficient management of forest resource data. In Cloud Computing storage applications, a large number of different types of storage devices in the network through the application of software together to work together to provide data storage and business functions. This is not limited by the specific geographical location, super strong scalability, high reliability and other features.

2) Provides the basis for intelligent decision through a reliable scheduling policy

One of the ultimate goals of forestry information is to provide people with the wisdom of the decision-making system to address the impact of human factors on the decision-making of forestry. Cloud Computing in the IaaS layer to provide a reliable scheduling strategy and is the key to the wisdom of Cloud Computing. Forestry tables model (including analysis forecast, forestry resource surveys, forest resources prediction model, analysis evaluation model and forestry resources assessment of disaster early warning model) using efficient and reliable Cloud Computing platform analysis and decision-making is the key to reliable scheduling in Cloud Computing. Efficient and reliable scheduling strategy can greatly improve the computational efficiency, reduce the computational cost, and provide the basis for the intelligent forestry decision.

3) Data mining of remote sensing satellite data

China's remote sensing satellite business is developing rapidly. Every day it produces huge amounts of remote sensing information. How to quickly deal with this massive data information for people's need to be able to use in different industries. Now it becomes an increasingly urgent demand. Serious ecological problems to a new period of forestry development like the demand for monitoring and evaluation of resources and without affecting the environment. There is a lack of information for describing the forest vertical structure and forest biomass. A precise calculation of the forest service resources and carbon sequestration data is required. Using the parallel processing technology of Cloud Computing to mine the intrinsic correlation of remote sensing data, laser radar data, forest parameters data and the continuous data of multi-spectrum are analyzed and processed in parallel. Using parallel processing technology of Cloud Computing can greatly reduce the data processing time.

References

胡传双,王婷,云虹,等,2009.工厂预制木结构建筑在中国的发展现状及展望[J].木材加工机械,20(S1):56-61,51.

胡勇庆,钱俊,2010.木材加工剩余板皮高效利用分析与展望[J].木材加工机械,21(03):49-51,18.

黄杰,赵京音,万常照,2008.RFID在农业中的应用与展望[J].农业网络信息(09):119-121.

刘姗姗,张绍文,2008.基于RFID技术的林木种质资源管理信息系统的开发构想[J].林业建设(05):27-31.

陆研,张绍文,2008.基于RFID技术的名木古树管理系统初探[J].山东林业科技(02):91-94.

孙琴,乔牡丹,沈洪霞,等,2009.林业信息化建设研究初探[J].内蒙古林业调查设计,32(06):88-89,102.

汤晓华,马岩,陈强,等,2005.圆与定压缩系数椭圆截面三点初步定心[J].木材加工机械(02):6-9.

王颖,周铁军,李阳,2010.物联网技术在林业信息化中的应用前景[J].湖北农业科学,49(10):2601-2604.

CHAPTER 2　Technical Foundation of Forest Management (Focus on General Principle)

The field of traditional tree measuring including visual estimation, angle gauge measuring, log scale technique, all these techniques are backward the measured data precision is low. With the rapid development of modern science and technology, more and more new technologies and equipment used in forest measurement. The use of these devices significantly improves the accuracy of the data which provide people a better grasp of the forest like stock, growth trends, biomass, volume, and other basic forestry data provides a solid database.

2.1　Principle of electronic theodolite / total station /CCD superstation

2.1.1　Introduction tototal station instrument

1) Total station overview

A total station is all-station electronic measuring instrument, (electronic total station) is a combination of optical, mechanical and electronic parts. A total station is one of the high-tech measuring instruments like the horizontal angle, vertical angles, distance (inclined, flat margins), and altitude difference measurement function of surveying instrument system. Due to the placement of the instrument in survey station, it is called total stations. Characteristics of total station instrument are multi-function, high efficiency and can be used in almost all areas of measurement. Total stations are applied during an angle measurement which plays a huge role in the surveying and mapping work.

2) Classification of the total station instrument

Total station adopts photoelectric scanning angle measuring system. Main types of the total station are encoding angle measuring system, raster angle measuring system and dynamic (raster) angle measuring system.

Structure composition of total station instrument is divided into combined (modular) total station instrument (ranging unit and electronic longitude instrument can both combined and separated by cable and interface device connection) and overall type (integrated) total station instrument.

Total station instrument by functions are divided into the general total station (used for the calculation of angle, distance, coordinates, height difference), intelligent total station (having software system and tools with built-in or expandable software, with automatic compensation device), automatic tracking total station and so on. More than last 10 years, with the development of technology manufacturing, microelectronics, and computer technology, the major manufacturers of measuring instruments in the world produces most of the total station instrument belongs to a new generation of the integrated intelligent total station.

3) The basic components of total station

The basic components of the electronic total station are the angle measuring systems, ranging power systems, data processing, a communication interface, display, keyboard and other components. The electronic total station is itself a computer control system with special features. Its computer processing unit is having the microprocessor, memory, input and output components. Inclined distance, horizontal angles, vertical angles, vertical axis tilt error of vertical dial, sight axis error, index error, prism constant, temperature, barometric pressure and some other information are dealt by the microprocessor, resulting in the correction of observational data and calculation of data. The measurement procedure is cured in the read-only memory of the instrument and the measuring procedure is completed by the program. Instrument design framework is shown in Figure 2-1.

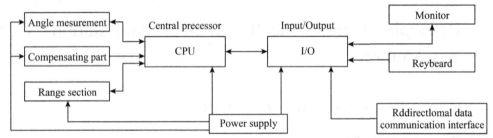

Figure 2-1 Total station design framework

Among:

(1) Power supply parts are rechargeable batteries, supply power to each part;

(2) Angular parts for electronic theodolite and can measure horizontal angles, vertical angles, set the azimuth;

(3) Part can be compensated instrument vertical axis tilt error of horizontal and vertical angle measurement of automatic compensation to correct;

(4) Locating parts for electro-optical distance measurement instrument can measure the distance between the two points;

(5) Accept input commands, controls a variety of observations to the CPU operating mode, data processing;

(6) Input and output including keyboard, screen, bi-directional data communication interface.

From an overall perspective, the composition of the total station instrument can be divided into two parts: one is to collect data and setting of equipment, mainly electronic angle measuring system, electronic distance measuring systems, and data storage systems. Secondly, the measurement process control equipment, mainly for the orderly realization of special equipment for each of these functions, includes a connection to the measured data peripherals and calculation, the microprocessor of the directive. The above two combinations of the total station can truly reflect both features of automated data collection and automatic processing of data, thus controls the whole measuring process.

4) The main features of computer total station

Computer total station is also known as intelligent type total station instrument, it has many features like double axis tilt compensation device, bilateral main and attached display, two-way transmission communications, large capacity of the memory or magnetic card and electronic record book and machine software. Thus measurement speed is fast and observation precision is high and operation is simple. This also has an advantage of its applicability in a large area, performance is stable. This instrument is deeply welcomed by general mapping technicians and became development direction of total station instrument mainstream since 1993.

Development of total station in early time was embodied in hardware devices only and thus had low quality and volume; medium-term development embodied in the software functions, such as conversion, automatic compensation of horizontal distance correction, additive constant, and multiplication constant correction. Today's development is comprehensive, fully automated and intelligent.

Therefore development status and prospects of the total station are moving towards automatic, intelligent and multifunctional. It will play an increasingly important role in topographic surveying, engineering surveying, industrial measurement, and construction surveying and deformation observations fields.

Development of total station has the following features:

(1) System of instrument: Total station from the 20th century, began to appear in the late 60s shows its systemic. Such as RegElta-14 from ZEISS in Germany and Geodimeter 700 in AGA in Sweden. They are equipped with recording and printing peripherals, so the total station is equipped with an RS-232C standard serial port for data output. Currently, this standard serial of development application not only can transform data from the instrument to record device i. e. electronic record book or electronic flat, that achieved data flows in one-way, and can enter data or program from computer to instrument. So as to update software on the instrument, even through the connection of computer and instrument, see the instrument as terminal, instrument real-time control operation by the program in a computer, that achieved data on two-way flow. A total station is no longer a single survey instrument but it is the combination of computer, software and communications equipment (such as telephones, fax machines, modems, etc) to form an intelligent mapping system.

(2) Double spindle automatic compensation correction: The instrumental error affects the measurement precision and is primarily from improper three-axis relationship of the instrument. In optical theodolite, it is mainly based on the relationship between Triaxial test corrections that reduce instrument error impact on angle measurement. In electronic instruments, it is mainly achieved through so-called 'three-axis-compensation'. Latest total station instrument has achieved 'three axis compensation' function (compensation device of effective work range General for ± 3'), that is the installation of compensation device in total station automatically detects errors due

to instrument vertical axis tilt and corrects the measuring angle. The error value is calculated out through detection results of instrument associate axis errors and horizontal errors, if necessary, observation angle is corrected by program built-instrument.

(3) Real-time auto tracking, processing and receiving computer control: In new structure of total station, the instrument is installed with the horizontally and vertically 360 telescope drive. This type of total station instrument can achieve unattended observation, automatic detection, automatic layout, three-axis error detecting, automatic search and track targets. Therefore, it will have an excellent advantage in dynamic positioning and application in some harmful environment.

(4) Easy to operate, functional: The development of total station makes it easy to operate and functional. As the instrument operation menu is often used English and in China, it is complicated to operate. Total-station instrument addresses this issue by using pictograms or mnemonics to help understand or use similar to windows-style interface to provide online help. In fact, it is not impossible to have the Chinese language. As the total station have LCD screen, some instruments provide a menu in several different languages, such as English, Japanese, French, German, and so on), although the number of the dot-matrix lines used in Chinese will be more, but the 'scroll bar' can solve this problem. Total station in full Chinese menu is not far away.

(5) Built-in more programs and standardization: In recent years a very important feature of the total station is the built-in procedures for the development of programs and increased standardization. The built-in program could provide observations in real time and calculate the final result. If the observer set the instrument in observation function and follow the steps correctly the observer will finish the survey work. But no program total stations can only provide observations value and values calculated by the observation value. In other words, the indoor work is directly finished in the industry through the program, the implementation process of the program is actually the implementation process of the instrument operation. At present, the equipment manufacturers have built-in practical programs i. e. degree positioning disc, lofting and coordinate measurement.

(6) Opening environment, users secondary development features: The biggest feature of the open environment is inclusive and flexible enough to meet different requirements on different occasions. With the progress of science and technology, that opens the environments throughout the entire development and applications. Users can only passively receive total station instrument functions provided. If some special requirements of the work are needed then the user can only use some alternative methods, instrumentation cannot take the initiative to command. Conditions in an open environment, users will be able to participate in the secondary development of instrument functions, which makes it really become instruments of the 'minds', the instrument in accordance with the wishes of the work of people.

(7) Instrument compatibility and standardization: Taking into account the interests of user's compatibility is essential. The compatible basis was first completed in the computer industry by IBM corporation, so that makes the rapid development in the computer industry. In the field of the electronic total station, users have already realized the drawbacks that are not compatible, such as the purchase of a set of instruments, after the years of use. If one of the key components is dam-

aged or have technical update problem due to incompatibilities between devices and other accessories in this instrument configuration cannot be used for other instruments, the user can only eliminate it as a whole. Currently, data logging equipment manufacturers are close to PCM-CIA card, but this is a small step in terms of compatibility. Taking into account the total station is a special industry of special instruments and the compatibility can only stay in the vision of manufacturer's own interests.

(8) Data achieve the ability to work together: Due to the instruments in real time operational requirements, 'industry' and 'fieldlization' is very necessary. In the past flow of work is from outdoor to indoor and to outdoor, now it is going be replaced by one-time field work. The improvement in the efficiency the instrument is based on data sharing of the instrument. This data sharing refers to the total station and other main types of instrument (such as a GPS receiver, a digital level) between data exchange. The exchange of data between different instruments can reduce the interface between the inner and outer raise the level of automation of measurement.

(9) High-precision: Accuracy is one of the most important parameters of the total station. The current highest precision total station, measuring accuracy is about ± 0.5, range accuracy ± (1mm+1μm). The emergence of the high precision instrument, resolve the issues relating to a series of a precise engineering survey, but in reality require higher precision in measurement engineering instruments to reduce the precision measurement of difficulty and workload, which is required for users, technology development requirements.

2.1.2 Electronic theodolite

1) Characteristics of electronic theodolite

Electronic theodolite is a combination of optical, mechanical and electronic technologies of the angle-measuring instrument. It can be designed into a complete instrument and electronic distance meter connected in blocks. Total station can also be designed as digital theodolite and electronic distance meter and combine to form an indivisible whole total station. But regardless of which form it can work independently.

2) Principle of electronic angle measurement

Electronic angle i. e. electronic, digital, automated angle measurement, the performance indirectly to the digital display angle measurement results. The essence is a set of angle conversion system to replace the traditional optical reading system, which is the use of photoelectric dial. The angular value of the scale into a photoelectric device can be recognized and received a specific signal, and then converted to the conventional angle value, thus achieving the reading of the digitization and automation.

Angle value and light signals of conversion are generally divided into two class: One is put degrees disc into the district, and ring for coding called coding degrees disc, it directly put angle conversion into II into the business code, so said absolute conversion system. Another is using grating degrees disc put units angle conversion into a pulse signal, then with computer cumulative changes of pulse number, obtained corresponding to angle value, so called incremental conversion system.

3) Compensator and electronic bubble

The compensator is generally divided into two types on the basis of working principle i. e., pendulum compensator and liquid compensator.

Pendulum compensator was more common in the early days of the electronic theodolite. The basic principle of pendulum compensator is during the tilt of instrument. The vertical displacement of the dial indicator cause the small swing of the pendulum and optical changes of vertical disc images, therefore point to the correct angle of the final output.

The liquid compensator is widely used by the most modern total station. The basic principle is that when the instrument is tilted, liquid tilt sensor mode cause beams of light with the displacement and displacement sensors in the array. Total station in the tilt of the microprocessor is calculated according to the size of the displacement of the instrument and caused by tilt correction and offered to the point output system.

Compensation is generally divided into two categories the single-axis compensation and two-axis compensation and three-axis compensation.

Single axis compensation: Single axis compensation is only vertical due to the vertical axis tilt disc reading errors.

Two-axis compensation: Can be caused due to the tilted vertical and horizontal disc reading errors.

Three-axis compensation: Compensation caused not only by the verticality of compensation caused by the vertical axis tilt and horizontal disc reading error but also due to the horizontal axis error and level of sight axis error and cause disc reading errors.

4) Compensation range of compensator

Compensation is a tilt compensator function can accept instrument levels i. e. when the tilt compensation within the scope of the instrument, will measure the amount of vertical and horizontal tilt. Compensator always limited the scope of the compensation; compensation for most of the total station instrument in the market range is generally about $3' \sim 5'$. Large compensation range allowing users with an increase in hostile environments (such as soft foundation or ground vibration belt) confidence in the operation and reliable results.

5) Compensation accuracy of compensator

Compensation accuracy compensator has two meanings: One is the amount of tilt measurement accuracy, it refers to compensation for the inclination measurement accuracy, usually the amount of tilt compensation accuracy is given in the user manual measurement accuracy. Other is vertical angle correction accuracy, which is the deviation in the vertical angle value within the range of the instrument nominal compensation effectively corrected vertical angle value of the instrument is in the correct position, the compensation accuracy of the calibration certificate is usually given vertical angle correction accuracy.

6) Electronic bubble

New total stations are equipped with electronic bubbles. Electronic bubbles are equipped with compensation display unit, it can directly display the instrument tilt.

The electronic bubble has two types, one is the digital type and other is the graphic type. In digital type the instrument is displaced directly on horizontal (x) and vertical (y) axis of tilt. Both are zero, for an instrument in the whole flat state; the another type is called graphics type, often with round points in a big circle in the of location to said, dang small round points are located in the big circle in the center an instrument for in a flat state.

In actual measurement when using electronic fast leveling bubble (simply leveled to compensate scope) or exact leveling, the instrument allows a user leveling the residual tilt compensated or not compensated. When the tilt is used to automatically correct the horizontal angle and vertical angle, even if the left or right half position measurement can obtain good accuracy, especially when large vertical angle automatically compensates for accuracy improvement is more obvious.

7) Three axis compensation

The old-fashioned single-axis compensator can only compensate for the tilt caused by the vertical axis of the vertical dial reading error, the biaxial compensator can be compensated by the vertical axis tilt caused by the vertical dial and horizontal dial reading error. Recently introduced the three-axis compensator not only compensates for errors in the vertical and horizontal dial readings caused by the tilt of the vertical axis, but also the reading errors of the horizontal dial due to the horizontal axis tilt and collimation error. In the triaxial compensation state, the factor causing the change in the horizontal dial reading (i. e., the factor used to correct the reading of the horizontal dial) is increased to three dimensions.

(1) Biaxial compensation: The so-called biaxial compensation is essentially a correction of the horizontal angle error caused by the vertical and horizontal tilt caused by the longitudinal inclination. For example, if the telescope is tilted up and down while the collimator is fixed, the horizontal dial reading will remain constant for uniaxial compensation, and the horizontal dial reading may be changing for biaxial compensation. It is on the horizontal tilt caused by the horizontal dial reading error correction.

(2) The horizontal axis error: The horizontal axis error, also known as the transverse axis error, generally denoted by means of its horizontal axis due to manufacturing, installation, and adjustment of imperfect cause ranging from the diameter of the horizontal axis or the support of the two brackets is not high due to tilt error. Due to the existence of a horizontal axis instrument error, when the whole instrument vertical and the horizontal axis is not level, this will lead to measurement error in the horizontal direction. If the horizontal axis tilt error of the horizontal effects of the observed readings, there are:

$$\Delta i = i \cdot \tan\alpha_T \tag{2-1}$$

Obviously, the size Δi proportional to the size and angle of only i, but also with the target vertical angle of about α_T.

(3) Collimation: Collimation also knew as the collimation error, this is is due to improper installation and adjustment of the telescope caused by the center of the crosshair from the correct position, resulting, collimation axis orthogonal to the horizontal axis. It is a fixed value; the outside

temperature changes can also cause changes in the collimation axis position, this change is not a fixed value. If the influence of the collimation error on the horizontal observation reading is:

$$\Delta c = c/\cos\alpha_T \tag{2-2}$$

Obviously, the effect of the collimation error on the horizontal reading is not only proportional to the collimation error, but also to the vertical angle of the target point. When the vertical angle is 0°, the collimation error is the same as the horizontal reading error caused by it. At this time can be left, right-hand side of the observed value of the difference to obtain, namely: $2c = L - R \pm 180°$.

(4) The basic idea of the three-axis compensation: The basic idea of the three-axis compensation is to compensate for the vertical axis tilt caused by the vertical and horizontal disc readings errors, the use of computer software to correct horizontal axis errors and collimation errors caused by the horizontal reading error.

Compensation for the uni-axial old total station, there is no correction function of the three, no matter how people see the rotating telescope, do not change the level of reading, it is not because the instrument is stable and reliable, in fact, because the instrument is not capable of this area corrections. For just a biaxial compensator total station, the only correct vertical and horizontal reading errors caused by the vertical axis is tilted so that when the compensator work, turn the telescope to change with the level readings; when the compensation is off after the telescope is rotated in any case, it does not change the level of reading.

2.1.3 CCD super-station instrument

1) Background

With the rapid development of photogrammetry and photographic measurement applications in different fields, the photogrammetric measurements and preliminary application of photogrammetry information is still limited to a single total station today, in engineering survey the presence of these limitations at different stages more and more obvious:

(1) Using traditional total station measurements only achieve individually for each point, heavy workload, and low efficiency;

(2) Using the traditional total station measurements, field measurements can only be saved text document, not easy after industry data during the inspection, verification;

(3) The use of engineering measurement mode of photogrammetry, the camera exterior orientation elements of photography is difficult to determine.

Therefore at present in the field of surveying and mapping and related fields, there are many limitations in total station measurements and preliminary application of photogrammetry. How to find a measurement method to overcome these deficiencies is particularly important.

2) Instrument Description

In order to overcome the deficiencies and limitations of existing measurement methods, make full use of precision measuring total stations and digital photogrammetry to their strengths, the purpose is to provide a new CCD photogrammetry total station equipment and new photogrammetric methods. It is achieved in particular:

(1) Firstly camera (CCD camera) calibration;

(2) According to the CCD camera field angle, determine the number of single-camera station photography;

(3) According to the CCD lens has a maximum depth of field length, taken between two stations to determine the baseline length B;

(4) Each camera site with a total station at least two feature points coordinates of the object side, and records relative to the outer bearing element CCD camera photography by total station readings repeat complete 360° panoramic photography;

(5) To set up the instrument on a coordinate system, which will determine by the upper observation point S1, repeat the above operation, complete panoramic photography and measurements and make relevant records;

(6) Repeat the above steps to complete the entire test measuring tape until it covers the entire area to be measured;

(7) Use within the industry supporting software, external industry photography and photo processing measurement data obtained the desired results.

The main purpose of instrument measurement processing software will achieve by using the VS2010 platform, C # language in the Net Framework 4.0 to achieve ground-chip, handle double-film measurement systems for analysis of forestry related field like even slopes there are chest high marking, no chest high marking for any terrain and other conditions (Figure 2-2).

Figure 2-2　CCD super-total station instrument

The measuring methods of the instrument have following advantages:

(1) In CCD total station engineering measurements point by point observations for each point with the total station are not required. Only two feature points are used to determined the coordinate and improve the efficiency;

(2) The combination of CCD and total station measurements retains the accurate measurement of total station and stores the image captured;

(3) The use of the combination of CCD and total station can use the advantages of both instruments and can determine exterior orientation element CCD camera photography by total station

measurements to facilitate post-calculation processing.

3) Field measurement principle and steps

To understand the basic use and operation of an ordinary total station we have to understand relevant knowledge of measurement (Figure 2-3):

(1) Explicit use of a CCD camera type and related parameters, rigorous camera calibration, obtained relative position of the camera center and the center of the total station can thereby determine the orientation of the elements within the CCD camera photography;

(2) According to CCD lens parameters, a clear field of view lens angle CCD size, according to the camera horizontal field angle is determined for each single base station Photography times $INT\left(\dfrac{360°}{\sigma°}\right)$;

(3) According to the maximum depth of field CCD camera Y max, to determine the best camera station baseline length B, but in actual production practice usually select $\left[\left(\dfrac{1}{10} \sim \dfrac{1}{20}\right)Y_{max}\right]$ Y_{max} as a camera stand baseline length is more appropriate;

(4) The above steps are working instrument factory before surveyors need to operate, only need to know the relevant parameter values;

(5) In the area tested strictly control point S_1 leveling combination instrument S_1 and known point coordinates relative to the object side (Xs_1, Ys_1, Zs_1). According to the baseline camera stands in their depth direction at first with a total station coordinate after orientation long B is determined as the other observation points on the object side S_2 coordinates (Xs_2, Ys_2, Zs_2), with the total station after accurately measure the object side relative coordinates measured object at least two feature points;

(6) Adjust the horizontal angle α in shooting and recording time and the vertical angle β, this time completing the photographing and measuring in one direction, then in front of the number determined in accordance with the photography n, each rotation corresponding horizontal angle

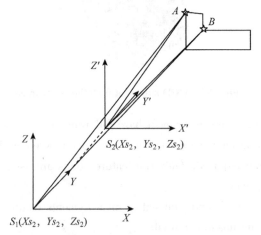

Figure 2-3 Measurement schematic

CHAPTER 2 Technical Foundation of Forest Management (Focus on General Principle) 59

$\left(\partial_o = \dfrac{360°}{n}\right)$, and then click the rotation n times, complete panorama panoramic photography and measurements;

(7) Combination instrument placed in the image of the other observation point S_2, and strict leveling rear-view observation point S_1 orientation, parameter setting and operating procedure was repeated observation point on the S_1 completed S_2 panoramic photography and panorama measuring point, corresponding records;

(8) Repeat the above steps, according to advance the design of measuring tape, the entire tape of the completion of the measurement, until it covers the entire area to be measured; and then, according to internal and external orientation element CCD camera photographing anti-collinear equation is calculated, using the corresponding record data into the collinear equation, like the film obtained at any point on the object side coordinates (X, Y, Z).

2.1.4 CCD superstition to be detected

In the use of CCD over-the-meter measurement of the target, if not in advance of comprehensive testing, a clear combination of the relationship between the two instrument center, the actual measurement is meaningless, so how to find a combination of instruments quickly and easily to achieve calibration technical methods is particularly important.

Detection steps:

(1) Shown in Figure 2-4, the production side length 20cm × 20cm square target, where the mid-point of a target cross hairs is characterized by the intersection point O, the four corners of the

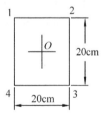

Figure 2-4   Schematic model of the target square

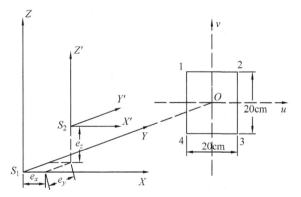

Figure 2-5   Showing the relationship between the coordinate system

target point are numbered by 1, 2, 3, 4 in a clockwise direction followed by the feature point.

(2) Select station and assume coordinates (0, 0, 0). At this station, place and adjust the CCD combination total station, first adjust the horizontal braking, then turn the total station vertical brake, keep the vertical angle of 90°, and aim at the square target, in which the characteristic point O of the square target is located at the center of the cross hair of the telescope.

(3) After aiming at the target point by the total station, take the CCD camera to take the square target and take the photograph.

(4) With a total station for four feature points as shown in Figure 2-4 on a square target relative coordinate measurement, measured relative coordinates of each feature point (X_i, Y_i, Z_i) ($i = 1, 2, 3, 4$).

(5) According to the principles of photogrammetry, the coordinates of the feature point on the image side of the front upright image by shooting on (u_i, v_i), the use of digital-analog

$$\begin{cases} (X'_i - X'_{i+1})^2 + (Y'_i - Y'_{i+1})^2 + (Z'_i - Z'_{i+1})^2 = a^2 \\ X'_i = \dfrac{Y'_i}{f} u_i \\ Y'_i = \dfrac{Y'_i}{f} v_i \end{cases}$$

($I = 1, 2, 3, 4$ and when $x = 4$, $i + 1 = 1$), again obtained four feature points corresponding to the object X_i', Y_i' Z_i' side relative coordinates ($i = 1, 2, 3, 4$), where, f is the focal length of the lens and the CCD is a known quantity.

(6) As shown in Figure 2-5, we know that there is a relationship between the rotation and offset total station coordinates and CCD camera coordinate system, to this end, using a mathematical model.

$$\begin{bmatrix} X_i \\ Y_i \\ Z_i \end{bmatrix} = \begin{bmatrix} X'_i \\ Y'_i \\ Z'_i \end{bmatrix} + \begin{bmatrix} 0 & -Z'_i & Y'_i \\ Z'_i & 0 & -X'_i \\ -Y'_i & X'_i & 0 \end{bmatrix} \begin{bmatrix} \varepsilon_x \\ \varepsilon_y \\ \varepsilon_z \end{bmatrix} + \begin{bmatrix} e_x \\ e_y \\ e_z \end{bmatrix}$$

Solver CCD camera center coordinates relative to the amount of displacement of the center of the total station (e_x, e_y, e_z) and ($\varepsilon_x, \varepsilon_y, \varepsilon_z$) the amount of rotation angle, relative to the total station namely CCD camera center stance offset parameter.

2.2 MINI total station / superstation instrument

2.2.1 MINI total station instrument

1) Hardware structure

Mini total station instrument hardware-structure including MEMS (Micro-Electro-Mechanical Systems) tilt sensor, laser ranging sensor, electronic compass, a central control unit, memory, LCD, micro keys, USB data communication interface and power supply, as shown in Figure 2-6, the hardware are highly dense, high reliability integration and according to the ergonomic design of the alu-

Figure 2-6　MINI total station instrument

minum alloy shell.

MEMS tilt sensor selection of three-axis acceleration sensor LIS331DLH design used to measure the tilt angle between the MINI total station and the measuring point. Laser ranging sensor based on an integrated circuit chip LMC6482, PL613, PL673 design for mini total station instrument to measure point distance, electronic compass used integrated circuit chip GY-26 for measuring point of the magnetic azimuth. Based on three measuring parameters the tilt angle, distance and the magnetic azimuth and the use of trigonometric function relationship between tree height, stand average height, density, and arbitrary diameter indirect measurement can be realized. The central control unit uses the single chip microcomputer C8051F410, mainly completes each kind of sensor signal acquisition, processing, and the output. LCD screen using the integrated circuit GY1606A4FSW6Q used to display the measurement parameters. USB data communication interface using the integrated circuit chip LPC2148 design used to measure the data export. C8051F410 chip flash used to store measurement data. The power supply module uses the integrated circuit TPS61020 design, which is used to supply power to the device. The keyboard is used for the operation of a variety of design functions. The rotatable mechanical paddles, rotating paddles can achieve 4 ratio selections(Figure 2-7 and Figure 2-8).

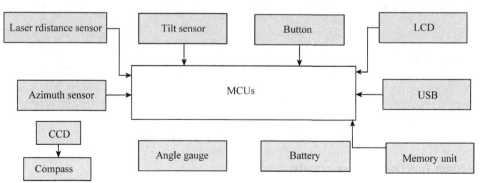

Figure 2-7　Hardware overall framework

Note: MCU: micro controller unit; LCD: liquid crystal display; USB: universal serial bus

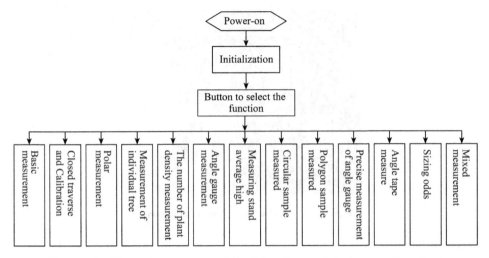

Figure 2-8   The main program and the 6 function module program flow chart

2) Software design

Software adopts modular structure design, including the following modules tree height measurement module, angle measuring module, stand average height measurement module, arbitrary diameter measurement module, measurement module of density are the basic measurement module. Figure 2-9 shows the program flow of each function module. The main measurement parameters such as the height of the tree, the number of the counted trees, the average height of the stands, the diameters at the diameter of the diameters, and the density of the trees are all displayed in real time, can be exported through the USB interface.

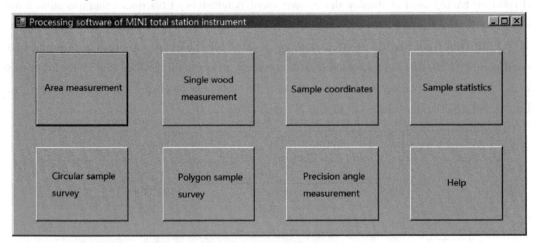

Figure 2-9   Industry processing software interface

3) Main functions and parameters

Mini total station instrument features include tree height measurement, the stand average height measurement, density measurement, angle gauge around test, random diameter measurement, basic measurement and other 13 functions. Distance measurement, the additional reflection film measuring a range of 0.5 ~ 100 m, the tree itself reflects the measurement range of 0.5 ~ 60 m, the smallest display unit is 1mm. Laser grade Ⅱ, a measurement error of 5 mm. Tilt angle measurement range of $-75° \sim 75°$, the measurement accuracy of 0.3°. Azimuth measurement range of $0 \sim 360°$, the measurement accuracy of 1°. With LCD display, key input, USB communication interface, lithium battery-powered by continuous working of 8h, the working environment temperature of $-20 \sim 50$ ℃.

2.2.2 MINI superstation instrument

1) Hardware structure

Mini superstation instrument hardware consists of a laser ranging sensor, MEMS (micro-electro-mechanical systems) tilt sensor, charge coupled synthetic device image sensor (CCD), the central processor, memory, liquid crystal display screen, power supply and miniature button etc. As shown in Figure 2-10. Through the integrated assembling of each component for the direct determination of tree, tree height and DBH factor accurate automatic measurement. The observation of the sample can be carried out on the bracket.

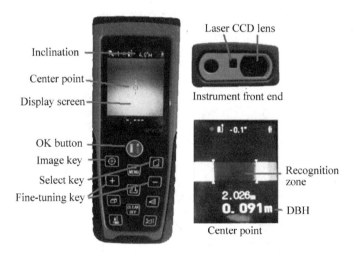

Figure 2-10　MINI super station schematic diagram

2) Function and parameter

The main functions of the MINI superstation instrument include tree height measurement and automatic measurement of diameter at breast height. Can realize the stand average height measurement based on the measurement of any other stem diameter, volume measurement, density measurement, basic measurement functions.

Instrument measurement distance range (range measuring) is $0.05 \sim 120$ m. Effective reflection distance $0.05 \sim 100$ m, minimum display value 0.001 m, measurement accuracy (precision measuring) is +1.5 mm. MEMS tilt sensor measurement range of $-90° \sim 90°$, effective measurement range $-75° \sim 75°$, measurement accuracy of 0.1°. The charge coupled device (CCD) image sensor is to calculate the accuracy of 1 mm. Operating ambient temperature $-10 \sim 50$. Instrument size 136 mm×52 mm×28 mm, continuous measurement time 5 h, normal working time 5000 s.

2.2.3 Principle of MINI total station instrument

1) Tree height measurement

The tree height is measured as shown in Figure 2-11.

(1) Measure the target station point A (X, Y, Z) to 1.3 m ($p_{1.3}$). The slope distance L and the inclination angle α, the horizontal distance S of the measured point to the measured tree T can be calculated by the formula (2-3).

(2) Calculate the tree height H, electronically display, and automatically according to the formula by measuring the inclination angle β of the point A (X, Y, Z) to the top of the tree tip P_v at the vertex (denoted as P_v) storage. In the left part of Figure 2-11, the horizontal position of the survey station is 1.3 m higher than that of the measured tree, and the horizontal position of the right half of the survey site is 1.3 m below the measured tree. Because of the built-in angle sensor can automatically measure the positive and negative tilt angle, without the need for manual interpretation, so both cases can be calculated using the same formula (2-3).

$$H = L \cdot \cos\alpha \cdot \tan\beta - L \cdot \cos\alpha \cdot \tan\alpha + 1.3 \tag{2-3}$$

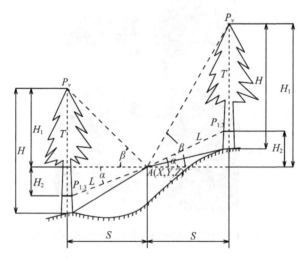

Figure 2-11 Tree height measurement principle diagram

Typeand l is the test site to $p_{1.3}$ slant distances (m); s test site to measure the horizontal distance of the tree T (m); alpha test site to $p_{1.3}$ tilt angle (DEG); beta test site to the tree vertex P_v tilt angle (DEG); H_1 is a station to be measured between tree treetop vertical distance (m); H_2 as the test site to $p_{1.3}$ between the vertical distance (m), h is the height of the tree (m).

2) The diameter measurement method of mobile

As seen by angle gauge principle, angle gauge openings between width l, angle gauge rod length L and fault area coefficient F_G into fixed proportion relation, as shown in Figure 2-12, eye from point O observation, O point and angle gauge opening ends (a, b) to form a fixed angle phi φ (Table 2-1). For the convenience of observation, only when measuring the opening measurement of *DBH* and basal area coefficient = 4 with an angle gauge.

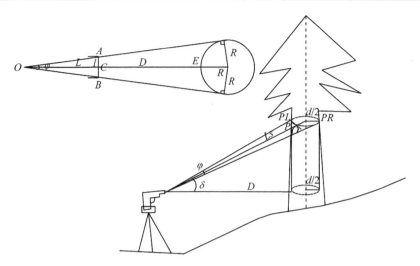

Figure 2-12 Schematic diagram of diameter measurement gauge method

Table 2-1 The diameter of formula corresponding to the four angle coefficient

F_g	d_i /cm
0.5	$\dfrac{2\sqrt{2}Ls\cos\delta}{2-0.014}$
1	$\dfrac{4Ls\cos\delta}{2-0.02}$
2	$\dfrac{4\sqrt{2}Ls\cos\delta}{2-0.028}$
4	$\dfrac{8Ls\cos\delta}{2-0.04}$

3) CCD diameter measurement

Focal length (f), field of view and the distance between the lens and the target to be taken to meet the following relationship:

$$\begin{cases} f = \dfrac{wL}{W} \\ f = \dfrac{hL}{H} \end{cases} \qquad (2\text{-}4)$$

In the formula (2-4), f denotes the focal length; w represents the width of the image, i.e. the subject on the imaging surface of the CCD target width, W indicates the width of the target subject; h denotes an image height, i.e. the subject in the CCD target surface image height, H represents the height of the subject. L represents the distance from the lens to the subject.

As we know F is a fixed value, L is known and only need to measure the w or h, you can calculate the W or H. That is, if the use of lens focal length f fixed L is known, then we can find K, $K = w/W$ for amplification.

Suppose $f = dL/D$, $D = dL/f$, where d is the image width, D is the diameter, and L is the distance measured by the laser rangefinder function of the MINI total station. F can be accurately ob-

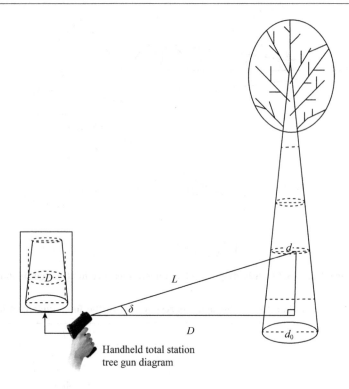

Figure 2-13 Schematic diagram of CCD diameter measurement

tained by calibration, but d and L there is a measurement error, in order to reduce the measurement error, first obtained in different measuring distance L corresponding to the value of $K = d/D$, and then use equation (2-5). The measurement principle is shown in Figure 2-13.

$$D = \frac{d}{K} \tag{2-5}$$

4) Single tree volume measurement

As shown in Figure 2-14, the calculation of single tree volume (2-6):

$$V = \frac{10^{-4}}{4}\pi \left[1.3 d_{1.3}^2 + \frac{1}{3}\sum_{i=1}^{3}(d_i^2 + d_{i+1}^2 + d_i d_{i+1})(h_{i+1} - h_i) + \frac{1}{3}d_4^2(H - h_4)\right] \tag{2-6}$$

Form in:

$$d_1 = d_{1.3};\ h_1 = 1.3;\ h_i = L_i \sin\delta_i$$

V for standing tree volume(m^3); L_i for measuring slope distance (m); the inclination angle (DEG); diameter at breast height (cm); h is the tree height (m); for the corresponding index of angle gauge and the tree of the tangent point height (m).

5) Closed traverse and calibration

Instrument measurement and recording elements are the followings inclination (δ), slope distance (S), magnetic azimuth (α) observe the inclination, slope distance and magnetic azimuth of the next target point (or measured wood) at the station. The starting point coordinates are (1000, 1000), and the names of the respective points are sequentially numbered. Coordinate incremental closure calculation and adjustment are as follows.

CHAPTER 2 Technical Foundation of Forest Management (Focus on General Principle)   67

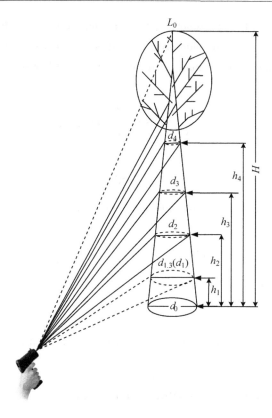

Figure 2-14 Single tree volume measurement

The coordinate increment is expressedas the calculation and adjustment of the coordinate increment closing error is as follows.

Coordinate increment:

$$D_{i,i+1} = L_{i,i+1}\cos\delta_{i,i+1} \tag{2-7}$$

$$\begin{cases} \Delta x_{i,i+1} = D_{i,i+1}\cos\alpha_{i,i+1} \\ \Delta y_{i,i+1} = D_{i,i+1}\sin\alpha_{i,i+1} \end{cases} \tag{2-8}$$

In the formula, L_i for the slant range, D_i for the flat pitch, δ_i for the dip angle.

Closed traverse coordinate increment closed f_x, f_y indicated difference respectively that there are:

$$\begin{cases} f_x = \sum \Delta x_{\text{measure}} \\ f_y = \sum \Delta y_{\text{measure}} \end{cases} \tag{2-9}$$

The total length of the wire is closed, with the said f_D, so there is:

$$f_D = \sqrt{f_x^2 + f_y^2} \tag{2-10}$$

The entire length error sum f_D, if the wire length $\sum D$, wire length is relatively closed for:

$$K = \frac{f_D}{\sum D} \tag{2-11}$$

According to the principle of proportional to the length error and edge length, the increment of coordinate closed f_x, f_y, signed by side grew up and is in direct proportion to the adjustment.

$V_{x_{i,i+1}}$, $V_{y_{i,i+1}}$: For the i to coordinate incremental tree to $i+1$ tree formed edge correction, there are:

$$\begin{cases} v_{x_{i,i+1}} = -\dfrac{f_x}{\sum D} D_{i,i+1} \\ v_{y_{i,i+1}} = -\dfrac{f_y}{\sum D} D_{i,i+1} \end{cases} \quad (2\text{-}12)$$

Finally, we can get the coordinate value after the closure error correction:

$$\begin{cases} X_{i+1} = x_i + \Delta x_{i,i+1} + V_{x_{i,i+1}} \\ Y_{i+1} = y_i + \Delta y_{i,i+1} + V_{y_{i,i+1}} \end{cases} \quad (2\text{-}13)$$

Closed traverse area (2-38) calculation:

$$S_n = \frac{1}{2} \sum_{i=1}^{n} x_i (y_{i+1} - y_{i-1}) = \frac{1}{2} \sum_{i=1}^{n} y_i (x_{i+1} - x_{i-1}) \quad (2\text{-}14)$$

In the formula X_i, Y_i is the coordinates of the i tree, n is equal to the number of points measured, when $i=n$, $i+n=1$.

2.2.4 MINI superstation instrument principle

2.2.4.1 Diameter measurement of standing tree at breast height

1) Measurement principle

When measuring the diameter at breast height of the standing tree under test the need is to point the laser range-finder spot on the tree. At this point, electric charge or image sensor device will automatically turn light to charge signal and through their own internal adc and the charge will be converted into digital signals and then through the digital signal processor (DSP) color correction the white balance processing into a visual image that appears on the screen. By adjusting the measure the position of the personnel to make the whole under test area (tree diameter at breast height) complete present within the screen. When measuring the automatic recording instrument measurement within the screen image, horizontal line at the center of the image again to baseline the upward and move down to 30 pixels, formation of top and bottom width of the 60-pixel rectangle processing recognition area. Using gray value $V = 0.299R + 0.587G + 0.114B$ calculation formula, type of R, G, B values for pixels, the recognition of all the pixel area gray, each pixel gray value and then to identify areas of each pixel traversal, calculate each column 60 pixel gray value of the mean, and mean value in the array.

The array of two intermediate values as a demarcation point, followed by traversing forward and backward, and to determine the demarcation point before and after the three parts of the mean. After the completion of traversal, find out three significant changes in the corresponding demarcation point (this step can be artificial fine-tuning to correct, to ensure that the identified area for the diameter of the diameter of the edge to be measured). According to the pixel coordinates of the cut-off point, calculate the pixel value N corresponding to the diameter to be measured.

When measuring the position of measurement point distance with a diameter at breast height measurement distance L can be recorded automatically by the laser range finder. The CCD focal length f is known, through the principle of CCD imaging (Figure 2-15), the image of the diameter at breast height under test calculation formula for calculating value R (1):

$$R = N \times \frac{L}{f} \qquad (2\text{-}15)$$

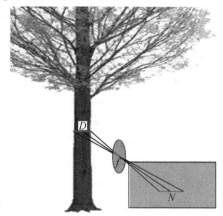

Figure 2-15 CCD imaging principle diagram

Type: N for DBH pixels under test (px), f for CCD composition is apart from the value (px), L is under test in diameter at breast height and the distance between the measuring point (m), R (m) for the image of the diameter at breast height under test value.

2) The DBH correction

While observing a single location on cylindrical objects, the light reflected by the diameter the cylindrical object is in the form of two tangent points of observation points and tangent cylinder wires, as shown in schematic diagram (Figure 2-16). If S_1 is the point of observation i. e. the distance of instrument and observing point, measured the value R is the length of AB. Laser rangefinder read automatically the measuring point distance L and DBH under test. Calculation formula of diameter D is obtained by calculation and analysis for the formul (2-16):

$$4LD^3 + (4L^2 - R^2)D^2 - 4LR^2D - 4L^2R^2 = 0 \qquad (2\text{-}16)$$

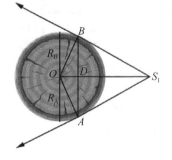

Figure 2-16 Diameter at breast height measurement schematic diagram

Use Shengjin Gong formula for solving a cubic equation, according to Shengjin Gong discriminant conditions, using the program to solve (2-16), we obtain the value of D calculated based on the actual relationship between D and R, whichever $D > R$, and close to the R value solver value as the diameter D.

3) The diameter at breast height position

Non-contact measurement of diameter at breast height measurement methods and principles are mentionedin many kinds of literature in China and abroad. In 2013, Jian-li wang et al. used a method combined with optical triangle, image processing and least square fitting method of buck diameter at breast height measurement; In 2015, Cao Zhong et al. using electronic theodolite DBH and tree height measurement and stocking and so on. But the vast majority did not mention how to determine the position of diameter at breast height, and how to judge the observation point is from the root collar, diameter at breast height, where 1. 3 m. It just puts forward a way of non-contact measurement but no real practical investigation of non-contact measurement.

In order to truly realize the function of the non-contact measurement in field investigation combined with the MINI-super station instrument and advantage of the characteristics of laser range finder and MEMS tilt sensor, automatic judgment DBH position.

The assumption that the root collar point to be measured is located above the observation point in the horizontal plane [Figure 2-17(a)], that the observation point, the range finder when spotted on tree root collar point inclination angle sensor readings alpha is greater than zero. The laser rangefinder will automatically record observation point the root collar L_0, H_0 the vertical distance between root collar point to a point which is at the same level with observation point, the horizontal distance between the observation points and target point L_S. Adjust the laser point on the tree diameter at breast height, which can measure and automatically calculates the numerical value of the diameter at breast height, the inclination angle sensor readings for angle beta calculated by the formula given below (2-17). In the second case the measured tree root collar located below the observation point in the horizontal plane [Figure 2-17(b)], alpha angle readings this time is less than zero. H_0 calculated value is negative, diameter at breast height is located above and below the plane of observation point decided the plus or minus value of beta, but had no effect in solving the formula. Through analysis and calculation, the calculation formula of angle beta for type (2-17).

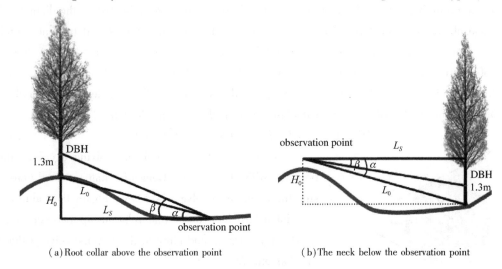

(a) Root collar above the observation point   (b) The neck below the observation point

Figure 2-17 DBH position determine the schematic diagram

$$\beta = \arctan\left(\frac{L_0\sin\alpha + 1.3}{L_0\cos\alpha}\right) \tag{2-17}$$

Where: L_0 is observation point to the root collar distance (m), alpha is laser the point on the root collar inclination of the readings (°), β is the laser spot on the diameter at breast height when the inclination of the reading (°).

2.2.4.2 Standing tree height measurement

When the tree height measurement the angle measuring function and MEMS angle sensor used for tree height measurement, which is the main feature of laser range finder. For tree height measurement first determine the suitable position for height measurement, make sure the whole standing tree is exposed in the line of sight range. Then the tree trunks were aligned, the root neck and the top of the tree three measurements were recorded to the trunk of the measuring point to the horizontal distance L_S, the measurement point to the root of the formation of the inclination angle α, measuring points to the top of the tree.

Assumptions for the standing tree the rootneck of the tree is below the observation point in the horizontal plane [Figure 2-18(a)], the inclination angle sensor readings alpha is less than zero, so the calculation formula for the type of tree height (2-18). When standing tree under observation the root neck of the tree is located above the water level [Figure 2-18(b)], the reading of the inclination angle alpha is greater than zero, and tree height calculation formula is:

$$H = L_S(\tan\beta - \tan\alpha) \tag{2-18}$$

Type: H is the height of the standing tree under test (m), L_S is the horizontal distance from observation point to the standing tree under test (m), beta is angle of measurement between tree top and horizontal point on tree with reference to observation point and α is the angle of measurement between the horizontal point with reference to the observation point to the root neck of the tree (°).

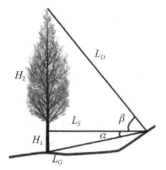 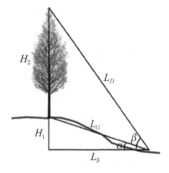

(a) Root collar below the observation point (b) The root collar above the observation point

Figure 2-18 Tree height measurement schematic diagrams

When using a laser range finder three measurements will be simultaneously recorded by the laser range-finder i. e. L_S the horizontal distance from an observing point on target tree, the distance of tree top from observation point L_D and distance of root collar point from observation point L_G. Only through analysis of these three distances we can get the tree height calculation formula for the type (2-19):

$$H = L_S\left\{\tan\left[\arccos\left(\frac{L_S}{L_D}\right)\right] \pm \tan\left[\arccos\left(\frac{L_S}{L_G}\right)\right]\right\} \tag{2-19}$$

Type: H is the height of the standing tree under test (m), L_S the horizontal distance from an observing point on target tree, L_D is the distance of tree top from the observation point, the distance of root collar point from observation point is L_G. When the value of the inclination angle α is less than 0°, the value of ± takes a positive value, and when the value of the inclination angle α is greater than 0°, the negative value is taken.

2.2.5 Application results and analysis

2.2.5.1 MINI total station instrument application result and analysis

The instrument has been used in Shangcheng County, Henan Province, Beijing Forestry University Jiufeng experimental forest, Yunnan Province Chuxiong Zixi mountain forest, and other places to test the working and functions.

1) Tree height measurement function test

In order to validate the accuracy and precision of the tree height measurement, twelve different height samples were selected from *Pinus tabuliformins*, *Ginkgo biloba* and *Sabina przewalskii*. Total station measurements were taken as the relative true values. Four observers respectively, the same strain of trees were observed 30 times, the measurement method is divided into observation with a tripod and handheld observation. Tree height measurement accuracy of 20~27 cm, the precision of 18~27 cm. The precision range is from 97.37% to 98.33%, which meets the requirement of 95% accuracy of the tree height measurement technology of National Forestry Forest Inventory (China State Forestry Administration, 2003). With tripod measurement and without stand measurement, the impact on accuracy is not obvious, indicating that the tree height measurement function, the instrument can handle the handheld operation.

2) Diameter measurement function test

The feasibility and accuracy of authenticating a mobile angle gauge method DBH measured, including sample trees 12 strains are observed the terrain is flat, which 6 strains of coniferous tree species and 6 strains of broadleaved species. 36 measurements were taken for each stand. The DBH relative to the true value D_0 of each plant sample was measured with a scale. The measurement results are shown in Table 2-2 and Table 2-3.

Table 2-2 Mobile measurement needle-leaved DBH and the accuracy analysis

S No.	Forest category	Varieties of trees	F_g coefficient	Diameter class	σ/cm	$\sigma_{中}$/cm	$\bar{x} - D_0$/cm
1	Needle	*Sabina chinensis*	0.5	10	0.320	5.508	-5.499
2	Needle	Chinese pine	0.5	18	0.989	9.546	-9.496
3	Needle	Chinese pine	0.5	26	0.634	12.772	-12.757
4	Needle	*Sabina chinensis*	4	38	2.054	7.369	-7.085
5	Needle	*Sabina chinensis*	4	48	1.595	8.169	-8.016
6	Needle	Chinese pine	4	58	1.738	15.014	-14.915

Table 2-3 Mobile measurement broad-leaved DBH data and the accuracy analysis

S No.	Forest category	Varieties of trees	F_g coefficient	Diameter class	σ/cm	$\sigma_{中}$/cm	$\bar{x} - D_0$/cm
1	Broad leaf	Chinese pagoda tree	1	10	0.380	3.986	-3.968
2	Broad leaf	Maple	0.5	22	0.972	11.770	-11.731
3	Broad leaf	Maple	0.5	26	0.374	13.520	-13.515
4	Broad leaf	Gingko	2	36	0.879	8.795	-8.752
5	Broad leaf	Chinese parasol tree	4	50	1.178	3.986	-12.846
6	Broad leaf	Chinese parasol tree	4	56	4.411	5.191	-2.834

Table 2-2 and Table 2-3 x for the sample mean, sample variance, Sigma is in error. The mean variance of coniferous tree species was between 0.320 and 2.054 cm, and the error was between 5.508~15.041 cm, and the difference between the mean value and the true value of the 36 measurements was between −14.915~−5.499 cm. Broad-leaved tree species measured mean variance between 0.374~4.411 cm, the error between 3.986~13.520 cm, the mean value of the 36 measurements and the difference between −13.515~−2.834 cm. The coniferous and broadleaf tree species had less influence on the method, but the mean value of the 36 measurements was smaller than the true value.

3) CCD diameter measure

In the Southwest Forestry University mountain experimental forest tree measurement of 465 strains, the measurement results shown in Figure 2-19. The first ruler with *DBH* measurement three times tree diameter, taking the average value as opposed to the true value, then in diameter well marked for another shot measurement, shooting, in order to convenient and high-efficient of standing trees were measured that doesn't presuppose angle and distance. The measurement results show that the mean diameter of diameter at breast height is 21.954 cm, the mean diameter at breast height is 21.642 cm, the absolute error is 3.2 mm, and the mean relative error is 1.42%.

465 standing trees were measured in the mountain forest farm of Southwest Forestry University. The measurement results are shown in Figure 2-19. First, measure the diameter of the tree 3 times, take the average as a relative truth value, and then do a good job at the *DBH* mark and then shooting measurements, shooting, in order to facilitate the rapid measurement of the standing trees, without preset angle and distance. The results showed that the mean *DBH* of 21.954 cm, 21.642 cm, 3.2 mm and 1.42%, respectively.

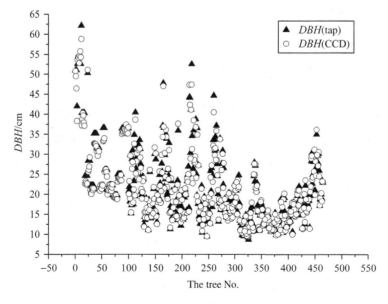

Figure 2-19 Scatter plots of diameter at breast height

4) Single tree volume measurement

The results showed that the error of measurement was larger than that of total station, the error was between ~46.39% and 4.6%, and the error was between −46.39% and 4.6%. The reason is that the use of mobile angle gaugediameter and diameter of the upper diameter of the larger error caused by temporarily unable to promote the use, however, verify the feasibility of the method, the need for further improvement.

5) Closed traverse and calibration

In order to verify the function and parameters, first select a certain measuring point in the sample. With the total station measuring coordinates calculate the area. Total station coordinate and area as a relative truth value (1000, 1000). Five points were selected in the experiment, the coordinates, and area of the five samples were measured by the total station. Then the four observers were used to analyze the coordinates of each sample point for four measurements. For each sample point, the area measured by the total station is compared with the average value of four times of total surveying station. The relative error of measuring total station is between 0.85% ~ 5.51%.

2.2.5.2 MINI superstation instrument application results and analysis

(1) After the boot press the image key, select the key to measuring the height of the tree diameter or the height of the tree.

(2) Measurement of diameter at breast height [Figure 2-20(a)], the instrument laser point at breast height [Figure 2-20(b)], click to determine the key to automatically display the diameter of the breast [Figure 2-20(c)].

(a) Tree alignment   (b) Aligned diameter at breast height   (c) Automatic calculation

Figure 2-20 Instrument interface at breast height measurement

(3) Tree height measurement. First, the laser point of the instrument is aligned with the standing tree, and the horizontal distance of the measured point to the standing tree is automatically recorded [Figure 2-21(a)]. Then, the laser point is aligned with the root of the standing tree to be measured, [Figure 2-21(b)]. Finally, the laser point is placed at the topof the tree to be measured. The inclination and distance of the measured point to the top of the stump are recorded automatically [Figure 2-21(c)]. At this point, the instrument automatically resolves the standing height [Figure 2-21(d)].

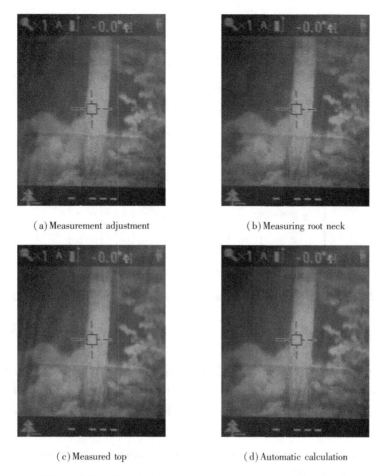

(a) Measurement adjustment (b) Measuring root neck

(c) Measured top (d) Automatic calculation

Figure 2-21 Tree height measurement instrument interface

(4) Measurement analysis of diameter at breast height. During the experiment, 10, 25, 25 measurements were made on three standing trees with a diameter of 7.5 cm, 12.3 cm, and 14.5 cm according to different measuring distance. The range of measuring distance was 2~42 m. The measurement results are shown in Figure 2-22.

The results of the experiments were analyzed, the absolute error range of the 60 measurements was −0.5~0.5 cm and the mean absolute error was 0.147 cm. Relative error range −4.07%~6.67%, mean relative error 1.26%. As is shown in Figure 2-22, when the measurement distance is reached 20 m, the curve fluctuates greatly, and the error is relatively large. After removing 14 measurements from a distance of more than 20 m, the average absolute error is 0.107 cm and the average relative error is 0.9%.

The reason of the CCD lens angle of view, the measurement of the shortest distance of different diameter at breast height also has the request, after the experimental verification, the measurement distance is satisfied:

$$L \geqslant 0.125[\text{int}(D) + 1] \tag{2-20}$$

Formula: L for measuring distance (m), D for breast height (cm), int (D) for the numerical value of the whole.

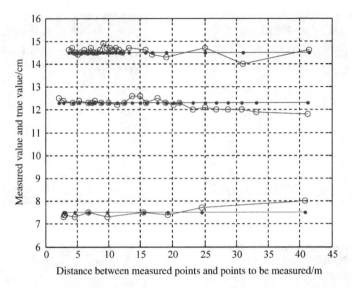

Figure 2-22 Measured values and true values under different measurement distances

After the experimental analysis, the measurement of diameter at breast height has the following conclusions:

①A single measurement of diameter at breast height, in order to improve the measurement accuracy, as far as possible to make the measurement distance is not more than 20 m. In order to make the entire diameter of breast height display in the display, the diameter of less than 20 cm, the shortest distance is greater than 2.6 m; diameter breast height is less than 40 cm, the shortest distance is greater than 5 m. For the measurement of tree diameter, the optimal measuring distance should be within the range of 5~20 m.

②In order to improve the measurement accuracy, the measurement of diameter at breast height, the bigger the diameter, the lower the relative error, the higher the measurement accuracy (Table 2-4).

③Experimental process, the image solution time 0.5~3 s, the closer the distance measurement, the faster the solution.

Table 2-4 Error analysis of calculating results of diameter at breast height

Parameter	Tree 1	Tree 2	Tree 3
Number of measurements	10	25	25
True value/cm	7.50	12.30	14.50
Average value/cm	7.48	12.25	14.56
Mean absolute error/cm	−0.02	−0.05	0.06
Mean relative error /%	−0.27	−0.39	0.44
Mean absolute error /cm	0.16	0.15	0.14
Mean relative error /%	2.13	1.24	0.94

(5) Tree height measurement analysis. During the experiment for three trees as high as 2.18 m, 3.6 m, 12.89 m tree according to the different measuring distance respectively for the measured 20 times, when measuring angle [the angle of inclination of the shadowed reduction to the root neck angle range (10° to 90°)]. The measurement results are shown in Figure 2-23.

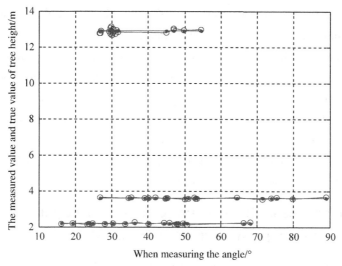

Figure 2-23　True and measured values of tree height measurements at different angles

The results of the experiments are analyzed, the relative error range of the 60 trees height measurement is −2.57% ~ 3.58%, and the average relative error is 1.13%. Through the experimental analysis, the measurement accuracy of the tree height can be greatly improved. In the measurable range, the higher the trees, the lower the relative error (Table 2-5), the higher the accuracy of the measurement.

Table 2-5　Tree height calculation results in error analysis

Parameter	Tree 4	Tree 5	Tree 6
Number of measurements	20	20	20
Ture value/cm	218.00	360.00	1289.00
Average value/cm	217.65	359.66	1287.28
Mean absolute error/cm	−0.35	−0.34	−1.72
Mean relative error/%	−0.16%	−0.09%	−0.13%
Mean absolute error /cm	3.43	3.42	11.02
Mean relative error /%	1.57%	0.95%	0.85%

In order to realize the digital, multi-function and precision measurement of the forest, the idea of digital, portable and multifunctional integrated measuring instrument, a kind of MINI total station / super-station instrument is developed. Instrument tree height measurement accuracy is 20 ~ 27 cm, the precision is 18 ~ 27 cm. The precision range is 97.37% ~ 98.33%, which meets the requirement of national forest resource continuous inventory technology, and the tree height measurement error is less than 3%. CCD diameter measurement precision between 96.11% ~ 98.44% while mobile angle gauge method DBH measurement and volume measurement error are larger, the need to further study, improve accuracy.

The closed traverse and the demarcation line between the 1.41% ~ 14.20% are closed, the relative error between the area measurement is in −2.52% ~ 2.32%, and the tree coordinate measurement and the automatic drawing of the distribution map of the sample are realized.

Handheld digital multifunctional mini total station instrument/super-station instrument development and test are under the background of conventional forestry to digital forestry. The aim is to explore the realization of a more efficient forest measurement method from the forest measurement equipment. Nevertheless, the product also needs further research and improvement, such as GIS, GPS and combined to increase the practicality of products, efficiency analysis of products, the most suitable range of sensitivity analysis, instrument operation point, and error analysis, various environmental test and analysis will be the focus of further study.

2.2.6 Estimating tree position, diameter at breast height, and tree height in real-time using a mobile phone with RGB-D SLAM

Forest ecosystems are considered important for the survival of both animals and human beings because of their environmental and socio-economic benefits. They not only provide services such as soil and water conservation, carbon storage, climate regulation, and biodiversity, but also provide food, wood, and energy. Forest resource information is an important solid basis for making decisions at different levels according to various needs such as timber harvesting, biomass, species diversity, watershed protection, and climate change impact evaluation, etc. Forest inventory primarily involves the collection of forest resource information, which aims to provide accurate estimates of the forest characteristics including the wood volume, biomass, or species diversity within the region of interest. These attributes are precisely estimated by models constructed using tree species, diameter at breast height (DBH), and tree height. The conventional instruments used to measure these properties are calipers and clinometers for the DBH and tree heights, respectively. Forest inventory has gradually improved with the advancement of remote sensing sensor technology, computing capabilities, and Internet of Things (IoT). New technologies with high levels of precision and accuracy in the field of forestry have been introduced in recent years. Terrestrial laser scanning (TLS), also known as ground-based Light Detection and Ranging (LiDAR), provides a solution for efficiently collecting the reference data. This technology has been used to extract various forestry attributes. However, the TLS has some operational and performance limitations, especially when it is used in large forests, as it is time consuming, laborious, and intensive to carry and mount; moreover, its data processing is also difficult. Mobile Laser Scanning (MLS) is a vehicle born system with an inertial navigation system (INS) and a laser scanner, which makes it possible to move and measure 6 degrees-of-freedom (DOF) of the laser scanner in forests. However, the poor GNSS coverage under the canopy could hinder the positioning accuracy, hence making it difficult to achieve a globally-consistent point cloud. Simultaneous localization and mapping

(SLAM) is a process that can simultaneously generate the map of unfamiliar environmental conditions, as well as locating the mobile platform. This technology is a potential solution that can use sensors such as cameras and lasers to perform relative positioning without GNSS signals through SLAM algorithms in real-time. Some previous studies initially applied SLAM to the forestry inventory process. However, the MLS system is relatively complex, heavy, and expensive.

The Time of Flight (TOF) camera is an alternative to LiDAR, and its basic principle is its consistency with LiDAR. The difference is that TOF cameras generally use infrared as the light source, and do not use point-by-point scanning, but rather, an integrated matrix of multiple TOF sensors to measure multiple distances simultaneously. TOF cameras are power efficient and smaller than LiDARs, and can even be integrated into a mobile phone. The combination of the TOF camera and RGB camera forms an RGB-D camera that can acquire the texture and depth information of the surrounding environment, and it is often used as input to the SLAM system which known as the RGB-D SLAM system. This sensor overcomes the shortcomings of the inability of a monocular SLAM system to obtain the scale and correspondence of a stereo SLAM. With the improvement of the SLAM algorithm and the advancement of chip computing capabilities, the RGB-D SLAM system can even run on a mobile phone. Compared to the traditional MLS system, the mobile phone has the advantages of being portable, inexpensive, and highly integrated. Previously, researchers mainly used the SLAM algorithm to improve the pose estimation accuracy of the MLS system offline. Some other researchers also used point clouds from RGB-D SLAM to obtain tree positions and DBHs offline instead of emphasizing real-time access to forest attributes. This paper aims to estimate forest attributes in real-time in a forest inventory using a mobile phone with RGB-D SLAM. We design algorithms for the mobile phone with RGB-D SLAM, to evaluate tree position, DBH, and tree height in forests online. In addition, AR technology is used to display the estimated results on the screen of a mobile phone, enabling the observer to visually evaluate their accuracy.

2.2.7 Theory and technology

2.2.7.1 SLAM

SLAM is the process by which a mobile platform builds a map of the surrounding environment and finds its location using the map in real-time, as shown in Figure 2-24.

The simultaneous localization and mapping (SLAM) problem. Here, x_k is the state vector describing the pose of the mobile platform at time k; u_k is the motion vector describing the movement of the platform from time $k-1$ to time k; m_j is the vector describing the position of the jth landmark.

A mobile platform moves in an unknown area and, at the same time, separately observes the surrounding unknown landmarks and records their relative motion through the mounted visual sensors and motion sensors. Then, the relative positions of the landmarks and the pose of the mobile platform are estimated in real time. In probabilistic form, data from motion sensors u_k are described as a motion model (a probability distribution):

$$P(x_k \mid x_{k-1}, u_k) \qquad (2\text{-}21)$$

Here, x_k, x_{k-1} are the platform poses at times k and $k-1$, respectively. Data from the vision

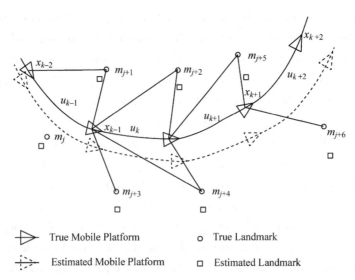

Figure 2-24 Simultaneous localization and mapping(SLAM) problem description

sensors z_k are described as an observation model (a probability distribution):

$$P(z_k \mid x_k, m) \qquad (2\text{-}22)$$

Here, m is the set of all landmarks. Generally, the SLAM problem is best described as a probabilistic Markov chain. That is, the joint posterior probability at time k, $P(x_k, m \mid Z_0:k, U_0:k, x_0)$, is solved using the posterior joint probability at time $k-1$, $P(x_{k-1}, m \mid Z_0:k-1, U_0:k-1, x_0)$, and some other conditions. In the SLAM algorithm, this is implemented using recursive steps, predictive update and observation update forms:

Predictive update:

$$P(x_k, m \mid Z_{0:k-1}, U_{0:k}, x_0) = P(x_k \mid x_{k-1}, u_k) P(x_{k-1}, m \mid Z_0:k-1, U_0:k-1, x_0) dx_{k-1} \qquad (2\text{-}23)$$

Observation update:

$$P(x_k, m \mid Z_{0:k}, U_{0:k}, x_0) = \frac{P(x_k \mid x_k, m) P(x_k, m \mid Z_0:k-1, U_0:k, x_0)}{P(z_k \mid Z_0:k-1, U_{0:k})} \qquad (2\text{-}24)$$

Obviously, the key to solving the SLAM problem is to give a reasonable expression of the motion model and the observation model, and then the predictive update and observation update are used to solve the target probability distribution. Three commonly-used methods for solving SLAM problems are extended Kalman filtering (EKF), sparse nonlinear optimization, and particle filtering.

2.2.7.2 The technology of a portable graph-SLAM device

SLAM algorithms have been implemented to track device positions in real-time to build maps of the surrounding three-dimensional (3D) world. Though most SLAM software today works only on high powered computers, Project Tango enables this technology to run on a portable mobile platform (Smartphone). Project Tango created a technology that enables mobile devices to acquire their poses in the three-dimensional world in real time using highly customized hardware and software. Three-dimensional maps can also be constructed by combining pose data with texture and depth information.

Figure 2-25 shows the smartphone with Google Tango sensors (Lenovo Phab 2 Pro) which was used in this paper. The mobile phone contains a combination of an RGB camera, a time of flight camera, and a motion-tracking camera called a vision sensor. The sensor is used to acquire texture and depth information of the 3D world so that the phone has a similar perspective of the real world. A 9-axis acceleration/gyroscope/compass sensor combined with the vision sensor can be used to implement a Visual-Inertial Odometer system using SLAM algorithms. Some special hardware, such as a computer-vision processor, is used to help speed up data processing so that the pose of the device can be acquired and adjusted in real-time.

Figure 2-25 Structure of a smart phone (Lenovo Phab 2 Pro) with RGB-D SLAM

The device uses a Visual-Inertial Odometer system for the front end and back end optimization algorithms implement adjustments to account for odometry drift. These adjustments are mainly done through pose graph non-linear optimization and loop closure optimization. These optimizations use visual features to identify previously-visited regions and then adjust the corresponding camera pose drift during this process.

2.2.7.3 Methods

An application was developed to enable the SLAM mobile phone to be used for forestry inventory. Figure 2-26 shows the workflow of our hardware and software system. The SLAM system uses an RGB-D camera and an inertial measurement unit (IMU) as inputs and produces RGB images, point clouds, poses, and time stamps. Our forest inventory system uses these data, and then interacts with the Android system to show the results on a screen or accept instructions from users. The forest inventory system includes mapping the plot ground and tree-by-tree estimation, as shown in Figure 2-27. The mapping process should provide a globally consistent map and the plot coordinate system for tree-by-tree estimation.

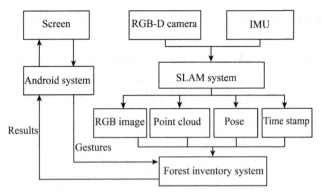

Figure 2-26　The workflow of our system

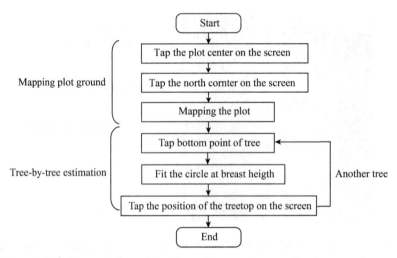

Figure 2-27　The workflow of the forest inventory system for data acquirement

2.2.7.4　Mapping of the plot ground

The plot ground was mapped before observing the trees. That mapping was used to correct the poses of the mobile phone when observing the trees through loop-closure detection and pose graph nonlinear optimization. To obtain a globally consistent plot ground map, the scan path was designed as shown in Figure 2-28(a). The plot center was used as the starting point of the mapping process towards the north of the plot, while the end was near to the north point. Figure 2-28(b) shows a typical instance of mapping path.

Before measuring the tree attributes, we built a new coordinate system during the mapping process to describe the position of each tree in the plot. This coordinate system, known as the Plot Coordinate System (PCS), has the center of the plot as its origin, and the horizontal eastward/ horizontal northward/vertical upward directions were defined as the $x/y/z$-axis directions. However, the mobile phone defined a coordinate system known as the Initial Coordinate System (ICS), in which the origin is the position of the device when it is initiated, the x-axis is towards the right side of the phone screen, and the y-axis is directed vertically upward, while the z-axis is towards

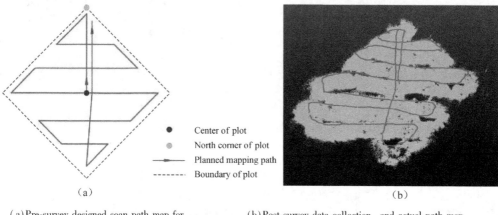

(a) Pre-survey designed scan path map for collecting data with mobile phone

(b) Post-survey data collection, and actual path map. Building the Plot Coordinate System

Figure 2-28　The mapping path of plot ground

the mobile phone screen. Conversion of the coordinate system was needed after the center and the north corner of the plot had been tapped on the screen during the mapping process, as shown in Figure 2-29(a) ~ (c). The center of the plot was tapped at the beginning of the mapping process, and the north corner was tapped at the end of the mapping process. After that, the transformation matrix between the PCS and the ICS was obtained. Then, the PCS was transformed into the Open-GL coordinate system and displayed as shown in Figure 2-29(d).

(a) Waiting to tap the plot center on the screen

(b) Waiting to move towards the north corner of the plot

(c) Waiting to tap the north corner on the screen

(d) Completion of the PCS building

(e) Waiting to tap a bottom point of a tree (f) Waiting to scan the breast height section of the tree

(g) Completion of fitting the circle at breast height (h) Waiting to tap the position of the treetop on the screen

(i) Completion of the estimation of the tree height (j) Finished estimating the attributes of the tree

Figure 2-29   Different states of the SLAM device during the observation process

2.2.7.5   Estimation of the stem position, DBH, and tree height

Estimation of the stem position and DBH After mapping the plot ground, we were able to observe each tree individually. While observing a standing tree, a point near the bottom of the tree was acquired to determine the breast height and the approximate stem position [see Figure 7(e)(f)]. After the bottom point had been taken, the breast height was displayed on the screen as shown in Figure 7(f). Then, the position of the stem center and DBH were calculated using the current camera pose and the point cloud (Figure 2-30).

For convenience of operation, an auxiliary coordinate system (Auxiliary Camera Coordinate System, ACCS) was established with a similar origin to that of the camera coordinate system (CCS), as the y-axis direction was same as the y-axis direction of the PCS (vertically upward), and the z-axis was in the plane formed by the new y-axis and the z-axis of the CCS. The point cloud data were constructed and transformed into the ACCS. The points belonging to the tree were

(a) The RGB image (b) The Depth image

Figure 2-30 The RGB image and depth image used to calculate the stem position and DBH

filtered according to the bottom point (x_{ac_bottom}). Figure 2-31 shows the scatter plot of these points on the plane $O_{ac}-x_{ac}-z_{ac}$, while the filter conditions were set to

$$\begin{cases} x_{ac\_bottom} - 1 < x_{ac} < x_{ac\_bottom} + 1 \\ z_{ac\_bottom} - 1 < z_{ac} < z_{ac\_bottom} + 1 \\ y_{ac\_bottom} - 1.25 < y_{ac} < y_{ac\_bottom} + 1.25 \end{cases} \quad (2\text{-}25)$$

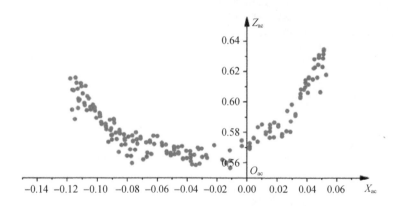

Figure 2-31 The scatter plot of all filtered points on the plane $O_{ac}-X_{ac}-Z_{ac}$

The marginal distribution of all filtered points in the x-axis direction was approximately uniformly distributed (Figure 2-32). When the mean (μ_x) and variance (σ_x) of the points were calculated, the range of the stem was ($\mu_x - \sqrt{3}\sigma_x$, $\mu_x + \sqrt{3}\sigma_x$) in the x-axis direction. RANSAC (Random Sample Consensus) was used to improve the robustness of the interval in this process (Figure 2-33).

The two edges of the stem were underestimated due to the assumption of uniform distribution. The stem edges were adjusted by linear fitting the points near to the stem edges, and the point on each fitted line that was farthest from the previous estimation interval was used as the boundary point (Figure 2-34); RANSAC was used in the process too.

The TOF camera uses the principle of perspective projection to obtain images. So, when the stem edge points were defined as T_1 and T_2, both were the tangent points of the stem circle and the

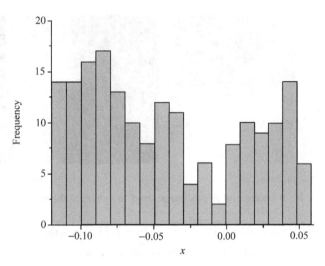

Figure 2-32 The edge distribution of all filtered points in the x-axis direction

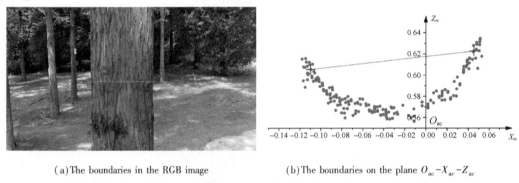

(a) The boundaries in the RGB image (b) The boundaries on the plane O_{ac}-X_{ac}-Z_{ac}

Figure 2-33 The stem boundary detection result by a uniform distribution

(a) The boundaries in the RGB image (b) The boundaries on the plane O_{ac}-X_{ac}-X_{ac}

Figure 2-34 The stem boundary detection result after linear fitting

lines T_1O_{ac} and T_2O_{ac}. The stem circle needed to meet the following conditions: §1 Line T_1O_{ac} and T_2O_{ac} had to be the tangents of the circle and points T_1, T_2 had to be the tangent points; and §2 The points between point T_1 and point T_1 had to be in the circle. Condition §3 could be expressed as the cost functions:

CHAPTER 2   Technical Foundation of Forest Management (Focus on General Principle)   87

$$\begin{cases} r = w \cdot zT_1(zc - zT_1) + xT_1(xc - xT_1) \\ r = w \cdot zT_2(zc - zT_2) + xT_2(xc - xT_2) \end{cases} \quad (2\text{-}26)$$

Here, r is the residual to be optimized; w is the weight of the cost function; $XT_1 = (xT_1, zT_1)$ and $XT_2 = (xT_2, zT_2)$ are the coordinates of points T_1 and T_2 in the plane $O_{ac} - x_{ac} - z_{ac}$; $X_c = (x_c, z_c)$ is the stem center coordinate; and d is the DBH of the tree. Each point in Condition §2 can be constructed as the cost function Here, $X_i = (x_i, z_i)$ is one of the points between points T_1 and T_2 in the circle. Because condition 1 is more important for the circle fitting than condition §2, the weight w should have a large value. In this paper, it was determined according to Condition §2; if the number of points in the optimization process was defined as N in condition §2, the weight w was equal to $N/4$. After the cost functions had been constructed, the Levenberg Marquardt algorithm was used to fit the stem circle. To increase the robustness of the optimization result, RANSAC was used in the fitting process. Figure 2-35 shows the fitting circle. Then, the stem position coordinate was converted into the PCS and projected into the OpenGL coordinate system, so the result could be viewed from the display [Figure 2-29(g)].

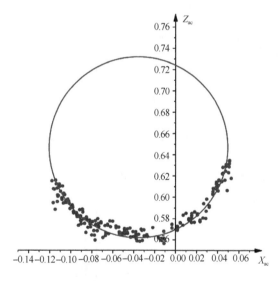

Figure 2-35   The fitted stem circle at breast height

Estimation of the tree height

To estimate the height of the tree, the device was kept at a position where the treetop could be observed simultaneously [Figure 2-29(h)]. After tapping the treetop position on the screen, the tree height was calculated in real time. As the treetop was on the connection line between the optical center of the camera and the tapped pixel (Figure 2-36), the coordinates of the treetop point were expressed as

$$r = -(x_c - x_i)^2 - (z_c - z_i)^2 \quad (2\text{-}27)$$

$$x_{p\_top} = t_{p\_c} + kR_{p\_c}K^{-1}u \quad (2\text{-}28)$$

Here, $x_{p\_top} = (x_{p\_top}, y_{p\_top}, z_{p\_top})$ is the coordinate of the treetop point in the PCS; tpc is the translation vector of the camera in the PCS; R_{p_c} is the rotation matrix of the camera in the PCS; K

is the intrinsic matrix of the camera; $u = (u, v, 1)T$ is the coordinate of the treetop pixel in the pixel coordinate system; k is an unknown factor. However, additional conditions were needed to determine k. If the standing tree was assumed to be vertical, the treetop would be on the vertical line passing through the center of the stem, but since a natural stem is not exactly vertical and the tapped treetop pixel showed deviation, the two lines were difficult to intersect or even impossible to intersect. To solve this problem, we assumed that the treetop was on the plane of the stem center, whose normal vector is denoted by

$$n = \begin{bmatrix} 1 & 0 & 0 \\ 0 & 0 & 0 \\ 0 & 0 & 1 \end{bmatrix} (t_{p\_c} - x_{p\_center}) \quad (2\text{-}29)$$

Here, $x_{p\_center} = (x_{p\_center}, y_{p\_center}, z_{p\_center})$ is the central coordinate of the stem. In this way, the treetop coordinate satisfied

$$(x_{p\_top} - x_{p\_center}) n = 0 \quad (2\text{-}30)$$

Then, after being calculated from (2-28) and (2-30), the treetop coordinate x_{p_top} was

$$x_{p\_top} = t_{p\_c} + \frac{(x_{p-conter} - t_{p-c}) n}{(R_{p\_c} K^{-1} u) n} \cdot (R_{p\_c} k^{-1} u) \quad (2\text{-}31)$$

Naturally, the tree height h equaled

$$h = y_{p\_top} - y_{p\_center} + 1.3 \quad (2\text{-}32)$$

As shown in Figure 2-29 i, j, the treetop point x_{p_top} was projected into the Open GL coordinate system and displayed on the screen.

O is the center of the stem at breast height; C is the optical center of the camera; T is the treetop; U is the treetop pixel on the image; n is the horizontal component vector of the OC vector; and P is the plane whose normal vector is n and on which O is a point.

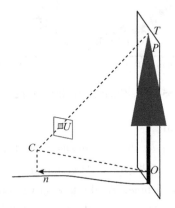

Figure 2-36 The geometric relationship between the points in the tree height solution

2.2.8 Conclusions

SLAM technology provides a solution for positioning without GNSS signals in places like forests. In this study, we measured the positions, DBHs, and heights of trees in real-time using the poses and dense point cloud data provided by a phone with RGB-D SLAM.

The tree positions were obtained from the device pose and a circle was fitted to the points around the breast height of each tree trunk. The drift of the phone pose mainly affected the accuracy of the tree positions. The method used to measure the height in this paper was similar to the traditional altimeter, which calculates the attribute by measuring the distance from the observation position to the tree and the inclinations of the tree bottom and treetop. The results showed a BIAS value of approximately -0.83 to 2.08 m and an RMSE value of $0.46 \sim 2.44$ m. The result also showed that the measurement results had high precision when the tree was not higher than 20 m, as shown in Figure 2-20, although AR technology was used, which enabled the observer to determine whether the displayed treetop position was appropriate; otherwise, it was adjusted as needed. This may be influenced by occlusion and the subjective decisions of observers. Reference evaluated a Laser-relascope, a classical traditional instrument; they reported a -0.016 m BIAS and a 0.190 m standard deviation. The limitation of the Laser-relascope is that it needs to be mounted on a fixed site, but our device did not need that. Reference used the height difference between the ground level and highest point on the point

In the SLAM algorithm, corners or blobs are often used as visual feature points. A good feature should have localized accuracy (both in position and scale), repeatability, robustness, etc. However, it is difficult to find good corners or blobs, especially in forests with complex ground conditions, such as shrubs. The plot data used in this article were taken from human accessible areas with fewer shrubs on the ground. However, most forests will not meet these special conditions. In addition, the RGB camera needs to search for feature points in an environment with sufficient illumination, while the TOF camera is susceptible to strong illumination because it uses infrared light as its light source. Therefore, the device is only suitable for use under the canopies during the day.

Although the device is a possible option for forestry surveys, it still has some limitations. For example, (1) although the device can estimate forest inventory parameters in real-time, it is less efficient than TLS, MLS, or the previous uses of SLAM because of the need to access each tree individually; (2) the taller tree accuracy of our tree height estimates was compromised a little, perhaps due to drift; and (3) the tree position coordinates obtained by the device do not use the geodetic coordinate system, but rather, the plot coordinate system of the origin at the center of the plot. In addition, the device uses the manual selection method to locate the bottom of the tree.

Unfortunately, the Google Tango Project has been terminated, i.e., is longer being developed or supported. However, ARKit, a monocular SLAM system, was released by Apple. Google has also introduced a similar solution—ARCore. These two technologies allow simultaneous localization and mapping for ordinary smartphones (without a TOF camera). Of course, due to the new features of these technologies, many aspects, such as how to get dense point clouds in real-time, still need to be investigated in the future.

It is recommended that future research studies are carried out to test the device under complex forest conditions, such as in areas with more shrubs, better or poorer light, different tree species, and in different aged forests. Future studies should also focus on extracting other forest inventory attributes, such as stem curves and crown diameters.

2.3 GNSS principle

GNSS is referred to the global navigation satellite system. It can also be translated into global satellite navigation system or global navigation satellite system. GNSS global satellite navigation system refers to the entire orbit satellite navigation system; it mainly included the Global Positioning System (GPS), the Russian 'GLONASS' System (GLONASS), European Galileo System (Galileo) and Chinese 'Beidou' System (BDS). This chapter primarily focuses the characteristics of above the world four big global satellite positioning system, composition and operation conditions were to be discussed.

2.3.1 Introduction of GPS system

1) GPS (Global Positioning System)

In the 70s of the 20th century, the armed forces of the United States jointly developed a new generation of military satellite navigation system (navigation by satellite timing and Ranging Global Positioning System), namely the NAVSTAR GPS, we called global positioning system (GPS system)

2) Characteristics of GPS system

The main features of the GPS system are widely used. US initial aim is to allow the Armed Forces to the GPS system to provide real-time and global navigation services, in order to gather information and monitor communications. GPS satellites can be used for vehicle navigation, air and sea navigation, missile guidance, engineering surveying, dynamic monitoring, and speed measurement.

The GPS system also has other characteristics like the observation taking is convenient, observation taking time is short and the instrument is easy to operate i.e. the relative static positioning within 20 km, only 15~20 minutes. In the process of observation measurement, the instrument only required to connect the cable and measuring the height of antenna then the monitoring instrument is in working state you can let other instruments, the observation task itself. After taking the measurement only need to turn off the power, in order to complete the data acquisition instrument.

The second one is a high positioning accuracy instrument. For example, the application of GPS in dynamic research, the use of GPS horizontal deformation monitoring precision has reached 1~2 mm/A, the vertical direction is 2~4 mm the relative accuracy of baseline / year, 1×10^{-9}.

The third one has good economic benefit; GPS satellite positioning technology compared to the conventional geodetic technology saves about 75% of the cost(Table 2-6).

Table 2-6 Parameters of GPS

Satellite parameters	GPS
Number of satellites (stars)	24
Number per orbit satellite	4
Number of satellites in orbit	6
Adjacent orbit satellite phase difference	40°
Orbit inclination angle	55°
Track long axis	26560 km
Track height	20200 km
Eccentricity ratio	<0.01
Operation period	11 hours and 58 minutes
Satellite lifetime	7.5~8 years

The main components of the GPS system are the space segment, the control segment and the segment of the user.

GPS space segment mainly consists of 24 satellites, including 21 satellites and three pieces of spare satellites, located above the surface at about 20200 km and operation cycle is about 12 hours, the satellite uniform distribution in the six orbital planes, orbit having an inclination angle of 55°(degrees).

A major part of the Control segment is the ground control system, which includes monitoring station. Master monitoring stations have a ground antenna, main control station in Colorado, Springfield. The ground control station is responsible for receiving satellite signals and returns a data were calculated and corrected(Table 2-7).

Table 2-7 Ground control section

Master control (control center)	Spring City, Colorado, USA
Injection along	The Atlantic: Amatsu Morimi military base
	India Ocean: Di Ge Garcia
	The Pacific: the U.S. military base at Kwajalein
	Transmit control signal to satellite via 3.5m antenna
Monitoring along	In addition to the main station, injection station plus Hawaii monitoring station receives the satellite signal, signal pre-processing; signal is sent to the central processing

User segment is the GPS signal receiver between the satellite and receiver. By the use of the receiver, it is divided into navigation receiver. According to the frequency carrier, it can be divided into single frequency receiver and dual frequency receivers. According to the receiver operating principles, it can be divided into code correlation receiver, square type receiver, hybrid receiver and interference receiver(Figure 2-37).

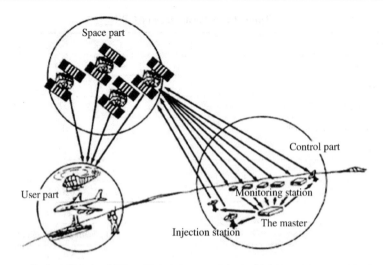

Figure 2-37 GPS components

2.3.2 GPS positioning principle

The 1970's the United States Army, Air Force jointly developed a new generation of the military satellite navigation system (Navigation By Satellite Timing And Ranging Global Positioning System), the NAVSTAR GPS, we call the Global Positioning System, referred to as GPS system.

GPS positioning principle is to use the spatial distribution of the satellite and the ground distance calculation of the ground position information and the use of space is the principle of the intersection.

GPS positioningis to solve the following main problems: Firstly, to observe the satellite position. Secondly to observed the distance between the test site and GPS satellite. In the ground, a location is needed to install the GPS receiver that can simultaneously receive signals from at least four satellites. Reuse Principle of Spatial Distance Intersection measures test site and the location of the receiver clock difference.

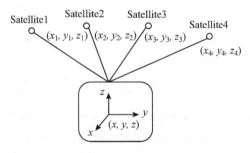

Figure 2-38 GPS positioning principle diagram

Shown in Figure 2-38, assume that at time t on the ground test point placement GPS receiver can determine the signal to reach the GPS receiver time Δt, plus receiver receives satellite ephemeris and other data can determine the following four equations:

$$\begin{cases} [(x_1 - x)^2 + (y_1 - y)^2 + (z_1 - z)^2]^{1/2} + c(v_{t_1} - v_{t_v}) = d_1 \\ [(x_2 - x)^2 + (y_2 - y)^2 + (z_2 - z)^2]^{1/2} + c(v_{t_2} - v_{t_v}) = d_1 \\ [(x_3 - x)^2 + (y_3 - y)^2 + (z_3 - z)^2]^{1/2} + c(v_{t_3} - v_{t_v}) = d_1 \\ [(x_4 - x)^2 + (y_4 - y)^2 + (z_4 - z)^2]^{1/2} + c(v_{t_4} - v_{t_v}) = d_1 \end{cases} \quad (2\text{-}33)$$

In the above four equations, x, y and z are the coordinates of the point to be measured, V_{t0} is the unknown clock parameter of the receiver, where $d_i = c\Delta t_i$, ($i = 1, 2, 3, 4$) T_i is the time that the signal of satellite i reaches the receiver. x_i, y_i, z_i is the space rectangular coordinate of satellite i at time t, V_{ti} is the clock difference of satellite clock, and c For the speed of light. From the above four equations to solve the coordinates of the measured point x, y, z and the receiver clock error V_{t0}.

According to the position of reference point, GPS can be divided into:

(1) Absolute positioning. Absolute positioning is also known as single point positioning, which uses receiver positioning method. In this type of positioning, absolute coordinates of the receiver antenna are used. This method of operation is simple, generally used in the application of the positioning accuracy is not high.

(2) Relative positioning. Relative positioning uses two or more receivers at the same time to observe the same set of satellites and to determine the position of the receiver antenna. GPS can also be divided according to the receiver in the operating state; the positioning method can be divided into:

Static position: When the point to be measured around the point of fixation without change of position or motion is quite slow and observation period cannot be detected. Only two times during the period of observation, the relative motion can reflect, making each of GPS data processing to be measuring points in the earth coordinate system protocol bits can be considered to be static. The point to be measured the amount known as static positioning.

Dynamic positioning: When the target point relative to the fixed point around in a possible perceived motion or apparent motion, determine the location of this dynamic test point is called dynamic positioning during observation. If in accordance with the principle of ranging, but also can be divided into test code pseudo range positioning, measuring phase pseudo range method positioning, positioning and so on.

2.3.2.1 Differential GPS positioning principle

1) Differential GPS principle

Differential GPS (DGPS, differential GPS-DGPS) is the use of precisely known 3D coordinates of the differential GPS base station. The correction of real time or afterward is sent to the user (GPS), revise the measurement data to the user to improve GPS positioning accuracy.

Different technology is used in test station for two stations on target measurements or at test station the is goal to view between measurements of seeking difference, thus eliminating public relations error, improves the positioning accuracy.

DGPS measurements require at least two GPS receivers (Figure 2-39). As were installed on known point coordinates i. e. ground points A and mobile carrier B. Two test station for the simultaneous measurement of navigation signals from the same GPS satellite (S_1, S_2, S_3, S_4), which measured station A receiver said as a reference receiver, measuring station B receiver known as a dynamic receiver. Reference receiver measured the coordinates and the known coordinates were compared, to obtain correct values and provide to the dynamic receiver, to eliminate or reduce the error.

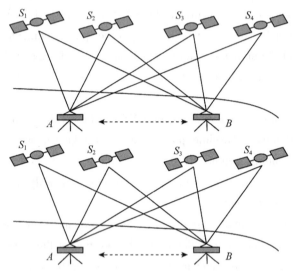

Figure 2-39 Differential GPS principle

2) Classification of differential GPS

According to the data transmitted by the differential GPS reference station, differential GPS can be divided into three categories, namely, position DGPS, pseudo range DGPS and phase DGPS.

(1) Position DGPS measurement. The DGPS data transmitted by the reference receiver to the dynamic user is ' the position correction value', which can correct the three-dimensional position of the user.

(2) Pseudo range DGPS measurement. The DGPS data transmitted by the reference receiver to the dynamic user is ' pseudo range correction value', which can correct the three-dimensional position of the user.

(3) Carrier phase DGPS measurement. The DGPS data transmitted by the reference receiver to the dynamic user is ' the carrier phase measurement correction value', which is to correct the carrier lag phase measured by the dynamic user and then solve the three-dimensional position of the user.

DGPS as compared to the normal GPS has (differential) corrected value. This change coincided with the improvement in the accuracy of GPS. The working principle of these three kinds of the differential mode is the same i. e. the base station transmits the corrected value, by the user receiving station and the measurement results are corrected to obtain accurate positioning results. The difference is that sending the correction value of the specific content is not the same; the differential positioning accuracy is also different.

CHAPTER 2 Technical Foundation of Forest Management (Focus on General Principle)

3) Pseudo range difference

The working principle of the pseudo range difference is described as follows.

At the base station A, the pseudo range of the reference receiver to the first GPS satellite is j:

$$\rho jA = \rho jAt + c(VTA - Vtj) + LjAtrop + LjAion \tag{2-34}$$

Form in:

ρjAt refers to real distance for t time A to satellite j;

ρjA refers to pseudo range for t time A to satellite j;

VTA refers to reference station receiver clock error;

Vtj refers to satellite clock error;

$LjAtrop$ refers to trop distance deviation caused by tropospheric delay at the reference station;

$LjAion$ refers to distance deviation for the ionospheric delay at the reference station;

C refers to electromagnetic wave speed,

The difference of the B point of the mobile station to the mobile station is corrected by the radio station:

$$\Delta\rho j = \rho jAt - \rho jA = - c(VTA - Vtj) - LjAtrop - LjAion \tag{2-35}$$

The pseudo range of B point observation is:

$$\rho jB = \rho jBt + c(VTB - Vtj) + LjBtrop + LjBion \tag{2-36}$$

The pseudo range correction (2-36) with type:

$$\rho jB + \rho jAt - \rho jA = \rho jBt + c(VTB - Vtj) + LjBtrop +$$
$$LjBion - c(VTA - Vtj) - LjAtrop - LjAion \tag{2-37}$$

Taking into account the conditions of the forest and the existing equipment communication capabilities, B, A points spacing is not more than 50 km, and thus

$LjBtrop \approx LjAtrop$, $LjBion \approx LjAion$, The above formula can eliminate the influence of satellite clock error, troposphere error, and ionosphere error. $c(VTB-VTA) = d$, The above formula can be expressed as:

$$\rho jAt + (\rho jB - \rho jA) = \rho jBt + d \tag{2-38}$$

Will (2-39) according to Taylor series expansion:

$$\rho jAt_0 + (\rho jB - \rho jA) + (- ljAtdXj - mjAtdYj - njAtdZj) =$$
$$\rho jBt_0 + d + (ljBtdXB + mjBtdYB + njBtdZB) + (- ljBtdXj - mjBtdYj - njBtdZj) \tag{2-39}$$

Because of A, B two-point distance is less than 50 km, so $ljAt \approx ljBt$, $mjAt \approx mjBt$, $njAt \approx njBt$, ρjAt_0、ρjBt_0 for ρjAt、ρjBt of the initial value, So according to the above formula can be eliminated most of the satellite ephemeris error, for further consolidation:

$$(ljBt\ mjBt\ njBt_1)(dXBdYBdZBd)T = (\rho jB - \rho jA) + \rho jAt_0 - \rho jBt_0 \tag{2-40}$$

When there are n ($n>4$) satellites, the pseudo range difference model is represented by the least square method:

$$G_{\delta_x} = L + V \tag{2-41}$$

$$\delta_x = (GTG) - 1GTL \tag{2-42}$$

2.3.2.2 GPS RTK principle

1) RTK overview

Real Time Kinematic(RTK) is a kind of real-time dynamic positioning technology based on carrier phase observation value. It can provide real-time test site in specifying the coordinates of three-dimensional positioning results, and to achieve centimeter-level accuracy. In RTK mode, base station link through the wireless data and its observations with the transfer station to the mobile station coordinate information. Mobile station not only receives carrier phase information from the base station but also received carrier phase information from the GPS, and the composition of the phase difference measuring the value of real-time positioning.

2) Composition and working principle of GPS RTK system

GPS RTK measurement systemis composed of three parts GPS receiving equipment, data transmission equipment and the software system. GPS receiving apparatus includes a base station and a mobile station GPS receiver. Data transmission system by transmitting station and mobile station to a receiving station, it is the key equipment to realize dynamic measurement; Software system the function of RTK real-time measurement technique calculates the three-dimensional coordinates of the mobile station is in the carrier phase observation time difference according to the GPS measurement technology. Its basic principle is a receiver as the base station, another station or several receivers as a mobile station, base station and mobile station simultaneously receives the same time signals from a GPS satellite radio transmission equipment. Through the data station observation received real-time reference sent to mobile station receiver; the observation station receives the base station flow transfer number. According to the future, according to the principle of relative positioning, a real-time solution for the whole week of ambiguity unknowns, and calculate the three-dimensional coordinates of the user station and the accuracy of the display.

Carrier Phase Differential GPS is divided into two categories: One is the base station that sends the carrier phase correction amount to the user station, to correct its carrier phase and then the solution coordinate; The other is the carrier phase reference station collects sent to the user to be evaluated poor solver coordinates. Whichever occurs RTK technology, the latter is true RTK technology (Figure 2-40). while the latter is a real RTK Technology.

In order to obtain the high accuracy real-time dynamic coordinate, RTK instruments now commonly used carrier phase difference measurement. Carrier phase measurement is divided into the key difference is requirements of integer ambiguity.

3) Principle of carrier phase measurement

Carrier phase measurement is responsible for GPS carrier wave in the propagation distance of the phase change value and determines the distance of signal propagation. From satellites, time t is required by phase for signal carrier (Figure 2-41), the propagation distance ρ to the receiver at K, signal phase φ_k, from the phase changes of the s k for ($\varphi_s - \varphi_k$).

($\varphi_s - \varphi_k$) including the whole week and less than a week of the fractional part, for the convenience of the carrier phase are in the number of weeks as a unit. If it can be measured ($\varphi_s - \varphi_k$), The distance ρ between the satellite and the receiver K is

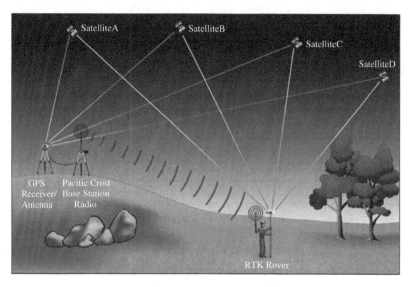

Figure 2-40 Schematic diagram of RTK measurement

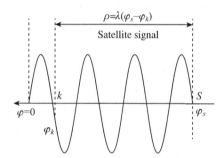

Figure 2-41 Schematic diagram of carrier phase measurement

$$\rho = \lambda \ (\varphi_s - \varphi_k) = \lambda (N_0 + \Delta\varphi) \tag{2-43}$$

Type, N_0 for carrier phase ($\varphi_s - \varphi_k$) (time t) of the whole week number part; $\Delta\varphi$ delta for less than a week the decimal part; lambda (λ) is the carrier wavelength is known values.

In the formula (2-43), we assume the same time both in the determination of T_i carrierphases satellites, determined by φ_k carrier phase at the receiver, the ($\varphi_s - \varphi_k$). But in fact, we cannot measure φ_s, so this method is not implemented. The solution to this problem is, if the receiver oscillator can generate a reference signal frequency and the initial phase of satellite carrier signal is exactly the same, at any time in the T_i phase reference signal receiver is equal to the satellite signal carrier phase. Therefore, as long as the determination of the receiver reference signal phase, the problem graphs (2-42) smoothly done or easily solved. Is the principle diagram the carrier phase measurement Figure (2-42) at any time in the satellite J T_i carrier phase $\varphi_s(t_j)$. Therefore, the determination of the phase reference signal and the received satellite carrier signal difference[$\varphi_k(t_j) - \varphi_k(t_i)$], which can be obtained by GPS signal transmission distance, $\rho = \lambda[\ \varphi_k(t_j) - \varphi_k(t_i)\ ]$.

Carrier phase measurements are received by GPS receiver satellite carrier signal and receiver local oscillator reference signal phase difference ($\varphi_{ki} t_k$ satellite carrier signal to represent K receivers clock time t_K to $k_i(t_K)$ received by the phase values, $\varphi_k(t_k)$ said K receivers in the clock time

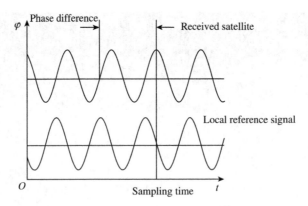

Figure 2-42 Principle of GPS carrier phase measurement

t_K of the local reference signal phase values, K receivers in the receiver clock time t_K observation satellite J to obtain the phase view measurement can be written as

$$\Phi_{ki}(t_k) = \varphi_k(t_k) - \varphi_{ki}(t_k) \tag{2-44}$$

The usual phase or phase difference measurements are measured in the phase of a week. In fact, if the count is carried out for a whole week, continuous phase measurements can be obtained from an initial sampling time (t_0).

At the initial time t_0, the measured less than a week of the phase difference is $\Delta\varphi_0$. The whole weeks N_{0j}, at this time the contains whole weeks phase observation value should be

$$\Phi k_i(t_0)v = \Delta\varphi_0 + N_{0j} = \varphi_{ki}(t_0) - \varphi_k(t_0) + N_{0j} = \varphi_k(t_0) - \varphi_{ki}(t_0) + N_{0j} \tag{2-45}$$

Receiver to track the satellite signal, continuous determination of less than a week of phase difference $\Delta\varphi(t)$ and the whole wave counter records from t_0 to time ti within the whole week number change amount of Int(φ), as long as the satellite S_j from t_0 to T_i between satellite signals without interruption, the initial time throughout the week fuzzy degree N_{0j} for a constant, so that any time t_i satellite SJ to K receivers of phase difference

$$\varphi_{ki}(t_i) = \varphi_k(t_i) - \varphi_{ki}(t_i) + N_{0j} + Int(\varphi) \tag{2-46}$$

From the above formula, from the first beginning, in the future observation, the concept of measurement includes the decimal part of the phase difference and the cumulative number of the whole week.

Carrier phase measurement, observation obtained whole weeks N_{0j} and lack of a week of phase difference $\Delta\varphi_0$ can be obtained the distance between satellite station, and interpret the coordinates of the receiver is calculated.

2.3.2.3 CORS system

1) The definition of CORS

The term 'CORS' is abbreviated as 'Continuously Operating Reference Stations' and also have abbreviation i. e. 'Continuously Operating Reference System'. The Chinese translation is continuous reference system. The two are essentially the same, refer to the user-oriented spatial location service infrastructure, the core is the long-term stability of the GNSS continuous operation of the reference station (also known as the base station), the control center.

2) The composition of CORS

CORS system is a typical distributed network system and is composed of 5 subsystems: continuous operation reference station system (reference station network), data communication subsystem, system control and data processing center (Center), customer service center and user application sub-system (user unit). In reference station system the whole system is composed of a plurality of data sources, with the regional distribution characteristics of the satellite tracking reference stations, for the realization of GNSS data, meteorological data, and information collection. The control system and data processing center is the core of the whole system. The server, work station, data storage device, communication equipment, the corresponding data processing & the database software, mainly with data processing, system management, information service and network management functions is responsible for control, monitoring, download, processing, distribution, and management of the reference station data, dynamic state data service function is to rely on wireless communication (such as GSM, RDS, GPRS, CDMA, etc.) achieve the structure of the CORS system as shown in Figure 2-43.

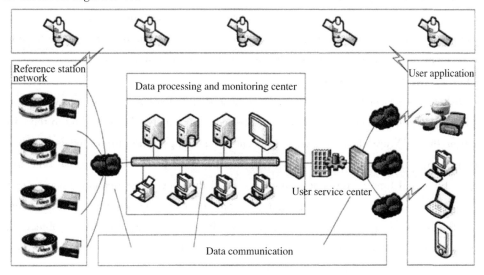

Figure 2-43 CORS basic system structure

3) The working principle of CORS

At present, the real-time dynamic positioning of CORS technology mainly includes virtual reference station (VRS), area correction technology (FKP), main and auxiliary technology (MAC) station.

(1) The virtual reference station (VRS). The virtual reference station isa station set up a certain number of references in a certain area, the accurate determination of base station location and the rate of change of the base station for receiving the satellite signal and transmits information to the information processing center. Data processing center and receiving station location information flow receiver outline send the data processing center according to the location of the mobile station, select a virtual reference station, then the base station according to the error information through

some mathematical model of interpolation error of the virtual station. The virtual mobile station to correct data broadcast for the mobile station, the virtual reference station location is usually a true position in the mobile station around the range of 5 m, ensure the correlation between virtual reference stations and error.

(2) Area correction technology (FKP). Area correction method using GPS reference station observation data, mainly phase observation value and pseudo range observations, and base station known coordinate data calculated the benchmark network with temporal or spatial correlation of the error correction model, ran after the measurement points of the approximate coordinates in measurement error corrections. Its application to observed values, thereby eliminating all space and time-related errors and improve the accuracy of positioning results.

(3) Main and auxiliary station (MAC) technology. Main and auxiliary station is to simplify the phase distance for a public all week unknown level, main and auxiliary method to send the main reference stations all positive and coordinate information, the auxiliary station only broadcast with respect to the primary reference station differential corrections and coordinate difference for reference, which can reduce reference station network data broadcast content. Using a single differential scattering and non-scattering phase correction of formation compression differential information, the regional error change positive model by mobile station equipment custom.

4) The classification of CORS

CORS has a variety of classification methods, according to the construction management units. Such as IGS, CORS of the national network and the cooperation network; according to the coverage area, such as the EPN area level, regional level; also according to the purpose of use, such as the reference network. In addition, it can also divide according to the service response time into two types, static type, real-time performance summary. From the above classification, resource distribution and management perspective we can make a comprehensive classification. Among them, according to the service response time from the performance point of view, the service object classification from the perspective of resources; according to the nature of builders, managers are divided from the above classification summary.

This classification method is the most important in all three levels of CORS i. e. the professional level, regional level, national level. CORS by the professional department of construction and management, from coverage, may belong to the region may also belong to the state, Such as China's CMONOC and RBN DGPS. Although they covered most parts of the country, but in the service object still belongs to the professional level. Regional CORS contains micro, city level, regional level (including the provincial, District, inter-provincial etc.) three. National contains two meanings, first from the perspective of the distribution of resources, should be the basic coverage of the national territory range. Secondly, from the perspective of management, it is by the central government or the competent administrative department of construction and management of the direct investment, in terms of management authority, service content, regional level, and professional level have an obvious difference. According to the service response time, CORS can be divided into two types: static CORS (Static CORS, STCORS) and real-time CORS (Real Time CORS, RTCORS).

RTCORS mainly uses real-time positioning technology such as DGPS, PPP, WADGPS, RTK, network RTK, etc., centimeter level real-time positioning means RTK, network RTK technology and in recent years the development of PPP technology. RTCORS is in the system of structure, management, information flow, etc. RTCORS has become one of the hotspots in the field of CORS construction because of the need of RTCORS to further meet the needs of geospatial information service in the service response (Figure 2-44).

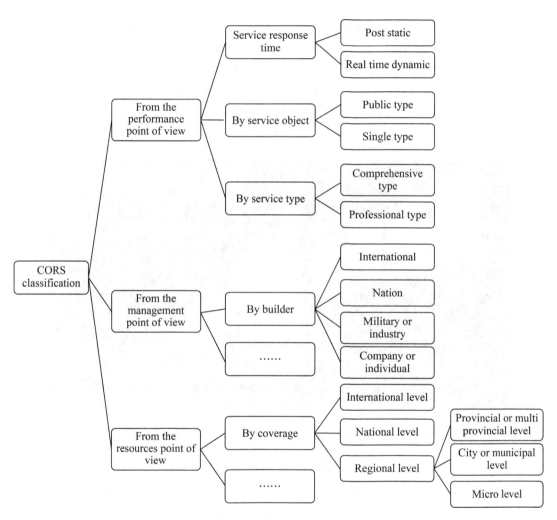

Figure 2-44　CORS classification structure

2.3.3　GALILEO system brief introduction

Galileo is the global navigation satellite system (GNSS) that is currently being created by the European Union (EU) and European Space Agency (ESA), the use of basic (low-precision) Galileo services is free and open to everyone. It is independent, global, controlled by Europeans, and with a satellite-based civilian navigation and positioning system.

1) Characteristics of GALILEO System

The GALILEO system is the first global satellite navigation system designed for civilian purposes in the world. It is more advanced, more reliable and more accurate than the GPS system. The main characteristics of the GALILEO system are independent systems, better compatibility, advanced and competitiveness and public cooperation in the world.

The orbit of Galileo satellite is at an elevation of 24 000 km and has higher accuracy than the GPS. The GPS system in the USA provides signals to other countries enough to find objects about 10 meters above the ground, and Galileo system can find 1 meter long targets.

2) The composition and operation of GALILEO system

From December 1999, Spain proposed the first set of solutions about establishing a civil positioning and navigation system controlled by the Europeans, after more than 1 year of study today the GALILEO system is being perfected from 15 solutions in 6 EU countries. The GALILEO system is composed of the constellation, the GALILEO satellite, and the ground part (Figure 2-45).

Figure 2-45 Simulation chart of GALILEO satellite

Satellite constellation of GALILEO system is made up of 30 medium altitude orbit satellites which are distributed in three orbits each of which has 10 satellites, 9 for normal work and 1 on standby (Table 2-8).

GALILEO satellite is composed of three parts, including satellite platform, navigation payload and search and rescue payload. Satellite platform has satellite power, thermal control system, data processing, other than that also have control system and orbit control system, propulsion system, electric control system, automatic measuring finally recording and tracking systems. Navigation payload includes satellite clock, antenna, and RF. Search and rescue effective load are mainly sophisticated communications equipment.

Table 2-8 Constellation parameters of GALILEO

Constellation parameters	Value
Satellites' number per orbit	10(9 normal work and 1 standby)
Number of satellite orbits	3
Number of satellites	30(27 normal work and 3 standby)
Orbit inclination angle	56°
Orbit height	23 616 km
Operation period	14 hours 4 minutes
Satellite life-span	20 years
Satellite weight	625 kg
Power supply	1.5 kW
Radio frequency	1202.025 MHz 1278.750 MHz 1561.098 MHz 1589.742 MHZ

2.3.4 A brief introduction gLONASS system

'GLONASS' in The Russian language is also known as 'ГЛОНАСС' which is the abbreviation of 'Глобальная навигационная спутниковая система', it means GPS.

1) Characteristics of GLONASS System

GLONASS system wasearlier controlled by the Russian military. In 1982 launched its first satellite and from this time to 1995 it had launched a total of 73 satellites. After the collapse of the Soviet Union to Russian Federation, due to economic reasons the last century on-orbit satellites were only nine. Only two satellite signals per day In December 2003, Russia developed a new generation of satellites delivered to the Federal Space Agency and the Department of Defense trial for the new GLONASS system. November 2011, Russian launched three 'GLONASS-M' global navigation satellites successfully into space. GLONASS system has 28 satellites, is expected in 2015 completed.

Despite the GLONASS's positioning accuracy is slightly lower than the GPS but its anti-interference ability is the strongest point. GLONASS can provide high precision 3D position in all-weather condition for land, sea, and air. Three-dimensional velocity and time information of GLONASS is like GPS. In addition, it also has applications in the surveying and mapping, geological survey, oil development, earthquake prediction, telecommunications and ground transport also and much more(Figure 2-46).

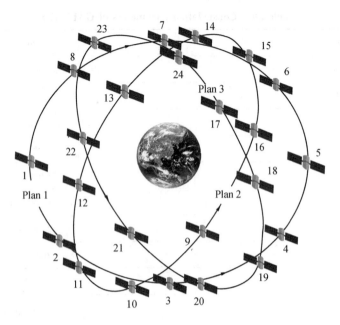

Figure 2-46 GLONASS Satellites

2) The composition and operation of GLONASS system

GLONASS system is composed of three parts: the GLONASS constellation, the ground supportsystem, and the user equipment.

GLONASS constellationis carried out by 21 satellites and three spare satellites, uniformly distributed in a nearly circular orbit, each orbital has eight satellites, orbit height is 1. 91 million kilometers, running period of 11 hours and 15 minutes, the orbital inclination angle 64. 8° (Table 2-9).

Table 2-9 Characteristics of GLONASS satellite orbit

Satellite parameters	GLONASS
Number of satellites (stars)	24
Satellites number per orbit	8
Number of satellite orbits	3
Adjacent orbit satellite phase difference	15°
Orbit inclination angle	64. 8°
Orbit long axis	25 510 km
Orbit height	19 100 km
Eccentricity	<0. 01
Operation period	11 hours 15 minutes and 44 sec
Satellite life-span	3~5 years

The ground support system is composed of the systém control center, the central synchronization device, the telemetry and remote control station and the field navigation control equipment (Table 2-10).

CHAPTER 2 Technical Foundation of Forest Management (Focus on General Principle)

Table 2-10 Ground support system

Satellite parameters	GLONASS
Master control (controlcenter)	Moscow
Injection along	Using the command tracking station in Russia to send control commands and navigation information to the satellite.
Monitoring along	There is the Call Trace System(CTS). It can distribute all satellite telemetry, range data processing, sent the signal to the master station in Russia

GLONASS user equipment is the receiver, providing both military and civilian services. It can receive the satellite navigation signal, measure thepseudo range and pseudo range rate of change, and from the satellite signal extraction, processing of navigation information, calculate the user's location, speed and time information.

2.3.5 Beidou system brief introduction

Beidou satellite navigation and positioning systems are China's active three-dimensional regional satellite positioning and communications systems, which is developed independently by China. After the US Global Positioning System and the Russian GLONASS, this is the third mature satellite navigation system. Beidou system operates in 2491.75 MHz frequency and can provide high-precision, real-time location services to regional users in all kind of weather. The accuracy of up to tens of nanoseconds, with considerable GPS precision, can play a substitute role of GPS(Figure 2-47).

Figure 2-47 Beidou satellites

1) The composition and operation of Beidou system

The Beidou satellite navigation system is composed of the space segment, the groundsegment, and the user segment. Beidou satellite navigation system space program consists of 35 satellites; including 27 are the earth's orbit satellite, 5 geostationary orbit satellites and 3 tilt synchronous orbit satellites. Five geostationary satellites are located at 58.75°, 80°, 110.5°, 140° and 160°. Medium earth orbit satellite is on the three track surface and the orbits are separated by 120 degrees. Until the end of 2012, Beidou has officially launched 16 satellites of which 14 networks and provide services, including 5 geostationary satellites, 5 tilt earth orbit satellites and 4 satellites in earth's orbit.

2) Working principle of Beidou satellite navigation system in China

Beidou satellite navigation and positioning systemare composed of three parts: navigation communication satellite, ground control center and the user terminal, as shown below (Figure 2-48).

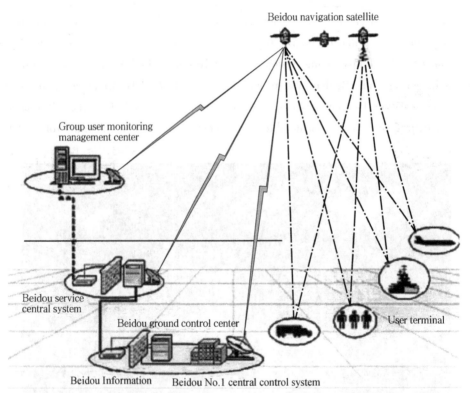

Figure 2-48 Composition sketch map of Beidou satellite navigation system constitution

The navigation communication satellite part is composed of three geostationary orbit satellites of which two satellites are operating satellites and a satellite is a spare satellite, satellite work is forward the task of execution ground control center and bi-directional radio signal when working. Ground control center includes a main control station, measuring rail station, altimeter station, correction station and computing center. It is mainly used to measure and correct parameters of navigation and positioning, in order to adjust the operation posture and pathway of satellite, at the same time the ground control center also prepares the ephemeris and completes the user position correction information and does positioning work for the user(Figure 2-48).

The user side is the device that is directly used by the user and it is mainly used to accepting the distance signal of ground control center forwarded by satellite. Beidou satellite navigation and positioning systems realize positioning based on ultimate principles of 'double-star positioning'. Two spherical surfaces are formed regarding the known coordinates on the track as the center of a circle and regarding the distance from the user terminal to two working satellites as radius and the user terminal is located on the arc of two spherical surface intersecting lines. Using mathematical methods to solve point of intersection of arc and earth surface can obtain the position of the user. Because positioning signal shall be sent to location satellite by user terminal when positioning and the user location shall be calculated according to the deviation of time from signal to positioning satellite, so it is called 'active location'.

Work flow of Chinese Beidou satellite navigation system is as follows:

As shownin Figure 2-49, the ground control center simultaneously sends inquiring signal to two working satellites and broadcasts to users in the service area via satellite transponder. The user responses to inquiring signal of one satellite and sends back response signal for two satellites, and the response signal forwards to control system of ground control center via satellite, control system of ground control center receives and demodulates the signal sent by the user and then the corresponding data processing is done according to the service content of user application for the location application of user. Central control system measures inquiring signal from the central control system and the inquiring signal is transmitted to the user via satellite. Then user emits positioning response signal, the response signal is transmitted to delay of central control system via the same satellite; the inquiring signal is given out from central control system and reaches the user via the same satellite, the user gives out a response signal, which is transmitted to two time delays of cen-

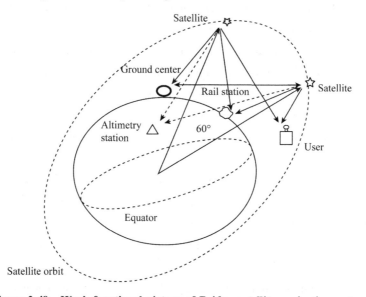

Figure 2-49 Work functional picture of Beidou satellite navigation system

tral control system via the other satellite.

Due to the location of ground control centers and two satellites are known, the distance from the user to the first satellite can be calculated by the above retardations and the sum of the distance from the user to two satellites. So it is known that the users are on a spherical surface regarding the first satellite as the center of the sphere and on the line of intersection between ellipsoids regarding two satellites as the focus. In addition, the central control system looks up to the elevation value of user from the digital topographical map stored in the computer and it is also known that users are on the ellipsoid parallel to earth criterion ellipsoid. Therefore, the central control system uses the numerical map to work out three-dimensional coordinates points for the presence of a user and sends the relevant information and contents of the communication to satellite as well as the relevant information and contents of the communication to the user or recipient via satellite transponder.

Some features of Chinese Beidou satellite navigation system are:

(1) The number of satellites is less, the structure is simple;

(2) Mainly focus on the key station of satellite ground;

(3) The user equipment is simple; the price is cheap and not expensive;

(4) Have the capabilities of a continuous positioning and provide two-way broadcast communication service.

The disadvantage is only two-dimensional positioning, high-dynamic positioning cannot be done, and a number of users are limited.

3) Beidou satellite navigation system has three main functions

(1) Rapid positioning: The system can provide high precision, fast real-time position service for the user in service area in all-weather conditions (can be completed within 1 second) and the positioning accuracy is 20~100 m.

(2) Message delivery: System user terminal has functions of two-way sending messages, depending on the different user rights. Registered users can send message to monitoring center by different time and can receive messages sent from monitoring center without time limit.

(3) Precision timing: The system has uni-directional and bi-directional timing functions. Depending on the different accuracy requirements, the timing terminal is used to complete the synchronization of time and frequency between user terminals and Beidou satellite navigation positioning system provides time synchronization precision of one-way timing 100 ns and two-way timing 20ns.

4) The advantages of the Beidou satellite navigation positioning system

(1) Compared with GPS and GLONASS, Beidou satellite navigation positioning system not only can do accurate positioning and timing services but also can complete the function of message sending under the condition of without support of other communication system, which is convenient for the communication of user side and supervision terminal.

(2) The coverage rate is better in China and there is basically no communication blackout. Beidou system not only covers China but also provide services to neighboring countries and re-

gions.

(3) Suitable for a wide range of remote monitoring, command, and management for group customer.

(4) The unique design of central node type positioning treatment and commanding user machine. It not only can make user know their location but also can tell others about their position, particularly can be suitable for the site which needs navigation and mobile data communication, such as transportation, dispatching, and commanding, search and rescue, GIS real-time inquiries.

(5) Positioning and timing system by independent research and development uses an encryption algorithm with seniority number, the system is stable and reliable the signal is safe, which is suitable for the domestic departmental applications with confidentiality and security.

(6) There is no need to lay ground station as a receiving terminal and the using cost and expenses are low. The potential is mainly reflected in the integrated fields of positioning communication. Currently, there are some users with positioning requirements and they do not have urgent need of using Beidou; for the users that not only need positioning but also need information transfer and involves the industries related to the security and confidentiality, applications, the application of Beidou satellite navigation and positioning system is very meaningful.

5) The characteristic comparison of Beidou satellite navigation positioning system and GPS system

(1) Different coverage area. Beidou satellite navigation positioning system mainly covers our local and surrounding countries and regions, it is a regional navigation positioning system. The coverage area is from 70° E to 140° E, 5° N to 55° N, however, GPS is GPS navigation and positioning system.

(2) Different positioning principle. Beidou satellite navigation positioning system is an active two-way ranging two-dimensional navigation, three-dimensional positioning data are provided to user after the resolving of control system of ground center; however, GPS is a passive one-way pseudo-code ranging three-dimensional navigation, and the three-dimensional positioning data of the user are independently solved by user device.

(3) Different orbital characteristics. Beidou satellite navigation positioning system is synchronous satellite, equatorial angle distance is approximately 60°, while the GPS is set 24 satellites on six orbital planes, equatorial inclination of pathway is 55°, equatorial angle distance of track surface is 60°.

(4) Different user capacity. Beidou satellitenavigation positioning system has active two-way ranging inquiry / response system. The user capacity depends on channel blocking rate, inquiring signal rate allowed by the user and the response rate of users, so user capacity of Beidou satellite navigation positioning system is limited. GPS is a one-way ranging system and the user can do range-based localization by receiving navigation message given by navigational satellite, so users capacity of GPS is unlimited.

(5) Positioning accuracy difference. three-dimensional positioning accuracy of Beidou satel-

lite navigation and positioning system is about 20 to 100 m, timing accuracy is about 100 ns. Three-dimensional positioning accuracy of the P code of GPS system can be up to P 6 m, while the three-dimensional positioning accuracy of C/A code is about 12 m, precision timing is about 20 ns.

2.4 3D laser scanning measurement principle

Laser scanning technology is also called 'real replication technology', is a high-technology appeared inthe mid-1990s and is a technological revolution after following GPS mapping technology. It breaks the traditional geodetic measurements such as triangulation and GPS single point measurement method. It breaks through the traditional geodetic measurements such as single point triangulation and GPS measurements by high-speed laser scanning, rapid access to a large area of high resolution spatial 3D point cloud data, measured object surface by automatically fitting real reflecting feature appearance with fast, non-contact, high penetration, dynamic, initiative, high-density, high-precision, digitization, automation and other features. This technology and GPS can combine in many complex environments together to achieve stronger and broader applications. In the field of Surveying and mapping instruments in common use are HDS series of Leica company production and production of Canadian company Optech ilris-3d 3D laser scanner, France for MENSI Corporation GS200, Austria RIEGL company produced VZ-4000. (For example precision, scanning speed, scanning distance, and laser archery divergence) may be a different quality of different products to get the data, but essentially the basic working principle of all is the same(Figure 2-50 and Figure 2-51).

Figure 2-50 Leica HDS8800 　　　　Figure 2-51 FARO focus-3d

2.4.1 Principle of 3D laser scanning technology

Three-dimensional laser scanning technology is a new technology in recent years, which has attracted more and more attention in the field of research in China. 3D laser scanning technology uses laser ranging principle by recording densely crowded space three-dimensional coordinates of the measured object surface, reflectance, texture information and the rapid reconstruction of the measured target 3D model of various maps data. Since the three-dimensional laser scanning system can get a large number of data points, so with respect to the traditional single-point measurement, 3D laser scanning technology is also known to be evolved from a single point measurement to a revolutionary breakthrough in technical measurements. The 3D laser scanning system includes the hardware part and the software part of the data collection.

The structure of the 3D laser scanner mainly includes the following parts: A high-speed precise laser ranging system, a set of laser reflection with a uniform angular velocity guide laser reflection and with a uniform angular velocity scanning reflection prism, horizontal deflection range controller, high deflection angle controller, data output processor (laptops), part of the instrument also has a built-in digital camera so that you can get the point cloud data directly to the target surface. In the software, the different manufacturers of a three-dimensional laser scanner with its own system software, such as the French FARO company's FARO Scene, Cyclone's Leica software, as well as Geomagic, Polywork software, etc.

Three-dimensional laser scanning principle mainly used the principle of laser ranging as shown in Figure 2-52. Pulse ranging is a kind of high-speed laser measuring technology. The principle of ranging system ranging range can reach several hundred meters to kilometers in distance. The laser ranging system is mainly composed of the laser transmitter, receiver, time counter and a microcomputer. Pulse ranging method mainly includes the following 4 processes:

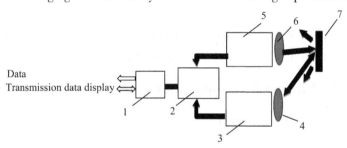

Figure 2-52 Schematic diagram of laser ranging principle

1. Microcomputer   2. Time range unit   3. Photodiode receiver
4. Receiver lens   5. Laser diode   6. Emitter lens   7. Target

(1) Laser Emitting. a laser pulse emitter under the action of the pulse trigger emits a very narrow speed laser pulses and turning toward the body by a scanning mirror, while the laser signal is obtained by sampling the main pulse;

(2) Laser detection. a laser echo signal that is reflected back from the ground through the same scanning mirror and telescope and converted into electrical signals.

(3) Delay estimation. to deal with the irregular echo signal, the time delay of target ranging is determined accurately and the echo pulse signal is generated;

(4) Time delay measurement. the time interval between the laser echo pulse and the main pulse of the laser is measured by a precision counter controlled by a precise atomic clock.

The 3D laser scanning measurement system is based on the specific coordinate system, which is based on the point cloud data acquired from the object with the spatial position information. This special coordinate system is called the instrument coordinate system and the coordinate axes of different instruments are not in the same direction. Is usually defined as the origin of coordinates in the laser beam i. e. Z axis is located within the instrument of vertical scanning plane upward positive; X axis is located perpendicular to the instrument of horizontal scanning plane and the Z axis; the Y axis is located on the instrument of horizontal scanning plane and the X axis vertical. At the same time, the Y axis is pointing in the direction of objects and with the X axis and Z axis together constitutes a right-handed coordinate system, as shown in Figure 2-53.

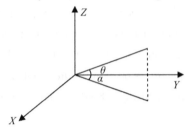

Figure 2-53 Positioning principle of ground 3D laser scanning system

Three-dimensional laser scanning point coordinates (X, Y, Z), the formula is:

$$\begin{cases} X = S\cos\theta\cos\alpha \\ Y = S\cos\theta\sin\alpha \\ Z = S\sin\theta \end{cases} \quad (2\text{-}47)$$

The vertical direction of the laserbeam said theta angle alpha said horizontal direction angle of the laser beam, S said the instrument to the oblique distance scanning point.

2.4.2 Function ofthree-dimensional laser scanning measuring instrument

1) Three-dimensional measurement

In traditional concepts of measurement, the final output of the measured data is the two-dimensional results (such as CAD drawings). The choice of measuring instruments like total station instrument, electronic theodolite, GPS is in the frontline in the development of 3D laser scanning technology today, 3D measurement gradually replace the 2D. 3D laser scanner for each measurement data contains not only the X, Y, Z point of information but also including R, G, B color information, along with information about the object of anti-color rate, such comprehensive information can give an object's visual effects on the computer that true representation of the general means of measurement cannot give.

2) Fast scan

In the conventional measurement methods, measurement of each point ranging from 2~5 seconds or even take a few minutes to measure the coordinates of a point but with the birth of 3D laser scanner change this situation. The initial measurement speed of 3D laser scanner about 1000 points per second give the world a new way and now the pulse scanner (scan station 2) has reached a maximum speed of 50 000 points per second. Phase scanner Surphaser 3D laser scanner with the highest speed has reached 1 million 200 thousand points per second, which is the basic guarantee of 3D laser scanner, a detailed description of the ancient objects, factory pipes, tunnels, and other complex terrain areas that were unmeasurable in the past.

2.4.3 Classification of 3D laser scanner

Three-dimensional laser scanners can be divided into different types on the basis of measurement i. e. pulse type, the phase difference and the principle of triangulation. Based generally on the principle of the phase difference of 3D laser scanner the measurement range is short only about one hundred meters and based on the principle of the pulse the range of the 3D laser scanner. Measurement process is longer, the measurement range reached as far as 6 km. In accordance with different carriers and different scanning platform 3D laser scanning system can be divided into airborne, vehicle-mounted, handheld and ground categories. Normally in accordance with the effective scanning distance of three-dimensional laser scanner they can be divided into the following.

1) Short range laser scanner

In short range laser scanner the longest scanning distance is less than 3m, generally the best scanning distance 0.6~1.2 m, usually this kind of scanner is suitable for measuring small molds, not only the scanning speed and high precision, accuracy up to a few tenths of millimeters i. e. Minolta vivid 910 high precision 3D laser scanner, handheld 3D data scanners fast Scan and so on.

2) Medium range laser scanner

The medium range 3D laser scanner is mainly applied in indoor places of the measured object or is still life measurement of large-scale generally for measuring three-dimensional statue of objects and the longest scanning distance is less than 30 m so its application mainly in small business.

3) Long range laser scanner

Three-dimensional laser scanner with scanning distance larger than 30m is a long distance 3D laser scanner, which is mainly used in the measurement of buildings, mines, dams, stations, large civil engineering and so on i. e. Austria Riegl, Switzerland Leica, the United States Faro and so belong to this type of scanner, the scanning distance of about 100~1000 m. LiDAR Terrestrial (TLiDAR), such as terrestrial 3D laser scanning system. Ground 3D laser scanning system (TLiDAR) is a 3D laser scanner based on the principle of the pulse. TLiDAR consists of the three-dimensional laser scanner, power supply, digital camera, registration ball, support and ancillary equipment, the use of non-contact laser measurement methods to obtain the surface of the target object point cloud data. The system to a certain sequence emits a narrow beam of the laser pulse and scanning the object surface, reflected by synchronous rotation and reflection receiving mirror received echo pulses, by a laser pulse transmitting and receiving time difference calculation of the distance and angle of the scanner and the object, while scanning control module for controlling and measuring, finally by the scanner's own coordinate system automatically calculates the relative 3D coordinates of the object to be measured.

4) Aviation 3D laser scanner

In aviation 3D laser scanning theScanning distance is usually greater than 1000 m and need to be equipped with precise navigation and positioning system that can be used for a wide range of terrain scanning measurement that is often referred to as the airborne lidar system. The system is mainly through the laser radar launch a pulse signal to the measured object, the reflected signal transmission and delivered to the lidar system and according to the background data obtained from the analysis of the target data measured. The system makes the measurement data acquisition and processing more efficient, so it can be widely used in many fields, such as shape mapping, environmental detection, 3D city modeling, earth science, planetary science and so on.

2.4.4 Application and prospect of three-dimensional laser scanning measuring instrument

In recent yearsthe three-dimensional laser scanning technology continues to develop and continuously maturing and 3D scanning equipment has gradually commercialized. The great advantage of 3D laser scanner is that it can quickly scan the object to be measured, without reflecting prism can directly obtain high precision scanning point cloud data. In this way, the real world can be effectively modeled and reproduced. Therefore, it has become one of the hot spots in the current research and has applications in the variety of fields like preservation of cultural relics, interior design, legal evidence, civil engineering, industrial measurement, plant transformation, traffic accident processing, ship design, pipeline design, mining, natural disaster survey, digital city terrain visualization, urban planning, tunnel, bridge reconstruction and military analysis.

1) The field of surveying and mapping engineering

3D laser scanning is used in the construction of dam and power station based topographic survey, highway surveying and mapping, surveying and mapping of railways, waterway surveying and bridge mapping, surveying and mapping of buildings foundation, tunnel detection and deformation monitoring, the dam deformation monitoring, underground engineering structure like tunnel, mine surveying and volume calculation.

2) Measurement structure

Bridge rebuilding extension project, bridges structure measurement, structure detection, monitoring, measurement of geometric size, spatial location conflict measurement, volume measurement, 3D high-fidelity modeling, offshore platform, measurement of ship-building factory, power plants, chemical plants and other large industrial enterprise interior equipment measurement; pipe, line measurement, all kinds of machinery manufacture and installation.

3) Protection of buildings and monuments measurements

Building interior and exterior measurement fidelity, monuments (ancient buildings, statues, etc.) the protection measure, restoration of cultural relics, ancient building surveying, data retention and other monuments protection, mapping sites, fake imaging, site virtual model, on-site protective image recording.

4) Emergency services

Counter-terrorism, land reconnaissance and attack, surveillance, mobile reconnaissance, disaster estimation, traffic accident orthophoto map, crime scene orthophoto map, forest fire monitoring, landslide debris flow warning, disaster early warning and on-site monitoring, nuclear leakage monitoring.

5) Entertainment industry

For the design of film products like the design of film actors and scenes, 3D game development, virtual museum, virtual tour guide, artificial imaging, virtual scenes.

6) Mining

In the open pit mine and metal mine operations, as well as some hazardous areas inconvenient to reach. For example the collapse areas, caves, cliff-side and another three-dimensional scanning.

In recent years, 3D laser scanner has been developed towards the direction of development, the most representative is onboard laser scanner and three-dimensional laser airborne radar(Figure 2-54 and Figure 2-55).

Figure 2-54　Optech vehicle vehicle 3D scanner

Figure 2-55　IP-S2 TOPCON laser mounted 3D laser scanner

The system sensor part of the vehicle mounted three-dimensional laser scanner is integrated into a transition plate which can be firmly connected with the common roof luggage rack or the custom component. The position of the bracket, position of the laser sensor head, digital camera, IMU and GPS antenna can be adjusted separately. The high strength structure ensures that the relationship between the relative attitude and position between the sensor head and the navigation device is stable and constant. Can be used in highway survey, maintenance and inspection, highway asset inventory (traffic signs, noise barrier, guardrail, drain, drainage ditch), road detection (cutting, road surface road deformation), highway geometric model (transverse and longitudinal profile analysis), structure analysis (overpass), flooding analysis and evaluation, GIS overlay analysis, analysis of landslide, hazard assessment (measurement and hazard analysis of landslide deformation, talc and water analysis), traffic flow analysis, safety assessment and environmental pollution assessment and earthwork analysis, driving vision and safety analysis and other aspects.

Airborne laser 3D radar system (light detection and ranging (LIDAR) is a set of laser scanner (scanner), global positioning system (GPS) and inertial navigation system (INS) and high resolution digital camera technology in a body of light mechanical and electrical integration system, used to obtain laser point cloud data and generate accurate digital elevation model (DEM), DSM (digital surface model), and access to the object DOM (digital orthophoto map information, through the laser point cloud data processing, can get the real 3D scene graph.

2.5 Space remote sensing platform

With the development of remote sensing technology, more and more high-resolution sensors, the remote sensing image information becomes more and more precise. The satellite image plays-more and more important roles in updating spatial imaging and topographic mapping field. Satellite remote sensing data has been widely used in making satellite images, thematic map making, update and revision of general maps, aeronautical chart, topographic map, and the production of shallow sea area charts.

IKONOS satellite is a milestone in the history of commercial satellite remote sensing resolution development. It is the first time in the civil field that the ground resolution of satellite-borne sensor reaches 1m. The new generation of 0.5 m satellite WorldView-1 and GoeEye-1 were successfully launched on September 18, 2007, and September 6th, 2008. GoeEye-1 satellite has currently the world's highest spatial resolution in commercial remote sensing satellites and its ground resolution is 0.41 m. IKONOS, QuickBird, GoeEye-1, WorldView- I / II provided the high-resolution image into the world of remote sensing image data market, greatly reduced the gap between the spatial resolution satellite images and aerial images. Break the situation that large-scale topographic map surveying and mapping can only rely on aerial photography. Table 2-11 lists the basics of several remote sensing satellites carrying high-resolution optical sensors.

Table 2-11 On-orbit high-resolution optical sensor

Satellite	Country	Launch date	Spatial resolution	Passage-way	Stereoscopic capability	Video overlay/km
SPOT-5	France	2002.05	2.5	1+4	Same/differentorbit stereo	60×60
IKONOS	USA	1999.09	1.0	1+4	Same/different orbit stereo	11×11
QuickBird	USA	2001.10	0.61	1+4	Stereo	16.5×16.5
TABLEOSAT-2	China	2004.05	2.0	1+4	Same / different orbit stereo	24×24
KOMPSAT-2	Korea	2006.07	1.0	1+4	Different orbit	15×15
Cartosat-1(IRS-P5)	India	2005.05	2.5	1	Same orbit	29/26
ALOS PRISM	Japan	2006.01	2.5	1+4	Same orbit	35,70
GeoEye-1	USA	2008.09	0.41	1+4	stereo	15.2
Resurs-DK1	Russia	2006.06	0.9	1+4		28×28
OrbView3	USA	2003.06	1.0	1+4		8×8
EROS-A	Israel	2000.12	1.1	1		9×9
EROS-B	Israel	2006.04	0.7	1	Same orbit	7×7
WorldView- I	USA	2007.09	0.5	1	Stereo	17.6
WorldView- II	USA	2009.10	0.46	1+8	Stereo	16.4
Day draw	China	2010.08	2.0	1+4		60
Pleiades-1	France	2011.12	0.5	1+4	Same/different orbit stereo	20
Resource 3	China	2012.01	2.5		Same/different orbit stereo	50
SPOT-6	France	2012.09	1.5	1+4	Same/different orbit stereo	60×60

The sensors of the commercial remote sensing satellites above adopt linear array CCD. According to the push-broom scanning imaging, the high resolution ground panchromatic and multi-spectral images can both be obtained. In order to shorten the revisit period, the satellites can side-look with a certain angle of the side in vertical to the running direction, so as to obtain an image of the adjacent tracks nadir (different orbit stereo). At the same time, in order to obtain the track with a stereo image pair, it can get foresight and backsight imaging along the track direction, Table in stereo pair with no significant time difference (same orbit stereo).

To obvious advantages of the high-resolution satellite is a high spatial resolution. It can acquire spatial data which is comparable to aerial data. It includes American IKONOS data and Quick Bird data, the France SPOT-5 satellite data, Israel EROS-A/B satellite data, as well as American GeoEye-1 satellite data and WorldView-Ⅰ/Ⅱ satellite data.

The following is the introduction of this high-resolution satellite in details:

1) IKONOS

The IKONOS satellite was launchedsuccessfully on September 24, 1999. It is the world's first commercial satellite to provide high-resolution satellite image. The successful launch of IKONOS satellite, can not only provide high definition and resolution satellite images of 1m but also develop a new more efficient, more economical way of obtaining a latest foundation of geographic inTableation. What's more, it established the standard of the new commercial satellite image(Table 2-12).

Table 2-12 IKONOS Satellite main parameters

Band	Wavelength/μm	Resolving power/m	Other
Blue	0.45~0.53	4	Track height: 681 km Track inclination: 98.1° Track running speed: 6.5~11.2 km/s Image acquisition time: 10:00 am~11:00 am Orbital period: 98 min +Track type: Solar synchronization Scanning band width: 11 km
Green	0.52~0.61	4	
Red	0.64~0.72	4	
Near IR	0.77~0.88	4	
Pan	0.45~0.90	1	

2) QuickBird

QuickBird is a high-resolution commercial remote sensing satellite launched by the United States on October 18, 2001. The spatial resolution of QuickBird is 0.61 m. It has significant advantages (main parameters are shown in Table 3-10) in multispectral imaging (a panchromatic band, a multispectral band), imaging width (16.5 km × 16.5 km) imaging pendulum angle. It can meet the more professional, more widely field of remote sensing users. It provides users with better, faster source of remote sensing(Table 2-13).

3) SPOT-5

The SPOT-5 satellite was launched in May 2002. The satellite carries two high-resolution collection devices. The same as the previous SPOT-1, 2, 3, 4 satellite, each detector can deflect at a certain angle, which makes the SPOT-5 satellite revisit the same locations in every 5 days.

Table 2-13　Quick Bird satellite main parameters

Band	Wavelength/μm	Resolving power/m	Other
Blue	0.45~0.52	2.44	Track height: 450 km
Green	0.52~0.60	2.44	Track inclination: 98° Revisit period: 16 days (70 cm resolution, depending on latitude)
Red	0.63~0.69	2.44	Quantization value: bits
Near IR	0.76~0.90	2.44	Ground imaging: : the along track / cross track direction(+/−25°) Irradiation width: to sub astral point as the center, around the 272 km
Pan	0.45~0.90	0.61	Imaging mode: single view 16.5 km × 16.5 km, Strip 16.5 km × 16.5 km

Enhanced ability makes the SPOT-5 satellite can obtain 120 km wide panchromatic and multispectral image. And outstanding star storage capacity makes data storage, record and playback have been optimized (Table 2-14).

Table 2-14　SPOT-5 satellite main parameters

Track		Sun-synchronous, highly 832 km, drop across the equator at 10: 30 local time		
Sensor		2 HRG high-resolution geometric imaging devices are the same, 1 HRS high-resolution stereo imaging device		
Scanning mode		Linear array push broom		
Band	Wavelength/μm	HRG/m	Vegetation imaging device	High-resolution stereo device HRS
PAN	0.49~0.69	2.5 or 5	—	10 m
B0	0.43~0.47	—	1 km	—
B1	0.49~0.61	10	—	—
B2	0.61~0.68	10	1 km	—
B3	0.78~0.89	10	1 km	—
SWIR	1.58~1.75	20	1 km	—
Width of cloth		60 km	2250 km	120 km

4) IRS-P5

Cartosat-1 satellite, also known as IRS-P5, is a remote sensing mapping satellite launched by India on May 5, 2005. It is equipped with two resolution 2.5 m panchromatic sensors which can continuous push broom to Table a same orbit stereo image pair. The sensors are located at 26° along the track direction and back angle 5°. At stereoscopic observing mode, through the quantitative adjustment of the satellite plan to compensate the rotation of the earth, the two cameras can get ground image in the same spot to stereo image pair.

The time difference between the two cameras for the same position is only 52 seconds, so the radiation effect of the two images is basically the same, which is conducive to the stereo observation and image matching. The effective width of the image pair is 26 km, and baseline height ratio (B/H) is 0.62. The data are mainly used for topographic mapping, elevation modeling, cadastral mapping, and resource survey (Table 2-15).

Table 2-15 IRS-P5 satellite main parameters

Track	Near polar solar synchronization
Track height	618 km
Total track number	1867
Long half shaft	6996.14 km
Eccentricity ratio	0.001
Dip angle	97.87°
Fall time	10:30 am
Time interval between adjacent tracks	11 days
Revisit period	5 days
Repetition period	126 days
Daily track number	14 individual
Orbital period	97 min

5) ALOS

ALOS is an earth observation satellite launched by Japan in January 24, 2006. It adopts advanced land observation technology. It can obtain the global high-resolution land observation data. The main applications are mapping, regional environmental monitoring, disaster monitoring, resource survey and other field. ALOS satellite contains three sensors: panchromatic remote sensing instrument for stereo mapping (PRISM), mainly used for digital elevation mapping; advanced visible and near infrared radiation meter (AVNIR-2), used for precise terrestrial observation; Phase L-band synthetic aperture radar (PALSAR), used for all-weather terrestrial observations at day and night(Table 2-16).

Table 2-16 PRISM main parameters

Band number	1(pan)
Wavelength	0.52~0.77 μm
Observation mirror	3 个(Ground imaging, forward-looking imaging, visual imaging)
Kigobi	1.0(Between forward-looking imaging and visual imaging)
Spatial resolution	2.5 m(Ground imaging)
Width of cloth	70 km(Ground imaging mode) 35 km(Joint imaging mode)
Direction angle	-1.5°~1.5°
Quantization length	8bit
activation	
Pattern 1	Substar, front, rear view (35 km)
Pattern 2	Substar (70 km) + rear view (35 km)
Pattern 3	Substar (70 km)
Pattern 4	Substar (35 km) + forward (35 km)

(continue)

Pattern 5	Substar (35 km) + rear view (35 km)
Pattern 6	Forward-looking after (35 km) + (35 km)
Pattern 7	Substar (35 km)
Pattern 8	Forward (35 km)
Pattern 9	Rear view (35 km)

6) WorldView- I / II

Worldview satellite system is Digital Globe company's next-generation commercial imaging satellite system. The system consists of satellite WorldView-I and satellite WorldView- II. The WorldView-I and WorldView- II satellite were launched in September 2007 and October 2009 (Table 2-17).

Table 2-17   WorldView- I / II Satellite parameters

Category	WorldView- I	WorldView- II
Launch date	2007.09.18	2009.10.8
Track	Height: 496 km Type: sun synchronous The descending node: place 10:30 am	Height: 770 Type: sun synchronous The descending node: place 10:30 am Cycle: 100 min
Band	Panchromatic, 400~900 nm	panchromatic and 8 multispectral bands 4 standard bands: red, green, blue, near infrared 4 new bands: red edge, coast, yellow, near infrared 2
Spatial resolution	Pan, nadir: 0.5 m	Pan: Ground: 0.46 m Deviation from the nadir of 200: 0.52 m Multi-spectrum: Ground: 1.8 m Deviation from the nadir of 200: 2.4 m
Dynamic range	11 bit	11 bit
Imaging bandwidth	Substar 17.6 km	Substar 16.4 km
Stereo mapping	Stereo mapping ability	Stereo mapping ability

7) GeoEye-1/2

The world's largest commercial satellite remote sensing company American GeoEye, successfully launched the most advanced technology on September 6, 2008, which is the most advanced and the highest resolution commercial satellite to date. The satellite has the highest resolution and measuring drawing ability is extremely strong. It has short revisit time. Vast numbers of users in the world pay much attention to it. The image of GeoEye-1 high-resolution satellite has broad application prospect. It has a great advantage in the realization of large area mapping project, the micro features of the interpretation and interpretation.

GeoEye-1 satellite mainly has the following characteristics:

(1) The real half meter satellite precision   Panchromatic images of 0.41 m resolution, resolution of the multispectral image 1.65 m positioning accuracy 3 m.

(2) Large-scale mapping capability   Collection of nearly 700 000 square kilometers of panchromatic image data every day or nearly 35 million square kilometers of the color image data fusion.

(3) Revisit period 3 days (or less) time to revisit the earth at any point of observation.

8) Pleiades-1/2

France launched a Pleiades-1 resolution optical satellite in December 17, 2011. The Pleiades is composed of two satellites constellation. The spatial resolution is 0.5 m, and width is 20 km. The satellite design lifetime is 5 years. Pleiades-1/2 is part of the 'Optical Radar Combined Earth Observation' (ORFEO) project. ORFEO is an advanced optical earth observation system sponsored jointly by the French space agency and the Italian Space Agency. The system can meet the dual-use demand of users in Europe, including map rendering, monitoring the volcano, geophysics and hydrology study, urban planning and other (Table 2-18).

Table 2-18   Pleiades-1/2 Satellite parameters

Impact product	Pan 50 cm 50 cm fusion color 2 m multi-spectrum Bundled products: 50 cm panchromatic andmultispectral 2 m	
Spectral band /nm	Pan: 480~830 Blue: 430~550 Green: 490~610 Red: 600~720 Near infrared: 750~950	
Image positioning accuracy	With ground control point	1 m
	No ground control point	3 m(CE90)
Covering ability	Double star single day coverage	1 000 000 km^2
	Stereo image pair	20 km×280 km

9) TianHui-1 satellite

TianHui-1 satellite is China's first single transmission type stereo mapping satellites. It provides quick access to 2 m panchromatic image, three ranges of 5 m line array stereoscopic image and red, green, blue, 10 m of near-infrared four-band multispectral images of the same area. The first satellite successfully was launched on August 24, 2010, and the second satellite successful was launched on May 6, 2012. The two satellites operate in a network, greatly enhancing the image acquisition capability. TianHui-1 satellite is completely designed according to mapping camera. Its main features are: more efficient to loads, high positioning accuracy, good data sets, and good ability to obtain data (Table 2-19).

Table 2-19 Sky painted one satellite main parameters

Daydraw the date of launch of the first stars		2010. 8. 24	
Daydraws the date of launch of the second stars		2012. 5. 6	
Track In Tableation	Track	Near circular orbit of the sun	
	Satellite altitude	500 km	
	Return period	58 day	
	Revisit period	9 day	
Sensor	Effective load	Spectral range/μm	Spatial resolution/m
	High resolution camera	0.51~0.69	2
	Three linear array camera	0.51~0.69	5
	Multispectral camera	0.43~0.52	10
		0.52~0.61	
		0.61~0.69	
		0.76~0.90	
The width of the cloth		60 km	
Lateral swing ability		±10°	

10) Resource-3 satellite

China's first self-developed civilian high-resolution 3D mapping satellite Resources-3 was successfully launched on January 9, 2012, which fills the gaps in China's civil surveying and mapping satellites. The satellite has the function of mapping and resources survey. It has the function of three-dimensional mapping. It has characteristics of high precision, high image data quantity, fast processing speed and wide application range.

Resources-3 satellite is loaded with 2.5 m resolution face color CCD camera, 4 m resolution before or after resolution camera and 10 m multispectral camera. It meets the need of resource remote sensing. At the same time, it has the abilities of 1 : 5 million scale map mapping and 1 : 2.5 million scale repairing function. It is a resource remote sensing satellite with a mapping function (Table 2-20).

Table 2-20 ZY-3 Satellite technical parameters

Project	ZY-3
Track	Type: sun-synchronous circular orbit The descending node: place 10: 30 am Height: 505.984 km Orbit inclination angle: 97.421° Regression period: 59 days Revisit period: 5 days

(continue)

Project		ZY-3	
Satellite weight		2630 kg	
Satellite lifetime		On orbit working life of 5 years	
Camera	Band	Spatial resolution/m	Wide in width/km
Face up to CCD	Pan	2.5	51
Forward looking CCD	Pan	4	52
After CCD	Pan	4	52
Multispectral camera /nm	Blue: 450~520	10	52
	Green: 520~590		
	Red: 630~690		
	Near infrared: 770~890		
Stereo mapping		Withstereo, mapping can stand	

2.6 Photography measurement principle

2.6.1 photo grammetry basic definition

2.6.1.1 Coordinate system commonly used in photogrammetry

The coordinate system commonly used in photogrammetry includes auxiliary plane coordinate system, the space coordinate system, and object space coordinate system.

1) Image plane coordinate system o-xy and fiducial mark o-uv

Image plane coordinate system is used to describe the pictures of the box as the location of the point on the image plane rectangular coordinate system, the general S point set as the center of photography, o point is the direction plane of photography and intersection point of the picture, which are the main points of the picture. At this point as shown in Figure 5-1, coordinate system o-xy is like a plane, o-uv for the standard coordinate system.

2) Like S-xyz space coordinate system

Like space coordinate system is a three-rectangular coordinate system, which is used to describe the position of image point in image space coordinate of pictures, the position of image point is decided by its coordinates on the pictures, in order to facilitate image point position and its corresponding points of the space position of conversion to each other, can introduce image plane coordinate system. The origin of the coordinate system as the main point o, sometimes photography center S as the origin, as shown in Figure 2-56, XYZ coordinate system S, center of photography as the main point of connection So that is the main optical axis as the Z axis, X, Y axis respectively with like u in a plane coordinate system, v parallel to the axis, the coordinate system can be easily converted from like space coordinate system and the like each image point in plane coordinate system in space coordinates all of the negative Z coordinates in the main from So. General writing-f.

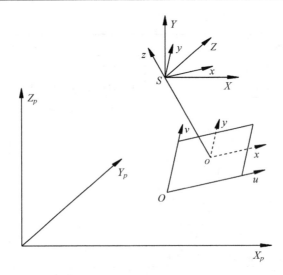

Figure 2-56 Photogrammetric coordinate system schematic

3) As auxiliary coordinate system S-XYZ space

As auxiliary space coordinate system is used to assistin the space and the calculation of coordinate system. The coordinate system uses the center of the camera as the coordinate origin. Generally, a certain vertical direction set is needed as the Z axis in the direction of the camera's shooting direction or flight direction as a circumferential direction, as shown in the figure above the S-XYZ coordinate system. Space-aided coordinate system is a typical relative coordinate system, that is the coordinates of each point in the coordinate system can only reflect the relative coordinate relationship. In the near-field photogrammetry, it is convenient to set up the space-aided coordinate system without the real value of the object point. The main research objectives of this paper are forest survey, specific to the DBH, tree height, volume, relative position, and these factors can be directly in the relative coordinate system that is, as the spatial coordinate system to obtain and taking into account the field survey is not convenient to obtain the coordinates of the object point of the true value, so the study in this chapter a large number of the use of space Auxiliary coordinate system.

4) The spatial coordinate system $S\text{-}X_p Y_p Z_p$

The spatial coordinate system used to determine the absolute coordinates of the object by the space rectangular coordinate system, general and commonly used in ground measurement coordinate system (geodetic coordinate system), as shown in figure of $S\text{-}X_p Y_p Z_p$, can through the object space coordinate of known points convert as auxiliary space coordinates to object space space coordinate system.

2.6.1.2 Internal and external orientation elements of photogrammetric images

Interior orientation elements and exterior orientation elements are the elements used to determine the position and orientation of the photograph in the object space coordinate system $S\text{-}X_p Y_p Z_p$. Interior orientation element is used to restore (photography) towards pictures beam and photo location factors and photograph defined exterior orientation element is used to make pictures beam in the object space coordinate system $S\text{-}X_p Y_p Z_p$ orientation and position of the element.

The shape will be shot by the inside azimuth element of images is a recovery factor, which is used to determine the center of photography (technically, the lens should be like square node) S and relative location of shooting pictures of P elements, through the inner orientation elements can be determined and the shape of the restoring moment beam shooting only.

Shown in the following Figure 2-57(a), as the main point is relative to the center of the image point o location (x_0, y_0) and center of photography to pictures of the vertical distance f this the inside azimuth element of three elements of the picture P.

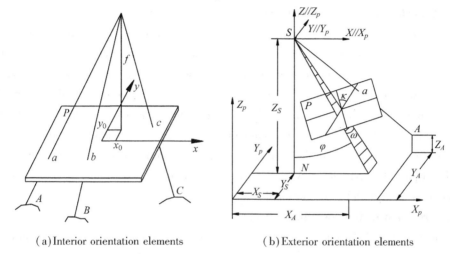

(a) Interior orientation elements   (b) Exterior orientation elements

Figure 2-57 Photo diagram of internal and external orientation elements

The inside azimuth element of the image under the condition of known, through the center of photography in the object space coordinate system S coordinates $(X_S$ and $Y_S, Z_S)$ and its spatial axis in the object space axis angular orientation elements, can only really set the image in the object space coordinates in space coordinate system.

The angular orientation elements generally described by three separate rotation angle. The three rotation angle will be because of the different choices have different ways of expression, but the final says the results are consistent. Three angular orientation element with $(X_S$ and $Y_S, Z_S)$ is known as the exterior orientation elements of the image.

Typically used to express the representation of image angular orientation elementsare A-v-kappa corner system, phi predominate-omega-kappa corner system predominate and omega ' - phi kappa' predominate the angle of the system.

Our usual corner system is phi-omega-kappa corner system, predominate the angle system used for double image photogrammetry, is also used in this article the angle of the system. As shown in Figure 2-57 (b), which was established in site taken as auxiliary coordinate system S-XYZ space, the axis of the XYZ respectively and object space coordinate system S-$X_p Y_p Z_p$ $X_p Y_p Z_p$ in parallel. Among them:

Phi: generally take flight direction or baseline direction for the X axis, the phi angle as the Y axis as the spindle rotation angle.

Omega: around the X axis rotation angle, but the X axis as the angle of rotation for phi with its initial direction.

Kappa: The rotation angle around the photographing direction, i. e., the Z-axis, coincides with the S_Z-axis at the start position.

Therefore, corresponding six exterior orientation elements which are respectively three line element X_S, Y_S, Z_S and three elements of angle phi, omega, kappa, predominate in Figure 2-57 (b) mark the image part P corresponding six exterior orientation elements, also known as absolute exterior orientation elements.

2.6.1.3 Coordinate rotation and translation

Rotation and translation can be transformed from one coordinate system to another. The rotation of coordinate system in photogrammetry generally refers to the transformation between the coordinates of image points in the image space coordinate system and the image space auxiliary coordinate system. Two coordinate systems of the same scale can always be converted to each other by rotation and translation. From the knowledge of higher mathematics, the rotation of rectangular space coordinate system is orthogonal transformation, a coordinate system can be rotated through a certain order in order to transform three angles into another coordinate system of the same origin, or through the origin. The parallel movement is transformed into a coordinate system in which the other three axes are parallel to the original direction.

1) The rotation of the coordinate system

The rotation of the coordinate system is illustrated by using the φ-ω-κ system as an example. When the image point in the image space coordinate system (x_a, y_a, z_a) is rotated from the starting position by φ, Turn the angle around the X axis, and finally turn the angle around the Z axis. From Figure 2-58, we can get the relationship between the coordinate (X, Y, Z) and the original coordinate of the rotated point a

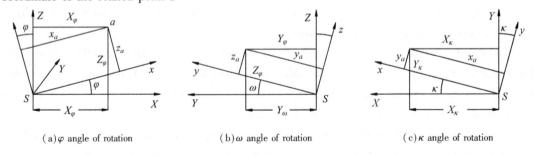

(a) φ angle of rotation　　(b) ω angle of rotation　　(c) κ angle of rotation

Figure 2-58　Rotating of coordinate system

$$X_a = R_\varphi R_\omega R_\kappa \begin{bmatrix} x_a \\ y_a \\ z_a \end{bmatrix} = \begin{bmatrix} \cos\varphi & 0 & -\sin\varphi \\ 0 & 1 & 0 \\ \sin\varphi & 0 & \cos\varphi \end{bmatrix} \begin{bmatrix} 1 & 0 & 0 \\ 0 & \cos\omega & -\sin\omega \\ 0 & \sin\omega & \cos\omega \end{bmatrix} \begin{bmatrix} \cos\kappa & -\sin\kappa & 0 \\ \sin\kappa & \cos\kappa & 0 \\ 0 & 0 & 1 \end{bmatrix} \begin{bmatrix} x_a \\ y_a \\ z_a \end{bmatrix}$$

$$= \begin{bmatrix} a_1 & a_2 & a_3 \\ b_1 & b_2 & b_3 \\ c_1 & c_2 & c_3 \end{bmatrix} \begin{bmatrix} x_a \\ y_a \\ z_a \end{bmatrix} \qquad (2\text{-}48)$$

Among them:

$a_1 = \cos\varphi\cos\kappa - \sin\varphi\sin\omega\sin\kappa$

$b_1 = \cos\omega\sin\kappa$

$c_1 = \sin\varphi\cos\kappa + \cos\varphi\sin\omega\sin\kappa$

$a_2 = -\cos\varphi\sin\kappa - \sin\varphi\sin\omega\cos\kappa$

$b_2 = \cos\omega\cos\kappa$

$c_2 = -\sin\varphi\sin\kappa + \cos\varphi\sin\omega\cos\kappa$

$a_3 = -\sin\varphi\cos\omega$

$b_3 = -\sin\varphi$

$c_3 = \cos\varphi\cos\omega$

Among them, R is the rotation matrix, the rotation matrix is very important in the photogrammetry. The result of the rotation matrix is based on φ-ω-κ corner system. If we choose other corner system, we can get the same method corresponding to the rotation matrix. It can be seen from the above formula, known to an image of the three angular orientation elements φ, ω, κ can be obtained after the trigonometric function obtained by the way the nine cosine parameters, which can be two origins of the same space coordinate system, and then rotate the matrix to achieve the mutual rotation between two spatial coordinate systems.

2) Coordinates translation

The translation of the coordinate system refers to the mutual conversion between the two coordinate systems by changing the coordinate increment in the case that the coordinate axes of the spatial coordinate systems are parallel to each other. It is also understood that the coordinate system conversion is performed by changing the origin coordinate in the case that the axis direction and the unit length of the coordinate axis do not change (Figure 2-59).

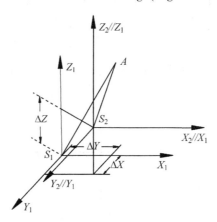

Figure 2-59 Schematic diagram of the translating coordinate system

Point A, as shown in the coordinate system S_1-$X_1Y_1Z_1$ as coordinates for the (X_A, Y_A, Z_A), coordinates in the coordinate system of S_2-$X_2Y_2Z_2$ as' (X'_A, Y'_A, Z'_A), and coordinate $X_1Y_1Z_1$ parallel to the axis S_2-$X_2Y_2Z_2$ respectively. Coordinate system S_1-$X_1Y_1Z_1$ and S_2-$X_2Y_2Z_2$ between the coordinates of the increment is as (as ΔX, ΔY and ΔZ), conversion can get the following formula:

$$\begin{bmatrix} X'_A \\ Y'_A \\ Z'_A \end{bmatrix} = \begin{bmatrix} X_A \\ Y_A \\ Z_A \end{bmatrix} + \begin{bmatrix} \Delta X \\ \Delta Y \\ \Delta Z \end{bmatrix} \qquad (2\text{-}49)$$

The above formula is the conversion formula of coordinate system translation, calculation of photogrammetry, therotation, and translation of coordinate system will be used in a large amount because the coordinate system chosen by the binocular camera in most cases in forest survey. Relative coordinate system, in the binocular camera between the two lenses and the binocular camera in the direction of rotation, through the rotation and translation of the coordinate system, can coordinate transformation of different coordinates to the same coordinate system, in order to obtain the correct investigation result.

2.6.2 Collinear equation and coplanar equation

1) Conditional collinear equation

Collinearity condition equation is central projection structure as the mathematical foundation (Figure 2-60), is used to describe the image point, center of photography and party at three points in a straight line on a condition equations. Photogrammetry is the basis of many methods in the collinear equation as a starting point, such as (like) space resection method, (or double) more space forward intersection method, beam method area net adjustment as well as the direct linear transformation (DLT) and so on. These methods are based on single structure as light as a research unit, at the same time with different to deal with problems, how to use the collinear equation and its concrete expression forms are also different.

$$\left. \begin{aligned} x - x_0 &= -f \frac{a_1(X_A - X_S) + b_1(Y_A - Y_S) + c_1(Z_A - Z_S)}{a_3(X_A - X_S) + b_3(Y_A - Y_S) + c_3(Z_A - Z_S)} \\ y - y_0 &= -f \frac{a_2(X_A - X_S) + b_2(Y_A - Y_S) + c_2(Z_A - Z_S)}{a_3(X_A - X_S) + b_3(Y_A - Y_S) + c_3(Z_A - Z_S)} \end{aligned} \right\} \qquad (2\text{-}50)$$

On type is the classic conditional collinear equation and the parameter type (a_i, b_i, c_i) and

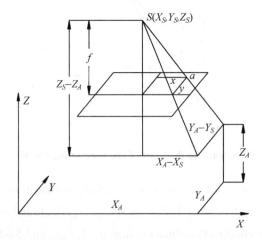

Figure 2-60 Collinear equation diagram (upright photography)

the meaning of the parameter value and type 5-2, it can be seen that type 5-4 is the 1 and 2 in the equation of a plane, it said it was the two intersecting line equation of plane, also is the center of photography, the image point and content of the PM. A single collinear equation can determine the direction of the object space points, and cannot determine the coordinates of object space points, but can be by different center of photography more than like the film of the same object space points collinear equation conditional place like space forward intersection simultaneous solution of object space coordinates, can also with different content on a single photo points collinearequation as space resection to solve the coordinates of center of photography.

2) conditional coplanar equation

In the basic theory of photogrammetry, the coplanar condition equation is also used to describe the pictures spatial relations(Figure 2-61). Coplanar condition equation is described pictures inside photography baseline and the light in the same plane with the same conditions ofan equation. Coplanar equation conditions from the points of light with the same name with the same shall be in the same plane (epic polar coplanar) the geometric relationships between the departure, through the coplanar equation can be directly stereo as internal orientation, relative orientation, and absolute orientation, thus under the condition of less object space points and photography Celiang Wang stereo pairs finished processing, multi-purpose at low precision photogrammetry.

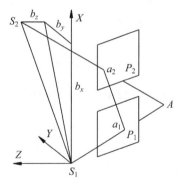

Figure 2-61   Schematic of coplanarity condition equation

As shown in the auxiliary coordinate system like space S-XYZ, S_1 and S_2 for the two points of photography, P_1 and P_2, stereo pairs constitute A set of baseline B for photography, A point for any item, a_1 and a_2 is point A on the P_1 and P_2 as some of the same name, is S_1S_2 and point A will form A triangle, namely S_1A, S_2A, S_1S_2 3 line, namely S_1A, S_2A, S_1S_2 3 lines coplanar, written in vector form the vector $\vec{b}$ ($\overrightarrow{S_1S_2}$), $\overrightarrow{S_1a_1}$, $\overrightarrow{S_2a_2}$ coplanar, from higher mathematics knowledge, 3 vectors coplanar condition is the mixed product is 0, that is:

$$F = \vec{b} \cdot (\overrightarrow{S_1a_1} \times \overrightarrow{S_2a_2}) = 0 \tag{2-51}$$

A written expression of determinant available:

$$F = \begin{vmatrix} b_x & b_y & b_z \\ u_1 & v_1 & w_1 \\ u_2 & v_2 & w_2 \end{vmatrix} = b_x \begin{vmatrix} v_1 & w_1 \\ v_2 & w_2 \end{vmatrix} - b_y \begin{vmatrix} u_1 & w_1 \\ u_2 & w_2 \end{vmatrix} + b_z \begin{vmatrix} u_1 & v_1 \\ u_2 & v_2 \end{vmatrix} = 0 \qquad (2\text{-}52)$$

Among them:

$$\begin{bmatrix} u_1 \\ v_1 \\ w_1 \end{bmatrix} = \begin{bmatrix} x_1 \\ y_1 \\ -f_1 \end{bmatrix} \qquad (2\text{-}53)$$

$$\begin{bmatrix} u_2 \\ v_2 \\ w_2 \end{bmatrix} = R_2 \begin{bmatrix} x_2 \\ y_2 \\ -f_2 \end{bmatrix} \qquad (2\text{-}54)$$

Type, R_2 is photo P_2 and pictures P_1 relative rotation matrix is made up of the relative orientation angle between the stereopair elements as the relative rotation angle between two slices of direction cosine predominate, and the relative orientation between the stereo pairs predominate horn line elements that the magnitude of the relative shift between two slices of b_x, b_y, b_z stereo pairs together enough to constitute a relatively exterior orientation elements, with the exterior orientation elements mentioned in the article. In the inside azimuth element of the known cases, relative exterior orientation elements can only decide the three-dimensional relative within the space the relative position of two pictures beam in the moment of photography, and absolute exterior orientation elements (X_S, Y_S, Z_S, φ, ω, κ) predominate is able to determine the three-dimensional relative within two pictures in the position of the instant photography and coordinates of the absolute position of space.

2.6.3 Ground photogrammetry general principles

Terrestrial photogrammetry generally used photographic equipment photography on the ground, including the measurement with the camera, the non-measurement with an ordinary digital camera, multi-axis ultra-station photographic apparatus, or asmartphone now.

In photogrammetry there are two basic measurements: Monolithic (single image) measurement and three-dimensional (double images) measurements. As a result, the ground can be divided into monolithic the single image photogrammetry and three-dimensional measurement. Due to the good efficiency of three-dimensional measurement, high accuracy, it has become the main method photogrammetry. But considering the particularity of forestry survey, the requirement of diameter at breast height and tree height in forestry survey monolithic and the three-dimensional measurement occupies the same proportion.

Terrestrial photogrammetry on the basis of the different state of photography object can be divided into two categories: static photography (pictured mountains, forests, and other animal body), dynamic photography (perturbation waves, water and other moving objects). Forestry survey belongs to still photography.

As terrestrial photogrammetry is photography performed on the ground, its efficiency is lower than aerial photography and is suitable for small area i.e. large scale measurement. But in forestry investigation, because of the need of diameter at breast height, tree height measurement, and the ground photogrammetry is more flexible, mobile high precision photogrammetry, So, as a result, the ground photogrammetry plays a more important than the aerial survey.

Terrestrial photogrammetry has the following features:

①In terrestrial photography small error can be restricted on the ground, operating a simple formula and simple equipment;

②Flexible, low cost and high precision;

③Easy to lead to measurement points;

④Is suitable for a small range of measurement;

⑤The principle is simple and easy to learn.

For single measurement, 3D ground photogrammetry optical axis horizontal photography, can be seen from Figure 2-62:

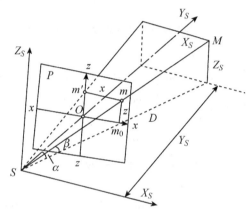

Figure 2-62 Coordinate relation diagram with the horizontal axis

$$\tan\alpha = \frac{x}{f} \tag{2-55}$$

$$\tan\beta = \frac{z}{Sm_0} = \frac{z \cdot \cos\alpha}{f} = \frac{z}{\sqrt{f^2 + x^2}} = \frac{z}{x} \cdot \sin\alpha \tag{2-56}$$

$$\frac{X_S}{x} = \frac{Z_S}{z} = \frac{Y_S}{y}$$

That is:

$$X_S = \frac{Y_S}{f} \cdot x \tag{2-57}$$

$$Z_S = \frac{Y_S}{f} \cdot z \tag{2-58}$$

The above formula shows that the use of monolithic (single) computing image point coordinates can be calculated to obtain the angle information of measurement points, two photographs taken from two angles can be solved through the intersection coordinates.

For oblique photography, because the main optical axis is tilted at an angle, we need to go through to get the image projection transformation to obtain image point coordinates (Figure 2-63).

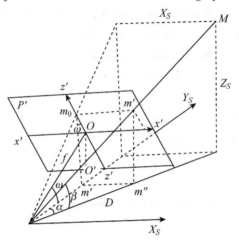

Figure 2-63　Coordinate relation diagram with the tilted axis

$$\tan\alpha = \frac{m'm''}{Sm'} = \frac{x'}{SO' - O'm'} = \frac{x'}{f \cdot \cos\omega - z' \cdot \sin\omega} \tag{2-59}$$

$$\tan\beta = \frac{mm'}{Sm''} = \frac{z' \cdot \cos\omega + f \cdot \sin\omega}{f \cdot \cos\omega - z' \cdot \sin\omega} \cdot \cos\alpha$$

$$= \frac{z' \cdot \cos\omega + f \cdot \sin\omega}{\sqrt{(f \cdot \cos\omega - z' \cdot \sin\omega)^2 + x'^2}} \tag{2-60}$$

$$\frac{X_S}{x'} = \frac{Y_S}{f \cdot \cos\omega - z' \cdot \sin\omega} = \frac{Z_S}{z' \cdot \cos\omega + f \cdot \sin\omega}$$

Then
$$X_S = \frac{Y_S}{f \cdot \cos\omega - z' \cdot \sin\omega} \cdot x' \tag{2-61}$$

$$Z_S = \frac{Y_S}{f \cdot \cos\omega - z' \cdot \sin\omega} \cdot (z' \cdot \cos\omega + f \cdot \sin\omega) \tag{2-62}$$

Comparison (2-58), (2-59) and (2-60), (2-61), the change (2-62) can be obtained:

$$\begin{cases} X_S = \dfrac{Y_S}{f'} \cdot x' \\ Z_S = \dfrac{Y_S}{f'} \cdot z'' \end{cases} \tag{2-63}$$

As can be seen from the formula, oblique photography is equivalent to a projection for the focal length changes and for Z_S changed not only the focal length, z' also conducted a projection transformation.

For double vane measurement in two different location photography, on the same target as required the two pictures. Such two pieces of pictures of the stereo pair, according to the two pictures to solve the two photography beams will need to know the location of the elements in the space, these elements are the main optical axis of the azimuth angle. The main optical axis angle, picture rotation angle and geodetic coordinates of the projection center decided six exterior orientation elements, each picture of exterior orientation elements have six, two pictures, a total of 12, work well the twelve elements of exterior orientation is solved the ground double stereo measurement.

Photo-coordinate rotation formula:

The $z'z'$ axis deviation from the main bar (zz axis) a kappa angle predominate, kappa angle is predominate in the figure.

Figure 2-64 Coordinate rotation diagram

By Figure 2-64 get:

$$\begin{cases} x = z' \cdot \sin\kappa + x' \cdot \cos\kappa \\ z = z' \cdot \cos\kappa - x' \cdot \sin\kappa \end{cases} \quad (2\text{-}64)$$

Placed at two different location in theodolite photography, photography on the same target according to certain requirements, we receive two pictures. The stereoscopic image of two pictures pairs has total 12 exterior orientation elements, 6 from each photograph. That is:

$$\begin{cases} \alpha_{01}, \omega_1, \kappa_1, X_1, Y_1, H_1 \\ \alpha_{02}, \omega_2, \kappa_2, X_2, Y_2, H_2 \end{cases} \quad (2\text{-}65)$$

The meaning of these elements for (see Figure 2-65):

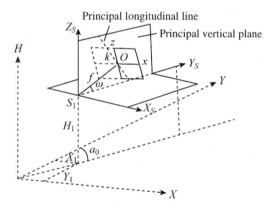

Figure 2-64 Exterior orientation elements diagram

α_0—The main optical axis direction angle of the earth.

ω —In the pictures of the main vertical surfaces when optical axis and the angle between the horizontal line. Omega is positive when elevation; When the dip angle is negative.

κ —Pictures of coordinates zz predominate shaft with pictures angle between the main bar. Zz shaft clockwise to the ordinate kappa negative predominate; zz to lord ordinate kappa to predominate is counter-clockwise.

X, Y, H, S geodetic coordinate projection center.

In practical engineering, the photo frame on the left piece of the main axis of the Y axis, in the left picture projection center S_1 as photography origin of coordinates, coordinates of photography. It needs to be the right piece projection center S_2 relative to the left of projection center S_1 as as geodetic coordinate as incremental X, Y, H coordinate system to switch to photography. Then transfer the photography coordinates into geodetic coordinate system.

The photography with the directional element coordinate system coordinate conversion of geodetic coordinate system coordinates can be obtained from Figure 2-65. Coordinate system diagram $Y_S S_1 X_S$ for photography, XOY for geodetic coordinate system.

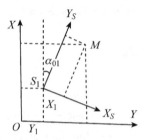

Figure 2-65 Transformation of the camera coordinate system and geodetic coordinate system diagram

Obtained by the coordinate transformation relationship:

$$\begin{cases} X = X_1 + Y_S \cdot \cos\alpha_{01} - X_S \cdot \sin\alpha_{01} \\ Y = Y_1 + X_S \cdot \cos\alpha_{01} + Y_S \cdot \sin\alpha_{01} \\ H = H_1 + Z_S + i \cdot V + (K + r) \end{cases} \quad (2\text{-}66)$$

Talked earlier about way of photography in a certain way, as shown in Figure 2-66 photography way to an arbitrary angle.

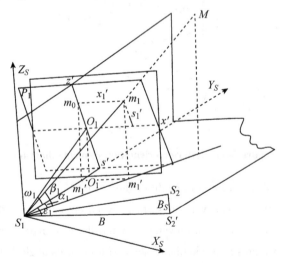

Figure 2-66 Arbitrary angle camera coordinate system diagram

Informed by Figure 2-67:

$$D_1 = \frac{B \cdot \sin(\varepsilon_2 + \alpha_2)}{\sin[(\varepsilon_1 + \varepsilon_2) + (\alpha_2 - \alpha_1)]} \quad (2\text{-}67)$$

$$Y_S = D_1 \cdot \cos\alpha_1 \quad \text{(a)}$$
$$X_S = D_1 \cdot \sin\alpha_1 \quad \text{(b)}$$
$$Z_S = D_1 \cdot \tan\beta_1 \quad \text{(c)}$$

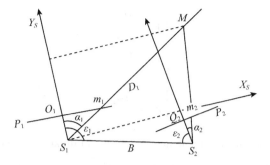

Figure 2-67 Arbitrary angle photography coordinate diagram

Introduction:
$$\sin(\varepsilon_1 + \varepsilon_2) = \sin[180° - (\varepsilon_1 + \varepsilon_2)] = \sin\gamma \tag{2-68}$$
$$\cos(\varepsilon_1 + \varepsilon_2) = -\cos[180° - (\varepsilon_1 + \varepsilon_2)] = -\cos\gamma \tag{2-69}$$

Then:
$$\begin{cases} Y_S = B \cdot (f \cdot \cos\omega_2\sin\varepsilon_2 - z'_2\sin\omega_2\sin\varepsilon_2 + x_2\cos\varepsilon_2) \\ \quad (f\cos\omega_1 - z'_1\sin\omega_1)/\{\sin\gamma \cdot (f^2\cos\omega_1 \cdot \cos\omega_2 - \\ \quad f \cdot \cos\omega_1 \cdot z'_2 \cdot \sin\omega_2 - f \cdot \cos\omega_2 \cdot z'_1 \cdot \sin\omega_1 + \\ \quad z'_1 z'_2 \sin\omega_1 \cdot \sin\omega_2 + x'_1 x'_2) - \cos\gamma \cdot [(f\cos\omega_1 - \\ \quad z'_1 \cdot \sin\omega_1)x'_2 - (f \cdot \cos\omega_2 - z'_2 \cdot \sin\omega_2)x'_1]\} \\ X_S = B \cdot (\sin\varepsilon_2 \cdot x'_1 \cdot f \cdot \cos\omega_2 - x'_1 \cdot z'_2 \cdot \sin\omega_2 \cdot \sin\varepsilon_2 + \\ \quad \cos\varepsilon_2 \cdot x'_1 x'_2)/\{\sin\gamma \cdot (f^2\cos\omega_1 \cdot \cos\omega_2 - \\ \quad f \cdot \cos\omega_1 \cdot z'_2 \cdot \sin\omega_2 - f \cdot \cos\omega_2 \cdot z'_1 \cdot \sin\omega_1 + \\ \quad z'_1 z'_2 \sin\omega_1 \sin\omega_2 + x'_1 x'_2) - \cos\gamma[(f\cos\omega_1 - \\ \quad z'_1 \cdot \sin\omega_1)x'_2 - (f\cos\omega_2 - z'_2\sin\omega_2) \cdot x'_1]\} \\ Z_S = B \cdot (f \cdot \cos\omega_2\sin\varepsilon_2 - z'_2\sin\omega_2\sin\varepsilon_2 + x_2\cos\varepsilon_2) \\ \quad (z'_1\sin\omega_1 + f\sin\omega_1)/\{\sin\gamma \cdot (f^2\cos\omega_1 \cdot \cos\omega_2 - \\ \quad f \cdot \cos\omega_1 \cdot z'_2 \cdot \sin\omega_2 - f \cdot \cos\omega_2 \cdot z'_1 \cdot \sin\omega_1 + \\ \quad z'_1 z'_2 \sin\omega_1 \cdot \sin\omega_2 + x'_1 x'_2) - \cos\gamma \cdot [(f\cos\omega_1 - \\ \quad z'_1 \cdot \sin\omega_1)x'_2 - (f \cdot \cos\omega_2 - z'_2 \cdot \sin\omega_2)x'_1]\} \end{cases} \tag{2-70}$$

The first type analysis is as follows:

(1) when $\kappa_1 = \kappa_2 = 0$; $\omega_1 = \omega_2 = \omega$; $\varepsilon_1 = 90° - \varphi$, $\varepsilon_2 = 90° + \varphi$ ($\varphi < 0$ is left, $\varphi > 0$ is right), namely $\gamma = 180° - (\varepsilon_1 + \varepsilon_2) = 0$, this is the way such as partial isoclinic photography.

(2) when $\kappa_1 = \kappa_2 = 0$; $\omega_1 = \omega_2 = \omega$; $\varepsilon_1 = \varepsilon_2 = 90°$, $\varphi = 0$, so $\gamma = 180° - (\varepsilon_1 + \varepsilon_2) = 0$, it's clearly a integrity isoclinic photography.

(3) when $\kappa_1 = \kappa_2 = 0$; $\omega_1 = \omega_2 = 0$; $\varepsilon_1 = 90° - \varphi$, $\varepsilon_2 = 90° + \varphi$ ($\varphi < 0$ is left, $\varphi > 0$ is right), namely $\gamma = 180° - (\varepsilon_1 + \varepsilon_2) = 0$, it's clearly a level such as partial photography.

(4) when $\kappa_1 = \kappa_2 = 0$; $\omega_1 = \omega_2 = 0$; $\varepsilon_1 = \varepsilon_2 = 90°$, the $\varphi = 0$, $\gamma = 0$, and it is level photography way integrity. Three original type will be:

$$\begin{cases} Y_S = \dfrac{B \cdot f}{p} \\ X_S = \dfrac{B \cdot x_1}{p} \\ Z_S = \dfrac{B \cdot z_1}{p} \end{cases} \qquad (2\text{-}71)$$

This is the most simple of ground photogrammetry coordinate calculation formula.

(5) when $\kappa_1 = \kappa_2 = 0$; $\omega_1 = \omega_2 = 0$; $\varepsilon_1 = 90° - \varphi$, $\varepsilon_2 = 90° + \varphi - \gamma$, it is left main axis is not perpendicular to the baseline of photography to photography.

(6) when $\kappa_1 = \kappa_2 = 0$; $\omega_1 = \omega_2 = 0$; $\varepsilon_1 = 90°$, $\varepsilon_2 = 90° - \gamma$, it is left main axis perpendicular to the photography baseline to photography.

2.6.4 Binocular camera vertical baseline photography

The reason for placing the camera up and downone upon another is to establish relative coordinate system more in line with normal cognition and is easy to rotation transformation; The second is the main body of forest survey objects are trees and trees are mostly irregular cylinder, if we choose to place camera like that (Figure 2-68).

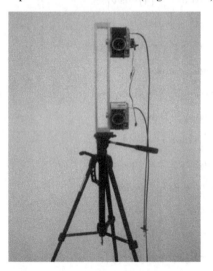

(a) Normal case photography   (b) Convergent photography

Figure 2-68 Schematic of binocular Camera

As shown in the Figure 2-69 (a), at this point in the camera to obtain stereo pair, same tree trunk edge points are not the same, but the camera photography beam and the tangent point of the trunk, as shown in Figure 2-69 in L_1 and R_1, the physical meaning of the collinear equation shows that the obtained object space coordinates are not on the edge of the trunk coordinates, but in the graph line RR_1 and LL_1 intersection of object space coordinates; And choose to put the camera up and down, it won't produce this kind of phenomenon, as shown in Figure 2-69(b) in.

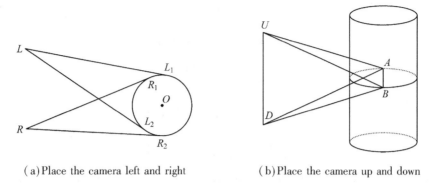

(a) Place the camera left and right (b) Place the camera up and down

Figure 2-69 Schematic of cameras placed

Under the condition of the vertical baseline, we set up as shown in Figure 2-70 as auxiliary space coordinate system, the attachment is at the center of the two camera photography baseline direction as the X-axis direction with a camera D center of photography as the origin, pass through the origin D and perpendicular to the plane of the baseline for YOZ plane. With the camera D photography direction in YOZ plane for the Z axis (photography direction is negative direction) coordinate system is established. At this point we can draw the following conclusion, in the two cameras through screws on the baseline, between two cameras $\cdot \omega$ 、$\cdot \kappa$ are 0, predominate as the only $\cdot \varphi$ relative rotation angle, and the baseline weight between the two center of photography b_y, b_z have 0, $b_x = B =$ baseline length, two center of photography of binocular camera relative coordinates were $D(0,0,0)$ and $U(B,0,0)$.

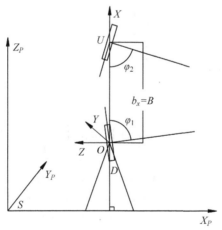

Figure 2-70 Schematic of binocular camera coordinate system

Is binocular camera above the $O-XYZ$ coordinate system is used to calculate the relative coordinates in forest investigation using the auxiliary coordinate system, because the baseline is the vertical state, so you can think of the X axis and geodetic coordinate system $S-X_P Y_P Z_P$, Z_P is parallel to the axis, but Y axis and Z axis does not necessarily with the earth coordinate system corresponding to the parallel axis, at this time only need to determine φ_1 and φ_2 angle between stereo pairs can determine the relative exterior orientation elements, by coplanar equation is available:

$$F = \begin{vmatrix} b_x & b_y & b_z \\ u_1 & v_1 & w_1 \\ u_2 & v_2 & w_2 \end{vmatrix} = 0 \qquad (2\text{-}72)$$

Then:

$$\begin{bmatrix} b_x \\ b_y \\ b_z \end{bmatrix} = \begin{bmatrix} B \\ 0 \\ 0 \end{bmatrix} \begin{bmatrix} u \\ v \\ w \end{bmatrix} = R \begin{bmatrix} x \\ y \\ -f \end{bmatrix} \qquad (2\text{-}73)$$

The R_1 corresponding three rotation angle is $(\varphi_1, 0, 0)$, R_2 corresponding three rotation angle is $(\varphi_2, 0, 0)$, in 2-73 by the direct solution of relative orientation from φ_1 and φ_2, was completed at this time the stereo pair 'absolute orientation' in the auxiliary coordinate system.

It is important to note from $\varphi_1 = \varphi_2 = 0$, namely integrity way of photography, commonly used in most cases single tree and sample observation, exterior orientation elements are known at this time, cannot do 'absolute orientation'; When the $\varphi_1, \varphi_2 \neq 0$, namely the convergent way of photography, is generally used in a few cases of tree height and volume measurement.

2.6.5 The ground any photography

Any photography to get a set of ground on the left as S_1 and S_2, right by the collinearity condition equation:

On the left are:

$$\begin{pmatrix} X_S \\ Y_S \\ Z_S \end{pmatrix} = \lambda_1 R_1 \begin{pmatrix} u_i \\ f \\ v_i \end{pmatrix} \qquad (2\text{-}74)$$

With the right piece of S_2:

$$\begin{pmatrix} X_S \\ Y_S \\ Z_S \end{pmatrix} = \begin{pmatrix} \Delta X \\ \Delta Y \\ \Delta Z \end{pmatrix} + \lambda_2 R_2 \begin{pmatrix} u_i' \\ f \\ v_i' \end{pmatrix} \qquad (2\text{-}75)$$

Type (2-74), (2-75), λ_1, λ_2 were left piece lambda S_1 and S_2 of right scaling coefficient; f, u_i, v_i, u'_i, v'_i were left S_1 and the inside azimuth element of the right piece of S_2 corrected; $\Delta X, Y$ and ΔZ is left piece of S_1 and S_2 of right three elements of exterior orientation linear X_S, Y_S and Z_S change; R_1, R_2, respectively is left patches S_1 and S_2 of rotation matrix right.

At this point, we to the left as a benchmark, the rotation matrix of the left piece is the unit matrix, namely: $R_1 = E$.

On the right,

$$R_2 = R_{\varphi_2} R_{\omega_2} R_{\kappa_2} = \begin{pmatrix} \cos\varphi_2\cos\kappa_2 - \sin\varphi_2\sin\omega_2\sin\kappa_2 & \sin\varphi_2\cos\omega_2 & -\cos\varphi_2\sin\kappa_2 - \sin\varphi_2\sin\omega_2\cos\kappa_2 \\ -\sin\varphi_2\cos\kappa_2 - \cos\varphi_2\sin\omega_2\sin\kappa_2 & \cos\varphi_2\cos\omega_2 & \sin\varphi_2\sin\kappa_2 - \cos\varphi_2\sin\omega_2\cos\kappa_2 \\ \cos\omega_2\sin\kappa_2 & \sin\omega_2 & \cos\omega_2\cos\kappa_2 \end{pmatrix}$$

Among them, $R_{\varphi 2}$, $R_{\omega 2}$, $R_{\kappa 2}$ for the right piece of S_2 predominate in like space coordinate system S -xyz around y axis, the x axis and z axis of rotation angle of rotation matrix

The simultaneous equation for calculating absolute exterior orientation elements:

$$\lambda \begin{pmatrix} u_i \\ f \\ v_i \end{pmatrix} = k_i \begin{pmatrix} \Delta X \\ \Delta Y \\ \Delta Z \end{pmatrix} + R_2 \begin{pmatrix} u'_i \\ f \\ v'_i \end{pmatrix}$$

Among them, the order:

$$\lambda = \frac{\lambda_1}{\lambda_2}, \quad k_i = \frac{1}{\lambda_2}$$

Type is nonlinear equations, must be the linearization to solve. Therefore, according to the Taylor series of type, too:

$$\begin{pmatrix} u_i \\ f \\ v_i \end{pmatrix} \lambda - \begin{pmatrix} \Delta X^0 \\ \Delta Y^0 \\ \Delta Z^0 \end{pmatrix} k_i - k_i^0 \begin{pmatrix} \Delta X \\ \Delta Y \\ \Delta Z \end{pmatrix} - \left(\frac{\partial R_2}{\partial \varphi_2} \varphi_2 + \frac{\partial R_2}{\partial \omega_2} \omega_2 + \frac{\partial R_2}{\partial \kappa_2} \kappa_2 \right) \begin{pmatrix} u'_i \\ f \\ v'_i \end{pmatrix}$$

$$+ k_i^0 \begin{pmatrix} \Delta X^0 \\ \Delta Y^0 \\ \Delta Z^0 \end{pmatrix} - R_2 \begin{pmatrix} u'_i \\ f \\ v'_i \end{pmatrix} + \left(\frac{\partial R_2}{\partial \varphi_2} \varphi_2^0 + \frac{\partial R_2}{\partial \omega_2} \omega_2^0 + \frac{\partial R_2}{\partial \kappa_2} \kappa_2^0 \right) \begin{pmatrix} u'_i \\ f \\ v'_i \end{pmatrix} = 0$$

Type, there are eight unknowns, λ, k_i, ΔX, ΔY, ΔZ, φ_2, ω_2, κ_2; The rest of the parameters are known. We just choose 6 of the same name as point ($i = 1, 2, 3, 4, 5, 6$), by the least squares iterative computation, can solve the right piece relatively left piece of relative exterior orientation elements.

2.7 Digital camera calibration

2.7.1 Interior and exterior orientation element solver

As for the calibration mode for ordinary and according to the basic principle of photogrammetry:

$$\begin{cases} x = -f \dfrac{a_1(X - X_S) + b_1(Y - Y_S) + c_1(Z - Z_S)}{a_3(X - X_S) + b_3(Y - Y_S) + c_3(Z - Z_S)} \\ y = -f \dfrac{a_2(X - X_S) + b_2(Y - Y_S) + c_2(Z - Z_S)}{a_3(X - X_S) + b_3(Y - Y_S) + c_3(Z - Z_S)} \end{cases} \quad (2\text{-}76)$$

In the formula above, (x, y) shows the image point in the image plane coordinate system, f represents a focal length photography. Let S Center of Photography and the ground point A coordinates, the ground photogrammetry coordinates are X_S, Y_S, Z_S and X_a, Y_a, Z_a, a_i, b_i, c_i is the direction cosine, represents the image side between the coordinate axes and the object-side cosine of the angle (is a function of exterior orientation elements):

$a_1 = \cos\varphi\cos\kappa - \sin\varphi\sin\omega\sin\kappa \approx 1$

$$a_2 = -\cos\varphi\sin\kappa - \sin\varphi\sin\omega\cos\kappa \approx -\kappa/\rho$$
$$a_3 = -\sin\varphi\cos\omega \approx -\varphi/\rho$$
$$b_1 = \cos\omega\sin\kappa \approx \kappa/\rho$$
$$b_2 = \cos\omega\cos\kappa \approx 1$$
$$b_3 = -\sin\omega \approx -\omega/\rho$$
$$c_1 = \sin\varphi\cos\kappa + \cos\varphi\sin\omega\sin\kappa \approx \varphi/\rho$$
$$c_2 = -\sin\varphi\sin\kappa + \cos\varphi\sin\omega\cos\kappa \approx \omega/\rho$$
$$c_3 = \cos\varphi\cos\omega \approx 1$$

Taking into account the camera radial distortion, eccentric distortion can be ignored, and the impact of other additional parameters, add direct lineartransformation of conditional equation in dot correction coordinate numbers, the formula above can change to:

$$\begin{cases} x - x_0 + \Delta x = -f\dfrac{a_1(X-X_s) + b_1(Y-Y_s) + c_1(Z-Z_s)}{a_3(X-X_s) + b_3(Y-Y_s) + c_3(Z-Z_s)} \\ y - y_0 + \Delta y = -f\dfrac{a_2(X-X_s) + b_2(Y-Y_s) + c_2(Z-Z_s)}{a_3(X-X_s) + b_3(Y-Y_s) + c_3(Z-Z_s)} \end{cases} \quad (2-77)$$

For the most digital camera lens system, the distortion can be accurately described by taking three coefficients for an ordinary camera with small cropping, we can also take only coefficients.

$$\begin{cases} \Delta x_i = (x_i - x_0)(k_1 r^2 + k_2 r^4 + k_3 r^6) \\ \Delta y_i = (y_i - y_0)(k_1 r^2 + k_2 r^4 + k_3 r^6) \\ r^2 = (x_i - x_0)^2 + (y_i - y_0)^2 \end{cases}$$

For camera calibration of 3D camera, we mainly used the piece monolithic disassembling from software to settle left and right piece. Use two-piece with the partial settlement left and right side of the baseline length B and angle.

When using the 3D camera calibration, try to place the level of 3D camera, we keep the focal length to maintain a minimum focusing distance of doing the processing, establish control points by the board upright photography, use software apart 3D photo into left and right piece of three-dimensional sheet formed relatively, and using the following formula to left and right sheet respectively sheet:

$$X_S = \dfrac{Y_S}{f \cdot \cos\omega - z \cdot \sin\omega} \cdot x$$

$$Z_S = \dfrac{Y_S}{f \cdot \cos\omega - z \cdot \sin\omega} \cdot (z \cdot \cos\omega + f \cdot \sin\omega) \quad (2-78)$$

Wherein, X_S, Y_S, Z_S is the point of photography geodetic coordinates, using the control network directly weighed, x_0, y_0, f, ω is the main camera distance, compared with the use of the formula Y eliminate variables into the control point coordinates can be calculated.

Use the formula:

$$x = z \cdot \sin k + x \cdot \cos k$$
$$z = z \cdot \sin k - x \cdot \cos k \quad (2-79)$$

CHAPTER 2 Technical Foundation of Forest Management (Focus on General Principle)

To compute element.

three-dimensional relative formula:

$$X_S = \frac{B \cdot x_1}{p} \cdot \left(\cos\varphi - \frac{x_2}{f}\sin\varphi \right)$$

$$Z_S = \frac{B \cdot z_1}{p} \cdot \left(\cos\varphi - \frac{z_2}{f}\sin\varphi \right) \quad (2\text{-}80)$$

To solve the sum of the baseline length B, we can calculate based elements to support tree measurement of the angle between the two 3D camera that is light photography, baseline length, and the two lens elements of interior orientation. When the angle $\leqslant 1°$ we can consider parallel light photography, oblique photography to simplify the calculation of the way.

2.7.2 Digital camera inspection and testing

Compared with the measuring camera type, the non-measuring camera has low prices, widely popular in use, but the big difference is in the lens optical distortion. There are no frame mark means so that the camera can not provide non-measurement orientation elements, so it is necessary to calculate orientation element by using certain means and methods.

2.7.2.1 Direct Linear Transformation Method principle

Direct linear transformation algorithm solution is to establish a direct linear relationship between image coordinates, the coordinates of the corresponding object point instrument and object space coordinates. Direct linear transformation solution proposed in 1971. The principle also came from the collinear equation interpretation.

Collinear equation:

$$\begin{cases} x - x_0 + \Delta x + f\dfrac{a_1(X-X_S) + b_1(Y-Y_S) + c_1(Z-Z_S)}{a_3(X-X_S) + b_3(Y-Y_S) + c_3(Z-Z_S)} \\ y - y_0 + \Delta y + f\dfrac{a_2(X-X_S) + b_2(Y-Y_S) + c_2(Z-Z_S)}{a_3(X-X_S) + b_3(Y-Y_S) + c_3(Z-Z_S)} \end{cases} \quad (2\text{-}81)$$

In the above formula systematic errors corrections (Δx and Δy) temporarily assume contains only non-perpendicular axis of varying scale $d\beta$ and linearity errors ds caused corrections section. Suppose x to scale without error, and the y scale factor $1+ds$. In this case, if x is distance to the main photo f_x, then y to the distance of the main photo f_y:

$$f_y = \frac{f_x}{1+ds} \quad (2\text{-}82)$$

The system error corrections Δx and Δy should be:

$$\begin{cases} \Delta x = ON_2 - OM_2 = |M_2P| \sin d\beta = (1+ds)(y-y_0)\sin d\beta \approx (y-y_0)\sin d\beta \\ \Delta y = ON_1 - OM_1 = |OM_1| \cos d\beta = OM'_1 = [(1+ds)\cos d\beta - 1](y-y_0) \approx (y-yt_0)ds \end{cases}$$
$$(2\text{-}83)$$

In this case, the collinearity equation form containing only linear error corrections are as follows:

$$\begin{cases} x - x_0 + (1 + ds)(y - y_0)\sin d\beta + f_x \dfrac{a_1(X - X_S) + b_1(Y - Y_S) + c_1(Z - Z_S)}{a_3(X - X_S) + b_3(Y - Y_S) + c_3(Z - Z_S)} = 0 \\ (1 + ds)(y - y_0)\cos d\beta + f_x \dfrac{a_2(X - X_S) + b_2(Y - Y_S) + c_2(Z - Z_S)}{a_3(X - X_S) + b_3(Y - Y_S) + c_3(Z - Z_S)} = 0 \end{cases}$$

$$(2\text{-}84)$$

Order r_1, r_2, r_3 is:

$$\begin{cases} r_1 = (a_1 X_S + b_1 Y_S + c_1 Z_S) \\ r_2 = -(a_2 X_S + b_2 Y_S + c_2 Z_S) \\ r_3 = -(a_3 X_S + b_3 Y_S + c_3 Z_S) \end{cases} \quad (2\text{-}85)$$

So substitute r_1, r_2, r_3 into the equation above, then:

$$\begin{cases} x - x_0 + (1 + ds)(y - y_0)\sin d\beta + f_x \dfrac{a_1 X + b_1 Y + c_1 Z + r_1}{a_3 X + b_3 Y + c_3 Z + r_3} = 0 & (2\text{-}86) \\ (1 + ds)(y - y_0)\cos d\beta + f_x \dfrac{a_2 X + b_2 Y + c_2 Z + r_2}{a_3 X + b_3 Y + c_3 Z + r_3} = 0 & (2\text{-}87) \end{cases}$$

Finish formula 2-86, then:

$$y + \dfrac{\dfrac{1}{r_3}\left[\dfrac{a_2 f_x}{(1+ds)\cos d\beta} - a_3 y_0\right] X}{\dfrac{a_3}{r_3}X + \dfrac{b_3}{r_3}Y + \dfrac{c_3}{r_3}Z + 1} + \dfrac{\dfrac{1}{r_3}\left[\dfrac{b_2 f_x}{(1+ds)\cos d\beta} - b_3 y_0\right] X}{\dfrac{a_3}{r_3}X + \dfrac{b_3}{r_3}Y + \dfrac{c_3}{r_3}Z + 1} +$$

$$\dfrac{\dfrac{1}{r_3}\left[\dfrac{c_2 f_x}{(1+ds)\cos d\beta} - c_3 y_0\right] X}{\dfrac{a_3}{r_3}X + \dfrac{b_3}{r_3}Y + \dfrac{c_3}{r_3}Z + 1} + \dfrac{\dfrac{1}{r_3}\left[\dfrac{r_2 f_x}{(1+ds)\cos d\beta} - r_3 y_0\right] X}{\dfrac{a_3}{r_3}X + \dfrac{b_3}{r_3}Y + \dfrac{c_3}{r_3}Z + 1}$$

The formula above change to:

$$y + \dfrac{l_5 X + l_6 Y + l_7 Z + l_8}{l_9 X + l_{10} Y + l_{11} Z + 1} = 0 \qquad (2\text{-}88)$$

Where in:

$$l_5 = \dfrac{1}{r_3}\left[\dfrac{a_2 f_x}{(1+ds)\cos d\beta} - a_3 y_0\right]$$

$$l_6 = \dfrac{1}{r_3}\left[\dfrac{b_2 f_x}{(1+ds)\cos d\beta} - b_3 y_0\right]$$

$$l_7 = \dfrac{1}{r_3}\left[\dfrac{c_2 f_x}{(1+ds)\cos d\beta} - c_3 y_0\right]$$

$$l_8 = \dfrac{1}{r_3}\left[\dfrac{r_2 f_x}{(1+ds)\cos d\beta} - r_3 y_0\right] = -(l_5 X_S + l_6 Y_S + l_7 X_S)$$

Finish formula 2-84, then: ①×$\cos d\beta$ − ②×$\cos d\beta$

$$x + \frac{\frac{1}{r_3}(a_1f_x - a_2f_x\tan\beta - a_3x_0)X}{\frac{a_3}{r_3}X + \frac{b_3}{r_3}Y + \frac{c_3}{r_3}Z + 1} + \frac{\frac{1}{r_3}(b_1f_x - b_2f_x\tan\beta - b_3x_0)Y}{\frac{a_3}{r_3}X + \frac{b_3}{r_3}Y + \frac{c_3}{r_3}Z + 1} +$$

$$\frac{\frac{1}{r_3}(c_1f_x - a_2f_x\tan\beta - c_3x_0)X}{\frac{a_3}{r_3}X + \frac{b_3}{r_3}Y + \frac{c_3}{r_3}Z + 1} + \frac{\frac{1}{r_3}(r_1f_x - r_2f_x\tan\beta - r_3x_0)X}{\frac{a_3}{r_3}X + \frac{b_3}{r_3}Y + \frac{c_3}{r_3}Z + 1} = 0 \quad (2\text{-}89)$$

So formula above change to:

$$x + \frac{l_1X + l_2Y + l_3Z + l_4}{l_9X + l_{10}Y + l_{11}Z + 1} \quad (2\text{-}90)$$

Where in:

$$l_4 = \frac{1}{r_3}(r_1f_x - r_2f_x\tan d\beta - r_3x_0) = -(l_1X_S + l_2Y_S + l_3Z_S) \quad (2\text{-}91)$$

According to the analysis above, the coefficients of $(l_1, l_2, l_3, \cdots, l_{11})$ are:

$$\begin{cases} l_1 = \frac{1}{r_3}(a_1f_x - a_2f_x\tan d\beta - a_3x_0) & l_5 = \frac{1}{r_3}\left[\frac{a_2f_x}{(1+ds)\cos d\beta} - a_3y_0\right] \\ l_9 = \frac{a_3}{r_3} \quad r_1 = -(a_1X_S + b_1Y_S + c_1Z_S) \\ l_2 = \frac{1}{r_3}(b_1f_x - b_2f_x\tan d\beta - b_3x_0) & l_6 = \frac{1}{r_3}\left[\frac{b_2f_x}{(1+ds)\cos d\beta} - b_3y_0\right] \\ l_9 = \frac{a_3}{r_3} \quad r_2 = -(a_2X_S + b_2Y_S + c_2Z_S) \\ l_3 = \frac{1}{r_3}(c_1f_x - c_2f_x\tan d\beta - c_3x_0) & l_5 = \frac{1}{r_3}\left[\frac{a_2f_x}{(1+ds)\cos d\beta} - a_3y_0\right] \\ l_9 = \frac{a_3}{r_3} \quad r_1 = -(a_1X_S + b_1Y_S + c_1Z_S) \end{cases} \quad (2\text{-}92)$$

So linear transformation of DLT:

$$\begin{cases} x + \dfrac{l_1X + l_2Y + l_3Z + l_4}{l_9X + l_{10}Y + l_{11} + 1} = 0 \\ y + \dfrac{l_5X + l_6Y + l_7Z + l_8}{l_9X + l_{10}Y + l_{11} + 1} = 0 \end{cases}$$

The linear equation into:

$$\begin{cases} l_1X + l_2Y + l_3Z + l_4 + 0 + 0 + 0 + 0 + xl_9X + xl_{10}y + xl_{11}Z + x = 0 \\ 0 + 0 + 0 + 0 + l_5X + l_6Y + l_7Z + l_7Z + l_8 + yl_9X + yl_{10}y + yl_{11}Z + y = 0 \end{cases} \quad (2\text{-}93)$$

In order to solve the 11 unknowns $(l_1, l_2, l_3, \cdots, l_{11})$, you need to select the six control point, and their spatial coordinates are known. According to this it can be listed at least 11 equations, so as to solve the approximation of $(l_1, l_2, l_3, \cdots, l_{11})$ coefficient

First, you need to establish a matrix: $\underset{(2n,11)}{A} \cdot \underset{(11,1)}{X} = -\underset{(2n,1)}{C}$

Expanding:

$$\begin{bmatrix} X_1 & Y_1 & Z_1 & 1 & 0 & 0 & 0 & 0 & x_1X_1 & x_1Y_1 & x_1Z_1 \\ 0 & 0 & 0 & 0 & X_1 & Y_1 & Z_1 & 1 & x_1X_1 & x_1Y_1 & x_1Z_1 \\ X_2 & Y_2 & Z_2 & 1 & 0 & 0 & 0 & 0 & x_2X_2 & x_2Y_2 & x_2Z_2 \\ 0 & 0 & 0 & 0 & X_2 & Y_2 & Z_2 & 1 & x_2X_2 & x_2Y_2 & x_2Z_2 \\ \vdots & \vdots & \vdots & \vdots & \vdots & \vdots & \vdots & \vdots & \vdots & \vdots & \vdots \\ X_n & Y_n & Z_n & 1 & 0 & 0 & 0 & 0 & x_nX_n & x_nY_n & x_nZ_n \\ 0 & 0 & 0 & 0 & X_n & Y_n & Z_n & 1 & x_nX_n & x_nY_n & x_nZ_n \end{bmatrix} \begin{bmatrix} l_1 \\ l_2 \\ l_3 \\ l_4 \\ l_5 \\ l_6 \\ l_7 \\ l_8 \\ l_9 \\ l_{10} \\ l_{11} \end{bmatrix} = \begin{bmatrix} -x_1 \\ -y_1 \\ -x_2 \\ -y_2 \\ \vdots \\ -x_n \\ -y_n \end{bmatrix} \quad (2\text{-}94)$$

By using least squares method to solve.

2.7.2.2 Determining the orientation of the inner and outer elements

1) Solve x_0 and y_0

Since the rotationmatrix, R is an orthogonal matrix.

So there are following basic formulae:

$$a_1^2 + b_1^2 + c_1^2 = 1, a_2^2 + b_2^2 + c_2^2 = 1, a_3^2 + b_3^2 + c_3^2 = 1 \quad (2\text{-}95)$$

$$a_1a_3 + b_1b_3 + c_1c_3 = 0, a_2a_3 + b_2b_3 + c_2c_3 = 0, a_1a_2 + b_1b_2 + c_1c_2 = 0 \quad (2\text{-}96)$$

Finish formula above,

$$l_9^2 + l_{10}^2 + l_{11}^2 = \frac{1}{r_3^2} \qquad r_3^2 = \frac{1}{l_9^2 + l_{10}^2 + l_{11}^2}$$

Then

$$l_1l_9 + l_2l_{10} + l_2l_{11} = -\frac{x_0}{r_3^2} \Rightarrow x_0 = \frac{l_1l_9 + l_2l_{10} + l_2l_{11}}{l_9^2 + l_{10}^2 + l_{11}^2}$$

$$l_5l_9 + l_6l_{10} + l_7l_{11} = -\frac{y_0}{r_3^2} \Rightarrow y_0 = \frac{l_5l_9 + l_6l_{10} + l_7l_{11}}{l_9^2 + l_{10}^2 + l_{11}^2} \quad (2\text{-}97)$$

Then solve x_0 and y_0 of the inner and outer elements.

2) solve f_x, f_y, ds and $d\beta$

$$l_1^2 + l_2^2 + l_3^2 = \frac{1}{r_3^2}(f_x^2 + f_x^2\tan^2 d\beta + x_0^2) = \frac{1}{r_3^2}\left(\frac{f_x^2}{\cos^2 d\beta} + x_0^2\right)$$

$$l_5^2 + l_6^2 + l_7^2 = \frac{1}{r_3^2}\left[\frac{f_x^2}{(1+ds)^2\cos^2 d\beta} + y_0^2\right]$$

$$l_1l_5 + l_2l_6 + l_3l_7 = \frac{1}{r_3^2}\left[x_0y_0 - \frac{f_x^2 \sin d\beta}{(1+ds)\cos^2 d\beta}\right]$$

$$r_3^2(l_1^2 + l_2^2 + l_3^2) - x_0^2 = \frac{f_x^2}{\cos^2 d\beta} = A$$

$$\Rightarrow r_3^2(l_5^2 + l_6^2 + l_7^2) - y_0^2 = \frac{f_x^2}{(1+ds)^2 \cos^2 d\beta} = B \quad (2\text{-}98)$$

$$r_3^2(l_1 l_5 + l_2 l_6 + l_3 l_7) - x_0 y_0 = -\frac{f_x^2 \sin d\beta}{(1+ds)\cos^2 d\beta} = C$$

$$\frac{A}{B} = (1+ds)^2$$

$$\frac{C}{B} = (-\sin d\beta)(1+ds) \Rightarrow$$

$$\frac{A}{B} \cdot \frac{B^2}{C^2} = \frac{1}{\sin d\beta}$$

$$ds = \sqrt{\frac{A}{B}} - 1 = \sqrt{\frac{r_3^2(l_1^2 + l_2^2 + l_3^2) - x_0^2}{r_3^2(l_5^2 + l_6^2 + l_7^2) - y_0^2}} - 1$$

$$f_x = \sqrt{A} \cos d\beta = \cos d\beta \sqrt{\frac{(l_1^2 + l_2^2 + l_3^2)}{(l_9^2 + l_{10}^2 + l_{11}^2)} - x_0^2} \quad (2\text{-}99)$$

$$d\beta = \arcsin \sqrt{\frac{C^2}{AB}} = \arcsin \sqrt{\frac{[r_3^2(l_1 l_5 + l_2 l_6 + l_3 l_7) - x_0 y_0]^2}{[r_3^2(l_1^2 + l_2^2 + l_3^2) - x_0^2][r_3^2(l_5^2 + l_6^2 + l_7^2) - y_0^2]}}$$

Where in:

$$f_y = \frac{fx}{1+ds}$$

3) Solving exterior orientation elements

$$l_9 X_S + l_{10} Y_S + l_{11} Z_S = \frac{a_3}{r_3} X_S + \frac{b_3}{r_3} Y_S + \frac{c_3}{r_3} Z_S = \frac{a_3 X_S + b_3 Y_S + c_3 Z_S}{a_3 X_S + b_3 Y_S + c_3 Z_S} = -1 \quad (2\text{-}100)$$

So we can establish three equations with formula l_4 and l_4, then we can solve $X_S Y_S Z_S$.

$$l_1 X_S + l_2 Y_S + l_3 Z X_S = -l_4$$
$$l_5 X_S + l_6 Y_S + l_7 Z X_S = -l_8$$
$$l_9 X_S + l_{10} Y_S + l_{11} Z X_S = -l \quad (2\text{-}101)$$

4) Azimuth outer element solver

$$a_3 = r_3 l_9 = \frac{l_9}{\sqrt{l_9^2 + l_{10} + l_{11}}}$$

$$b_3 = r_3 l_{10} = \frac{l_9}{\sqrt{l_9^2 + l_{10} + l_{11}}}$$

$$c_3 = r_3 l_{11} = \frac{l_9}{\sqrt{l_9^2 + l_{10} + l_{11}}}$$

$$b_2 = \frac{(l_6 r_3 + b_3 y_0)(1 + ds)\cos d\beta}{f_x}$$

$$b_1 = \frac{l_2 r_3 + b_3 x_0 + b_2 f_x \tan d\beta}{f_x} \quad (2\text{-}102)$$

According rotation matrix, we can get:

$$\tan\varphi = -\frac{a_3}{c_3} \qquad \varphi = \arctan\left(-\frac{a_3}{c_3}\right)$$

$$\sin\varphi = -b_3 \quad \Rightarrow \quad \omega = \arctan(-b_3) \quad (2\text{-}103)$$

$$\tan\kappa = -\frac{b_1}{b_2} \qquad \kappa = \arctan\left(-\frac{b_1}{b_2}\right)$$

Thus, a total of nine elements of interior and exterior orientation on the amount of the settlement is completed.

2.7.2.3 Digital Camera Calibration Based on the principle of DLT

An image of the calibration field is acquired using a camera each control point on the image coordinates of (u_i, v_i), corresponding to the coordinates of space point (X_i, Y_i, Z_i) After DLT and optical distortion linear transformation formula $L = (M^T M)^{-1} M^T W$ to finish simplification, iterative formula iteration Where in:

$$L = (l_1 \quad l_2 \quad l_3 \quad l_4 \quad l_5 \quad l_6 \quad l_7 \quad l_8 \quad l_9 \quad l_{10} \quad l_{11} \quad k_1 \quad k_2 \quad p_1 \quad p_2 \quad \alpha \quad \beta)^T$$

$$W = \left[-\frac{x_i}{A} \quad -\frac{y_i}{A}\right]$$

$$M = -[E \quad F]$$

$$E = \begin{bmatrix} \frac{X_i}{A} & \frac{Y_i}{A} & \frac{Z_i}{A} & \frac{1}{A} & 0 & 0 & 0 & 0 & \frac{x_i X_i}{A} & \frac{x_i Y_i}{A} & \frac{x_i Z_i}{A} \\ 0 & 0 & 0 & 0 & \frac{X_i}{A} & \frac{Y_i}{A} & \frac{Z_i}{A} & \frac{1}{A} & \frac{y_i X_i}{A} & \frac{y_i Y_i}{A} & \frac{y_i Z_i}{A} \end{bmatrix}$$

$$F = \begin{bmatrix} (x_i - x_0)r^2 & (x_i - x_0)r^4 & [r^2 + 2(x_i - x_0)^2] & 2(x_i - x_0)(y_i - y_0) & (x_i - x_0) & (y_i - y_0) \\ (y_i - y_0)r^2 & (y_i - y_0)r^4 & 2(x_i - x_0)(y_i - y_0) & [r^2 + 2(y_i - y_0)^2] & 0 & 0 \end{bmatrix}$$

$$r = \sqrt{(x_i - x_0)^2 + (y_i - y_0)^2}$$

$$A = l_9 X_i + l_{10} Y_i + l_{11} Z_i + 1$$

Get DLT iterative linear transformation coefficients and optical distortioncorrection parameters L.

Use modulus $\begin{cases} l_1 X_S + l_2 Y_S + l_3 Z_S = -l_4 \\ l_5 X_S + l_6 Y_S + l_7 X_S = -l_8 \\ l_9 X_S + l_{10} Y_S + l_{11} Z_S = -1 \end{cases}$ The instrument can be solved exterior orientation elements (X_S, Y_S, Z_S).

By using modulus

$$\varphi = \arctan\left(-\frac{l_9}{l_{11}}\right)$$

CHAPTER 2 Technical Foundation of Forest Management (Focus on General Principle)

$$\omega = \arcsin\left(-\frac{l_{10}}{\sqrt{l_9^2 + l_{10} + l_{11}}}\right)$$

And

$$\kappa = \arctan\left(\frac{\left(l_2 r_3 + \frac{l_{10}}{\sqrt{l_9^2 + l_{10} + l_{11}}} x_0\right) + \left(l_6 r_3 + \frac{l_{10}}{\sqrt{l_9^2 + l_{10} + l_{11}}} y_0\right)(1 + \mathrm{d}s)\cos\mathrm{d}\beta f_x \tan\mathrm{d}\beta}{\left(l_6 r_3 + \frac{l_{10}}{\sqrt{l_9^2 + l_{10} + l_{11}}} y_0\right)(1 + \mathrm{d}s)\cos\mathrm{d}\beta}\right)$$

(2-104)

The instrument can beused as exterior orientation elements solver $(\omega, \varphi, \kappa)$.

Distortion corrected parameters obtained by the DLT linear change as follows (Table 2-21).

Table 2-21 Camera calibration results

Calibration Description	Symbol	Camera 1(X)/pixels	camera 2(J)/pixels
Principal point of abscissa	x_0	2458.0477	2452.4848
Principal point ordinate	y_0	1661.9577	1665.0902
Camera focal length	F	4961.4577	4962.2832
Radial distortion coefficients 1	K_1	5.457578e-09	5.842168e-09
Radial distortion coefficients 2	K_2	-4.0678e-16	-4.218370e-16
Eccentric distortion factor 1	P_1	-8.774575e-08	-2.085813e-08
Eccentric distortion factor 2	P_2	-1.442552e-07	-5.467009e-08
Non-square scaling factor	A	4.267730e-05	-8.437985e-05
Non-orthogonal aberration coefficient	B	-8.754646e-06	2.743436e-05

2.7.2.4 Camera motion detection

Product squareis shown below three to five randomly targets were placed in a space large enough before, during and after the vertical plane, forming a non-indoor venue with no fixed control points, known coordinates and landmarks movable camera calibration and dynamic detection technology (Figure 2-71).

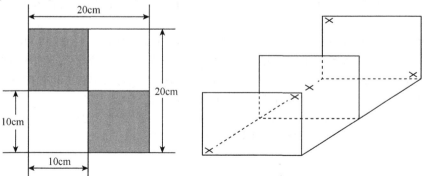

Figure 2-71 Square target schematic

The non-linear change of coordinates formallinearized operations, the following formula:

$$\begin{cases} \cdot u = (u_i - u_0)(k_1 r^2 + k_2 r^4) + P_1[r^2 + 2(u_i - u_0)^2] + 2P_2(u_i - u_0)(v_i - v_0) \\ \cdot v = (v_i - v_0)(k_1 r^2 + k_2 r^4) + P_2[r^2 + 2(v_i - v_0)^2] + 2P_1(u_i - u_0)(v_i - v_0) \end{cases}$$

In the formula, $r^2 = (u_i - u_0)^2 + (v_i - v_0)^2$;

$\cdot u, \cdot v$ —Denotes an image point coordinates nonlinear corrections;

(u_0, v_0)—Is the principal point coordinates;

k_1, k_2—Radial distortion coefficient;

P_1, P_2—Eccentric distortion factor:

(u_i, v_i)—Is an arbitrary image point coordinates.

Linearized equationrepresented as a matrix form of:

$$V = W \cdot X + B$$

In the formula, $V = \begin{pmatrix} \cdot u \\ \cdot v \end{pmatrix}$;

W—The coefficient matrix of the linearized matrix;

$X = \begin{pmatrix} u_0 & v_0 & k_1 & k_2 & P_1 & P_2 \end{pmatrix}^T$;

B—Constant the term.

Use more than five on target 45 points and 60 length value, combine with the top of simultaneous equations, substituting seeking to calculate the value of each calibration element, repeated iteration until the difference between the two operations to meet the minimum criteria are met.

References

范海英,杨伦,邢志辉,等,2004. Cyra 三维激光扫描系统的工程应用研究[J]. 矿山测量(03):16-18,4.

巩垠熙,何诚,冯仲科,等,2013. 基于改进 Delaunay 算法的树冠三维重构单木因子提取[J]. 农业机械学报,44(02):192-199.

胡传双,王婷,云虹,等,2009. 工厂预制木结构建筑在中国的发展现状及展望[J]. 木材加工机械,20(S1):6-61,51.

胡勇庆,钱俊,2010. 木材加工剩余板皮高效利用分析与展望[J]. 木材加工机械,21(03):49-51+18.

黄杰,赵京音,万常照,2008. RFID 在农业中的应用与展望[J]. 农业网络信息(09):119-121.

刘昌霖,2014. 三维激光扫描测量技术探究及应用[J]. 科技信息(05):61,35.

刘姗姗,张绍文,2008. 基于 RFID 技术的林木种质资源管理信息系统的开发构想[J]. 林业建设(05):27-31.

陆研,张绍文,2008. 基于 RFID 技术的名木古树管理系统初探[J]. 山东林业科技(02):91-94.

孙琴,乔牡丹,沈洪霞,等,2009. 林业信息化建设研究初探[J]. 内蒙古林业调查设计,32(06):88-89,102.

汤晓华,马岩,陈强,等,2005. 圆与定压缩系数椭圆截面三点初步定心[J]. 木材加工机械(02):6-9.

唐雪海,2011. 北京市城六区三维绿量估算与分析研究[D]. 北京:北京林业大学.

王颖,周铁军,李阳,2010. 物联网技术在林业信息化中的应用前景[J]. 湖北农业科学,49(10):2601-2604.

张会霞,朱文博,2012. 三维激光扫描数据处理理论及应用[M]. 北京:电子工业出版社.

CHAPTER 3 Forest Manager's Basic Numerical Tables and Model

3.1 The summary of forest numerical table

In the survey of forest resource and forest management, a variety of management tables have been constructed, most commonly used are the standing timber volume table, timber Products Rate Table and site ranking table.

3.1.1 Standing timber volume table

In the forestry production practice, we usually determine the volume by stock volume table, depending upon species. Volume table is prepared according to the function of stem volume, DBH and tree height are main factors. The most commonly used are ordinary volume table and binary volume table. Ordinary volume table is constructed by using DBH as the main factor and the table which is constructed by using the relationship of tree height and diameter (DBH) for volume estimation is said to be as a binary volume table.

Ordinary volume table can give us good accuracy in the application area. Since ordinary volume table only considers the relationship between volume and diameter, but in the same area and same diameter trees have different tree height and volume differences may occur from small scale to large scale, so ordinary volume table cannot be used in the large area. If you didn't understand the applicability of ordinary volume table within the region we should select samples of wood randomly and testing the applicability of the measured material for a volume table, such as the error exceeds 5%, to be recompiling.

It is convenient to measure the accumulation amount by ordinary volume table, only needed the single tree survey. statistically, the number of trees species in each size class, check the corresponding individual volume of tree species by ordinary volume table, multiplying the volume of each diameter class with the accumulation of species. This will give the total volume of each species, by this way we can get the entire stock volume stands.

Binary volume table takes into accountby having two factors DBH and tree height. Its scope, and accuracy is much higher than ordinary volume table. Binary volume table is usually used within a wide range such as in provinces.

Using binary volume table in the forest stand survey, the first step is to measure each diameter class and tree height and then tree height curve drawing, isolated by the curve on the average height of each size class. Then according to the respective diameter class and the corresponding average height value is obtained. Binary volume table give each diameter class individual volume, multiplied by the number of trees for each diameter class, to give the corresponding volume for each diameter class. From that, we can get the total species the volume the volume of each species and can obtain stock volume stands.

Whether we use the ordinary volume table by analytical method or binary volume table, there is a volume formula which is generally marked in the volume table. Therefore as this expression programmed into a computer application, when calculating the stand volume will be more convenient.

3.1.2 Timber products rate table

The volume tables can be used to determine the amount of wood in standing timber products. Timber products rate table can be divided into vertical timber rate table and stand timber products rate table.

1) Vertical timber rate table

Unitary merchantable volume tables are based ondiameter factor to determine the quantity material of the wood species. It is the same as unitary merchantable volume table, which is applicable to small region, and also the local rate table. First volume of each tree species is obtain in order the accumulation and then multiplied by the single-lumber species out of the material on the table to find the diameter of the wood species of the material rate, so we can get the diameter of the various wood species of the amount of material in order of the timber species of wood volume respectively, output can be obtained from the stand and the wood species of the different wood material.

Binary tree yield rate table is used to determine the rate of timber products based on two factors DBH and tree height of stumpage. The accuracy of binary volume tables is higher than unitary volume tables. The range and use is also in large scale. Usually this type of production the number of tables in several advanced trees (site level) prepared for the convenience of the application. In timber rate table we can also measure the corresponding tree height and diameter of bark volume.

2) Stand timber products rate table

According to the diameter distribution of forest elements, the timber yield table is based on the timber material diameter, DBH, average tree height and the output rate of each wood species is listed.

It is convenient to use the forest timber material output scale in forest manager investigation and planning and design. In the field survey do not have to carry out a huge workload of each wooden feet, only by visual survey or other means to obtain the average small diameter of small classes and the average tree height table look out for wood species, accumulation, that is, the amount of forest timber production.

3.1.3 Site ratings list

Site quality is an important factor in forest inventory as well as the use of forest land survey and also from an evaluation point of view it is an extremely important factor. Commonly used tables in the forest site survey are ranking table, standing scale, index table, site type table and a number of site index tables.

1) Class table

Class table is compiled on the basis of stand height taller than the average age. The average correlation between forest productivity is a reflection of the number of tables. Tree height is a reflection of the most sensitive site productivity factor applied to determine the average high site quality stands, easy to use and usually divided into five level positions expressed in Roman numerals.

When the table is applied, the position level of the small class can be identified from the class table by using the age and average tree height of the small class trees obtained from the survey. Since the average stand height is highly affected by the stand density, it will appear that the site grade of the stand is higher than that of the less dense stand under the same site condition. In addition, the differences in stand dominant tree species will also result in different quality of forest land under the same site conditions, especially in the case of planted forests.

2) Index table

Many scientists believethat average forest growth can't reflect the quality of forest, but it reflects the quality of the site. Site quality table shows the relationship between the dominant tree stand high and the average age of the trees this is called index table.

The advantage site index table usually refers to the height of the highest tree per 100 m^2, and the rank of the position index is expressed as the advantage of the tree in the base year. The base year is when the tree height of the tree species grows relatively stable of age. In the southern China fir and masson pine, the base year is defined as 20 years. For example, if the position index of *Pinus massoniana* is 20, it means that the advantage of 20-year-old *Pinus massoniana* can grow to 20 m.

3) Land condition type table

The above two types of site quality grading table must be used to check the tree height, therefore the evaluation of forest land must be a forest and the growth of young trees stage height is also very unstable. Thus the two types of tables cannot be used for evaluation of the site quality of non-forested, un-forested and young stands. In addition, the trees of different ages are very different, and the growing process is different from the same age forest, so they cannot use these two kinds of tables.

In order to evaluate the site quality of non-forestland and heterogeneous forest, environmental factors are often used in the production to evaluate the quality of the site, which is called the site condition table. For example, Fujian Province, the topography selected terrain, soil factors, the site of the three factors will be divided into the site level of fertile type, more fertile type, general type, barren type, the table is more convenient to use.

4) Quantify the site index table

In order to overcome these shortcomings some foresters in the 20th century since the use of quantitative theory, the various types of qualitative environmental factors and the advantages of trees height growth process established a quantitative model of high forest age, and compiled a quantitative site index table. In the application of the table as long as the factors found in the score value, the cumulative number of status can be obtained by the small index.

3.1.4 Number of tables from the forest to forest model

The forest data table described above is widely used in the forest resources survey and forest management for a long period of time, even at this stage. However, the forest parameters of this numerical method are very important for the description ofmacro forest data, quantification of forest form. There are some shortcomings in expression and future morphological prediction, such as heavy workload, error-prone, and unfavorable description of forest morphology. Therefore the laboratory of precision forestry in Beijing Forestry University is practicing the production and accumulation of a large number of data observations in the forest in order to achieve the exact parameters of expression and the ultimate goals. With massive data established on the basis of the relevant model, the realization of state forests and forest parameters precise description, to explore the relationship between the factors and forest growth that provides the basis for precise management of forest, greatly simplifying forest survey workload and improve the efficiency of forestry work.

3.2 Tree volume built research model

3.2.1 Field observation

Field observation as shown in Figure 3-1.

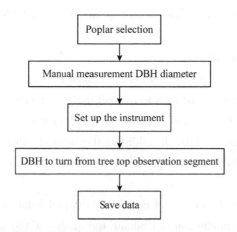

Figure 3-1 Field observation technology roadmap

1) Electronic theodolite tree measurement

First DBH is measured by using tape i. e. at 1.3 meters, followed by the use of electronic theodolite to find trunk distance from the theodolite pointed at DBH (1.3 meters). The horizontal level distance from theodolite to the trunk and tree top from theodolite, follow these steps:

① Measure the ground diameterwith d_0 and the diameter $d_{1.3}$ at the height 1.3 by dia tape;

② Set the electronic theodolite first and from the height of 1.3 meters to the top of the tree and obtain the horizontal angle $\alpha_1, \alpha_2, \alpha_3, \cdots, \alpha_n$ of the sight line of the electronic theodolite telescope and zenith distance $\gamma_1, \gamma_2, \gamma_3, \cdots, \gamma_n$. as shown in Figure 3-2.

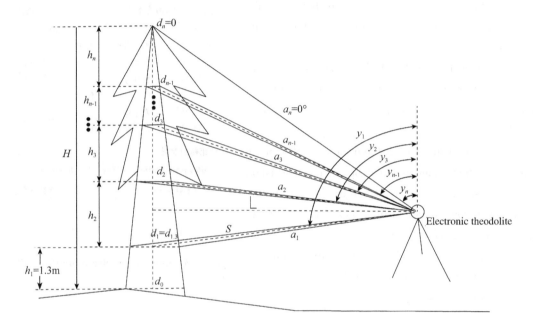

Figure 3-2 Electronic theodolite precision measuring principle

2) Total station tree measurement

Firstly by using measurement tape, measure stumpage diameter at 1.3 meters DBH in height, followed by the use of total station for the measurement of horizontal angle and zenith distance, and then use the total station ranging function measuring 1.3 m trunk center to the instrument center Slope distance. This step is followed by measuring 1.3 meters above the tree trunk to the diameter, horizontal angle, and finally the use of differential quadrature method to calculate the standing timber volume. Specific steps are as follows:

(1) Use ruler to measure DBH diameter d_0, height 1.3 diameter $d_{1.3}$;

(2) Place the total station, aiming at the edge of the trunk 1.3 meters to obtain the total station telescope sight line of sight left and right along the horizontal angle α_1 and aiming at the right edge of the zenith distance γ_1, and then aiming $\alpha_{1/2}$, After that measure slope distance S and save;

(3) From the tree height of 1.3 meters to the top of the tree in order. The segments edge of the telescope to obtain total station telescope line of sight left and right along the horizontal angle $\alpha_2, \alpha_3, \cdots, \alpha_n$ and the zenith distance when aimed at the right edge of the stump $\gamma_2, \gamma_3, \cdots, \gamma_n$;

(4) Click Finish to use the built-in software to get the results directly.

3.2.2 Data transmission

Total station data results can bestored directly into electronically readable and removable SD card. After using super CCD electronic theodolite station apparatus and the raw data need to be calculated using the post-processing software, electronic theodolite for example specific methods are as follows:

(1) First, by using the serial cable electronic theodolite connected to the computer and then turns the instrument interface to adjust the data transmission interface;

(2) Open the South CASS software on the computer and after that open, the menu data, (read total station data) appear under the interface shown in Figure 3-3;

(3) The instrument columnselects E500 South handbook, other communication parameters consistent with the instrument settings. Click on the send button to send electronic data and then press Enter to accept the data on your computer;

(4) After the transfer is successful a different sound will be heard and software data will appear as shown in Figure 3-4;

Figure 3-3 Interface of electronic theodolite

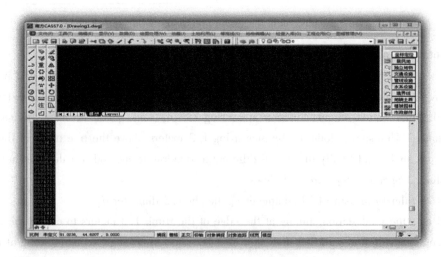

Figure 3-4 Data of software

(5) Finally, the establishment of TXT files copy and paste the data in the past by date, instrument type collated as shown in Figure 3-5.

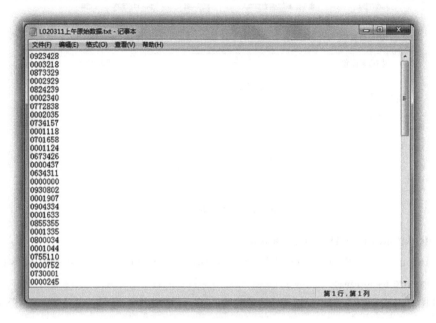

Figure 3-5　TXT documents

3.2.3　Internal calculation and error analysis

3.2.3.1　Internal calculation

The internal calculation refers to the electronic theodolite and CCD superstation data processing system, all of data is processed and calculated in the field quickly after data collection.

1) Electronic theodolite within the computing industry

The tree is measured by non-destructive precision measurement and modeling software (FSMV1.0) to complete the specific steps as follows:

(1) Data import (Figure 3-6).

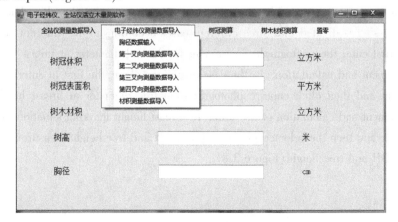

Figure 3-6　Data import

(2) Click on the button to calculate (Figure 3-7).

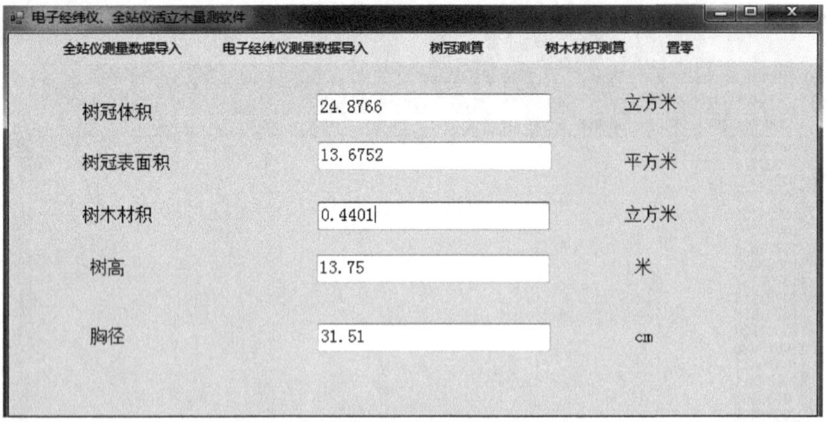

Figure 3-7 Computing interface

2) Calculation of CCD super station

Calculation is done through the CCD super-station instrument data processing software and its specific steps are as follows:

(1) First click on the file, calibration data into the camera, using the data format TXT format; Content as:

X02458.0477	Description: main point X0
Y01661.9577	Description: main point Y0
f 4961.4577	Description: focal length f
K15.45757850e-09	Description: radial distortion coefficient K1
K2-4.0678e-16	Description: radial distortion coefficient K2
P1-8.774575e-08	Description: eccentric distortion coefficient P1
P2-1.44255243e-07	Description: eccentric distortion coefficient P2
α4.26773e-05	Description: square ratio α
β-8.75464676e-06	Description: orthogonality of distortion coefficient β

(2) Then the image captured by the open test tree photographs, the chip that is the target shooting; Processing is divided into three different cases according to the different conditions: for flat condition first enter the photography instrument height and diameter at breast height and tree height measurement and calculation; In the case of uniform slope, the first to enter a value for the average gradient and then click capture photographs of tree diameter at breast height, then tree height measurement and calculation of its value; for breast height marking situation the first to enter camera angle and then the selected photograph of DBH and tree height measurement and calculation of the DBH and tree height (Figure 3-8);

CHAPTER 3 Forest Manager's Basic Numerical Tables and Model

Figure 3-8 Processing interface

(3) At the end the software process is completed and a large number of samples are obtained (Figure 3-9).

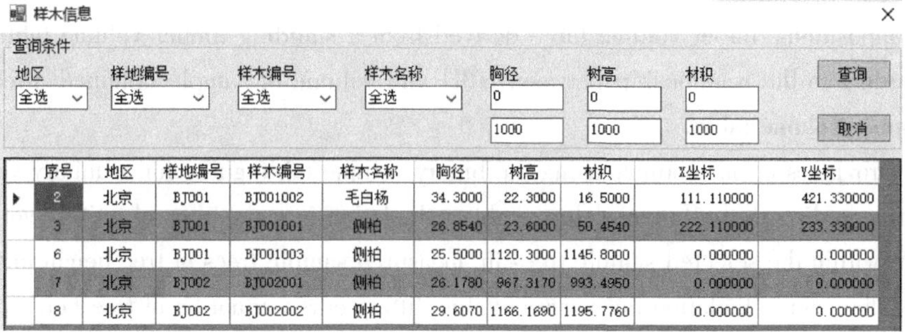

Figure 3-9 Sample wood information

3.2.3.2 Error analysis

Through the error propagation theory of the electronic theodolite measuring tree principle, the error different sample's tree height and volume are recorded. The actual measurement of volume data of the poplar tree in Beijing as the verification truth value is used. To provide a scientific basis and recommendations for the non-destructive preparation of standing timber volume table technology.

Table 3-1 can be obtained from the electronic theodolite, different specifications of sample tree; the change in error with tree size is slightly different. Among them the error increases as timber size increases and vice versa range from max value $\pm 7.1868 \times 10^{-2}$ m to min value of $\pm 4.5716 \times 10^{-2}$ m. Phase error increased from 0.29% to 0.89%; the error in volume reduced from $\pm 4.449 \times 10^{-3}$ m^3 to a minimum of $\pm 5.050 \times 10^{-4}$ m^3, the relative error increased from 0.50% to 2.33%.

Table 3-1 Error analysis of electronic theodolite measurement with different specifications

Sample size	Electronic theodolite angle measuring/°	Tree height			Tree volume		
		σ_H/m	H/m	σ'_H/%	σ_V/m^3	H/m^3	σ'_V/%
Big	±2.0	±7.1858×10^{-2}	24.79	0.29	±4.449×10^{-3}	0.8852	0.50
Middle	±2.0	±6.5867×10^{-2}	14.92	0.44	±4.567×10^{-3}	0.2653	0.76
Small	±2.0	±4.5716×10^{-2}	5.13	0.89	±5.050×10^{-3}	0.0212	2.23

The results show that the errors of the same precision electronic theodolite are different with the change of the tree size. The error of the tree height and volume is smaller as the tree size becomes smaller. This method is ideal for large and medium sized trees. For smaller trees, because of their height and small volume, they are more affected by fixed errors. They should be used rationally according to the purpose of investigation.

3.2.4 Establishment of forest volume model

1) Variable volume model

According the main technical regulations of our country's forest professional survey a standing tree volume model establishment has two methods. One is to build the tree height-diameter model, the binary oppositions timber volume table derived from a standing timber volume table; the second is according to the relationship between DBH and volume of sample obtained directly from a standing timber volume table.

In the provinces of our country using the binary volume table guide for a unitary volume table compilation according to the binary volume table calculation is main steps of unitary volume table, coding table within the selected sample trees in, measured sample trees of tree height and DBH and respectively diameter calculation out average value, fitting curve equation of tree height, the substitution of duality standing tree volume model, can be indirectly obtained volume diameter unitary volume model. The key lies in the choice of tree height curve equation.

According to past practice of a standing timber volume table, the common tree height curve equation:

$$H = a_0 + a_1 \lg D$$
$$H = a_0 - a_1 D + a_2 D^2$$
$$H = a_0 D^{a_1}$$
$$H = a_0 + a_1/(D + K)$$
$$H = a_0 e^{-a_1/D}$$
$$H = a_0 + a_1/D$$
$$H = a D^{2/3}$$
$$H = a + b \ln D$$
$$\ln H = a + b/D$$

In the above formulas, a_0, a_1, a_2 are the equation parameters, H is the height of the tree, D is the diameter and K is a constant. Reasonable selection of tree height curve is helpful to improve the applicability of the unitary volume model. But as the fitting curve equation of height unable to control the fitting effect of volume estimation, so this method of compiling a local volume table has very limited accuracy. In order to improve the method, Zeng Wei-sheng proposed the second-order regression estimation method based on the tree-height curve equation that is fitting the height curve equation into the binary tree volume model. The data of the samples are then fitted to the model again to re-estimate the parameters of the model. This will be directly used to carry out the fitting volume; can be very easy to fit the volume model control accuracy.

In addition, we can also use directly according to the volume and diameter at breast height (DBH) for the preparation of a timber volume table, a common element volume regression equation as shown below.

$$V = a_0 + a_1 D^2 \tag{3-1}$$

$$V = a_0 D + a_1 D^2 \tag{3-2}$$

$$V = a_0 + a_1 D + a_2 D^2 \tag{3-3}$$

$$V = a_0 D^{a_1} \tag{3-4}$$

$$\lg V = a_0 + a_1 \lg D + a_2 \frac{1}{D} \tag{3-5}$$

$$V = a_0 \frac{D^3}{D + 1} \tag{3-6}$$

$$V = a_0 D^{a_1} a_2^{D} \tag{3-7}$$

2) Two factors volume model

There are two main methods mainly used in China and abroad to establish volume equations. First is the standing timber volume equation, which is based on the principle of mathematical statistics and the direct estimation of the average standing timber volume condition. The second is the general indirect estimation of standing timber volume equation, that is established in a certain region within the different estimates of the overall volume of standing timber equation. In our country, the indirect equations of standing timber volume are used. It is necessary to estimate the deviation of the estimation result by the indirect method. Therefore, the errors will be controlled to a certain extent by the selection of the samples and the test of the volume table model.

Since the advent of the binary tree volume table in 1846 more than a hundred years, foresters have been seeking ways to improve the volume model, proposed dozens of commonly used binary tree volume model. As follows:

$$V = a_0 + a_1 D + a_2 D^2 + a_3 DH + a_4 D^2 H \tag{3-8}$$

$$V = a_0 + a_1 D^2 + a_2 D^2 H + a_3 H + a_4 DH^2 \tag{3-9}$$

$$V = a_0 + a_1 D^2 + a_2 D^2 H + a_3 H^2 + a_4 DH^2 \tag{3-10}$$

$$V = a_0 D^{a_1} e^{a_2 H - \frac{a_3}{H}} \tag{3-11}$$

$$V = a_0 D^{a_1} H^{a_2} \tag{3-12}$$

$$V = a_0 (D + 1)^{a_2} H^{a_3} \tag{3-13}$$

$$V = a_0 + a_1 D^2 H \tag{3-14}$$

$$V = D^2(a_0 + a_1 H) \tag{3-15}$$

$$V = \frac{D^2 H}{a_0 + a_1 D} \tag{3-16}$$

$$V = a_0 D^{a_1} H^{3-a_2} \tag{3-17}$$

$$V = a_0 (DH)^{a_1} \tag{3-18}$$

$$V = a_0 + a_1 D^2 + a_2 D^2 H + a_3 H \tag{3-19}$$

$$V = a_0 D^2 H \tag{3-20}$$

$$V = a_0 D^2 e^{a_1 - \frac{a_2}{H}} \tag{3-21}$$

$$\lg V = a_0 + a_1 \lg D + a_2 (\lg D)^2 + a_3 \lg H + a_4 (\lg H)^2 \tag{3-22}$$

$$V = a_0 D^2 H + a_1 D^3 H + a_2 D^2 \lg D \tag{3-23}$$

In the formula, V is tree volume; D is DBH; H is tree height; lg is logarithm; a_1, a_2, a_3, a_4, a_5 are the parameters of the model.

3) The method and steps

(1) Sort the data. In order to judge the existence of abnormal data, the scatter plot of DBH-volume, height-volume, DBH-height, $D^2 H$-volume were calculated respectively, and by using the method of triple standard deviation and double standard deviation and the abnormality of the data is judged.

(2) Data according to the proportion of each diameter 20% extracted with tree species used as volume model conforms to the accuracy test, the remaining 80% of the sample data for modeling.

(3) The use of software for a volume model fitting, the fitting model of the precision test, select the optimal model.

(4) Use software to make the binary volume model fitting, the fitting model of the precision test, select the optimal model.

4) Accuracy verification method

After the model is set up the model can be used to determine whether the model can be used in practice or not. For this purpose, it is necessary to establish a test and evaluation criterion which can correctly describe the characteristics of the model.

In the statistical literature, common evaluation of the regression model of statistical indexes and criteria are residual distribution, the residual sum of squares Q, the coefficient of parameter variation, parameter interpretation, surplus standard deviations, complex correlation coefficient R and model of extrapolation performance etc. Including the distribution of random parameter stability and residual to determine that the built model is robust and whether they fully meet the performance of the main basis.

The model of forestry meter test often considers several indicators like the total average relative error of the RS, the average relative error E, RMA, the forecast precision of absolute value of relative error P, etc. In addition to the precision of the model test, according to the requirements of the forestry survey technology when compiling it should also separately collected another set of test samples for the applicability of the model test. In addition, as the general model application in the production of some of the independent variable values is beyond the scope of modeling sample. If the model misconduct, extrapolating ability is poor this often leads to absurd forecast results. So we should also analyze behavior model, the final comprehensive evaluation conclusion is given. Among them, the commonly used model accuracy test and evaluation index calculation formula are as follows:

In addition as a general model, some of the independent variables in the production application will exceed the scope of modeling samples. If the model behavior is improper or poor extrapolation this often leads to absurd estimates. Therefore, we should conduct behavior analysis on the model, and finally give the comprehensive evaluation conclusion. Among them, the commonly used model accuracy test and evaluation index calculation formula is as follows:

Correlation index (R^2):

$$R^2 = 1 - \frac{SSE}{SST} \tag{3-24}$$

The value of R^2 max decreases as the increase of the number of independent variables in the model, at the same time R^2 is rarely selected on the optimal equation basis. The basis of choosing the best model should be: (i) all the variables are in as little as possible; (ii) The R^2 value is substantially not less than R^2 max (the maximum value of R^2). If the variables contained in the largest model also exist in other models then it is common to map the value of R^2 to the value of p. This typical plot reflects the fact that the R^2 value approaches the upper asymptote of R^2 max as the p-value decreases as the p-value is large. However, there is a point, which is the starting point for the sharp decline in the value of R^2, which corresponds to the p-value of the corresponding model, is often defined as the final model.

Mean square error is also known as the standard error ($RMSE$), Which is the observed value deviates from true value of the square and the square root of the ratio of a number of observations.

SSE is the sum of the squares of each residual and is called the sum of squares of residuals, which indicates the effect of random errors. The smaller the squares sum of the residuals the better the fitting degree of the model(Figure 3-10).

$$\sum_{i=1}^{n} (y'_i - \bar{y}')^2 \tag{3-25}$$

5) The software control steps (in a volume model for example)

(1) To import data, click the data import button in the pop-up window and select the EXCEL data to import and open(Figure 3-11).

The data stored in the EXCEL data format, as shown below (the first column volume, as in table second DBH) (Figure 3-12).

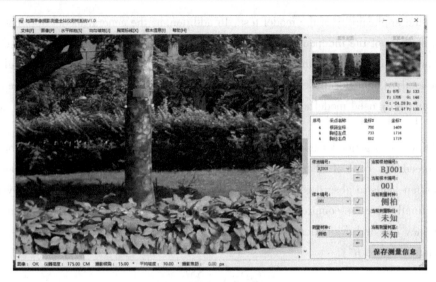

Figure 3-10 A volume table

样木信息

查询条件

地区 全选 | 样地编号 全选 | 样木编号 全选 | 样木名称 全选 | 胸径 0 1000 | 树高 0 1000 | 材积 0 1000 | 查询 | 取消

序号	地区	样地编号	样木编号	样木名称	胸径	树高	材积	X坐标	Y坐标
2	北京	BJ001	BJ001002	毛白杨	34.3000	22.3000	16.5000	111.110000	421.330000
3	北京	BJ001	BJ001001	侧柏	26.8540	23.6000	50.4540	222.110000	233.330000
6	北京	BJ001	BJ001003	侧柏	25.5000	1120.3000	1145.8000	0.000000	0.000000
7	北京	BJ002	BJ002001	侧柏	26.1780	967.3170	993.4950	0.000000	0.000000
8	北京	BJ002	BJ002002	侧柏	29.6070	1166.1690	1195.7760	0.000000	0.000000

Figure 3-11 Import data

一元材积数据 [兼容模式] - Microsoft Excel

	A	B
1	0.0098	6.1
2	0.0124	6.8
3	0.0137	6.6
4	0.0094	5.9
5	0.0124	6.9
6	0.0125	6.5
7	0.0145	6.4
8	0.0135	6.7
9	0.0108	6.7
10	0.0057	5.3
11	0.0166	8
12	0.0178	7.9
13	0.0233	8.3
14	0.0257	8.7
15	0.0224	8.9
16	0.0143	7.8
17	0.0225	8.9
18	0.0449	8.7
19	0.0239	7.7
20	0.0255	8.9
21	0.0238	8.5
22	0.0303	9.1
23	0.03	9.2
24	0.0256	9.3
25	0.0277	10.3
26	0.037	10.6
27	0.0416	10.5
28	0.0332	10
29	0.0406	9.9

Figure 3-12 EXCEL table

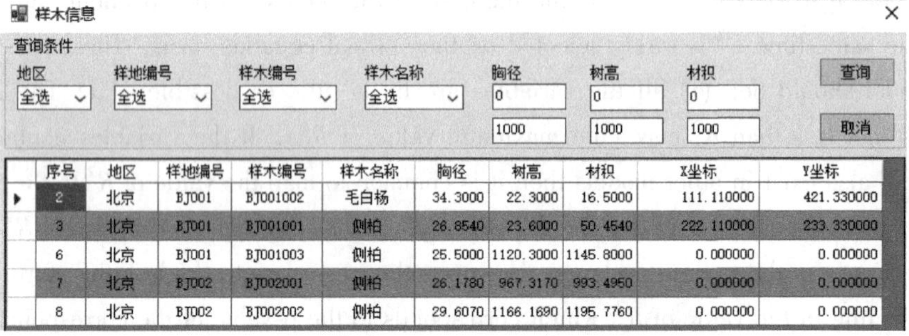

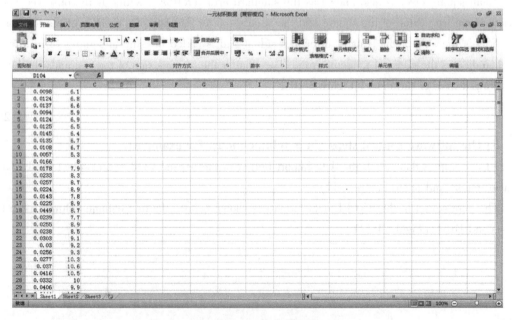

(2) The fitting volume model, select a volume model, click on the volume table generation calculation fitting model equation (Figure 3-13).

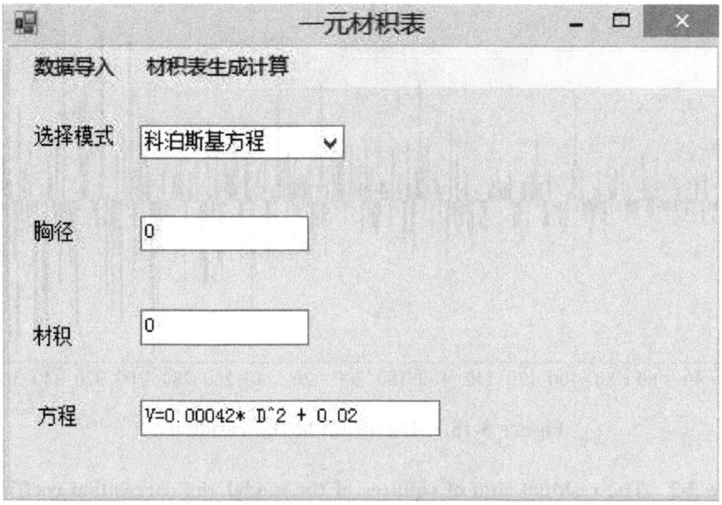

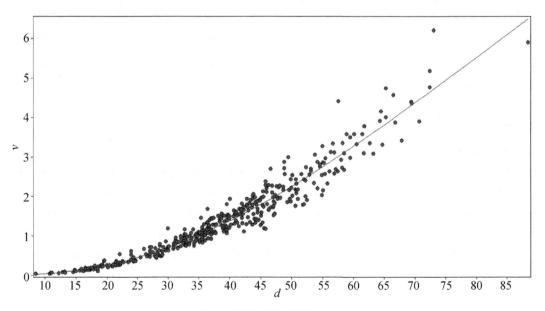

Figure 3-13 Volume table

(3) Model accuracy comparison, through the residual square sum of each model, correlation coefficient R, square R^2 and standard deviation and so on the comparison of the optimal model (Figure 3-14, Figure 3-15 and Table 3-2).

Figure 3-14 Model fit curve

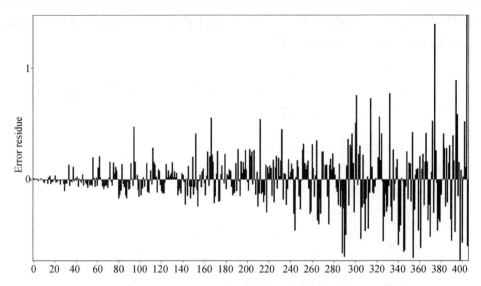

Figure 3-15 Histogram of the residuals

Table 3-2 The residual sum of squares of the model, the correlation coefficient
R and the square of the standard deviation of R^2

Model	a_1	a_2	a_3	SSE/Sum of squared residuals	R^2/ Fixed system	RMSE/ Mean square error
5-2	-0.0955125	0.0009242		26.6291	0.938436	0.256103
5-3	-0.0041970	0.0009671		26.90891	0.937858	0.257445
5-4	-0.2097012	0.0060472	0.0008526	26.49655	0.938743	0.255465
5-5	0.0006574	2.0748204		27.26988	0.937442	0.259166
5-6	-2.9680168	2.0051600	-4.5820200	2.009177	0.999971	0.070347
5-7	0.0009015			27.43295	0.938334	0.25994
5-8	0.0000710	2.8359632	0.9855240	25.85823	0.940219	0.252369

6) Example

To study as an example ' Beijing poplar stocking model' the project was completed from January 2013 to August 2013 in all districts of Beijing (except for the Mentougou District) (Figure 3-16). Electronic theodolite was used for non-destructive standing timber product measurement precision measurement of 1680 poplar trees. Which includes tree height and volume in fast-growing poplar area of Beijing, *Populus tomentosa* is used for development binary volume model and the preparation of the volume table is done by using the above method, *Populus tomentosa* a best binary volume model as shown in the diagram.

Table 3-3 Beijing *Populus tomentosa* a best binary volume model

Species	Best binary volume model
Populus tomentosn	$\lg V = -3.2163034 + 2.1656449\lg(d) - 2.4686708(1/d)$ $\lg(V) = -3.407038758 + 1.97600079\lg(D) - 0.027922921(\lg(D))^2 - 0.5622167911\lg(H) + 0.628400863(\lg(H))^2$
Populus tomentosa	$\lg V = -3.8090281 + 2.4817193\lg(d) + 0.6922169(1/d)$ $\lg(V) = -4.240849889 + 1.65478874\lg(D) + 0.075470681(\lg(D))^2 + 1.156009433\lg(H) - 0.054278676(\lg(H))^2$
Populus × canadensis	$\lg V = -2.9680168 + 2.0051600\lg(d) - 4.5820200(1/d)$ $\lg(V) = -4.661253304 + 2.183383911\lg(D) + 0.064042591(\lg(D))^2 + 1.19450357\lg(H) - 0.106679992(\lg(H))^2$

3.2.5 Stumpage non-destructive analytical curve

Electronic theodolite non-destructive measuring method is a manual measuring method of diameter at breast height. The horizontal angle and zenith distance of the corresponding position of the standing tree trunk are measured by using the electronic theodolite, according to the approximation method of analytic wood. Using the principle of triangle elevation computation tree height is measured. In accordance with the round table tired add to simulate the standing tree average section distinguish method for calculating the volume quadrature method, as shown in Figure 3-16, Figure 3-16 is the angle from instrument to the trunk; v for the observation of the zenith distance; D for the trunk diameter, m; h for the sub-height, m; H for the height of the tree, S_0 is the slant distance from the instrument center to DBH at m.

Known measurement of the horizontal angle β_0, diameter at breast height $D_{1.3}$, the calculation method for the oblique distance $S_0 = \dfrac{\frac{1}{2}D_{1.3}}{\sin\dfrac{\beta_0}{2}}$, because the horizontal angle β_0 is generally small, so

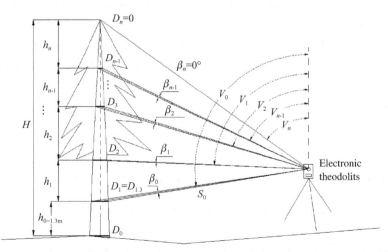

Figure 3-16 Electronic theodolite / total station

as to be considered $\sin\dfrac{\beta_0}{2} = \dfrac{\beta_0}{2}$, then the calculation formula can be simplified as

$$S_0 = \dfrac{D_{1.3}}{\beta_0} \qquad (3\text{-}26)$$

The tree height H is the slope distance from the instrument to the center of the tree trunk and is calculated from the zenith distance of the instrument center relative to the tree trunk and the tree top, calculated as

$$H = S_0 \dfrac{\sin(v_0 - v_n)}{\sin v_n} + h_0 \qquad (3\text{-}27)$$

In the formula, the section height of the trunk h_0 is 1.3m. The formula for calculating the diameter of any sub-section, when the section is too far, i.e. the measurement of the horizontal angle as shown in Figure 3-17.

$$D_i = S_0 \beta_i (i = 1, 2, \cdots, n - 1) \qquad (3\text{-}28)$$

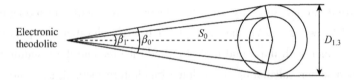

Figure 3-17 Calculation principle of slant distance

Where diameter D_0 is ground diameter and diameter at breast height D_1 ($D_1 = D_{1.3}$) to be measured using diameter tape, so $i > 1$.

Similarly when measuring different segments of high zenith distance is $v_0, v_1, \cdots, v_n$, arbitrarily segmented h_i formula for:

$$h_i = S_0 \left[\dfrac{\sin(v_0 - v_i)}{\sin v_i} - \dfrac{\sin(v_0 - v_{i-1})}{\sin v_{i-1}} \right] (i = 1, 2, \cdots, n) \qquad (3\text{-}29)$$

Among them, the segment height h_0 was 1.3 m, and the measurement of diameter at breast height was needed $i \geq 1$.

As shown in figure 1 the volume V of the trunk is according to the simulation of the frustum of a cone accumulation method. The trees as a cone, the trunk as a frustum of a cone, the calculation method for the trunk volume V is

$$V = \dfrac{\pi}{12} h_1 (D_0^2 + D_0 D_{1.3} + D_{1.3}^2) + \cdots + \dfrac{\pi}{12} h_i (D_{i-1}^2 + D_{i-1} D_i + D_i^2) + \cdots + \dfrac{\pi}{12} h_n D_{n-1}^2 \qquad (3\text{-}30)$$

3.2.6 National, provincial and municipal, unitary and binary model

The whole volume measurement of standing tree can be completed by the number of different unitary, binary and multiple volume table. The calculated volume at different positions of the trees can be divided into the trunk, the whole tree (only contains dry, branch), stems, roots, bark, no explanation is often referred to as the trunk volume with the bark.

Methods of compiling thetable can be divided into the graphic method, shape method, and volume method. The graphical method has taper map and volume curve, volume line and volume line diagram and other method which is widely used in volume regression equation method (mathematical model). Representative compiling basic data, accuracy, and consistency and choosing the best mathematical model is the key to ensuring the accuracy of tables.

Using the mathematical model of the listed below (D for diameter at breast height, H for tree height):

A volume type:
$$V = aD^b$$
$$V = b_0 + b_1 D^2$$

The binary volume type:
$$V = aD^b H^c$$

Binary volume table gives a fixed volume of standing timber stumpage. Binary volume table is based on two factors DBH and tree height for preparation and investigation of volume. Because the stem form is closely related to the diameter at breast height and tree height, stem form factor as an independent variable cannot effect the table directly and error is not large.

So by the rapid development of the binary volume table, it becomes nation's most common basic volume table. China's national binary volume model is in the form of $V = aD^b H^c$ is all over the country. 180 species, 197000 trees based on the data compiled, a total of 56 binary Standing volume table, of which thirty-five conifers, broad-leaved trees twenty-one, table system error is generally within plus or minus one percent, a few in the positive and negative 1% to 2%.

In unitary volume table, only one-factor DBH is used. Unitary volume table has the advantage of easy to use, but as it does not consider tree height and stem form of change, so it does not have the geographical scope and restricted to local areas, i. e. a local volume table. In 1878 French forester A. Wagner, put forward the following table which was modified in applications by the Swiss forester H. Biollie, known as the 'tower' table. 'Tower' etymology, Arabic for 'indicator'. So called 'Tarif' table on the form and content are very different. In order to improve the precision, we also developed a method to control the height change of tree height with tree height, status level, age group, forest type and so on, and compile the one volume table separately in different grades. According to the height, position classification is also expressed in a variety of form.

Volume table in a certain area using a common average volume curve and grade (class), this volume table can give several volume curves, has the characteristics of binary volume table. But the accuracy of the measurement volume checked by trial and error, choose from the original article bottom diameter of 2.5 meters in diameter for gauging diameter and material two factors to compile the article Sugihara volume table, as the national standards promulgated in 1984. China's provinces, municipalities, and autonomous regions are applicable to a volume table in the form of generally $V = aD^b$, forests Woodland also applies to own a volume table.

3.3 Bamboo model to study

Bamboo forest is made up of single dominant species of bamboo plants, its distribution range is very wide and there are many types of bamboo like bamboo for chicken feed, fresh bamboo etc. The largest distribution area is the moso bamboo forest, generally throughout the hilly altitude of 900 meters. Generally, the plant is 10 to 12 meters high, the plant stem diameter 6 to 13 cm and coverage are 75% ~ 95%. In some places, bamboo mixed with other tree species the commonly mixed species are fir, *Liquidambar formosana* and *Pinus massoniana*, continous niche and so on.

Under-story shrub layer common types include zhang, wild chloranthus Ji wood, leather, taper the eight immortals, tofu sticks, even the core tea, zijin cattle, indocalamus, etc. Herb layer there are deer species common umbrella, point grass, *Herba lophatheri*, ferns, etc., because of the upper coverage, vegetation layer usually less development.

Our country has the world's most abundant bamboo resources. Cultivation and processing of bamboo started very long ago in China. The bamboo forest area is about 7 million hectares in China, accounting for one-third of the world's bamboo forest area, the total volume of timber bamboo of about 97 million tons. China in terms of Bamboo and bamboo products production is no1 in the world. Bamboo shoots and bamboo fungus is a natural health food, also is the important resource of export.

Bamboo widely distributed in the south of China, it is fast growing, regeneration ability and has the very high carbon sequestration function. But the lack of mechanistic model hindered the study of the bamboo forest ecosystem. This study aimed at Jiangsu Yixing bamboo forest ecosystem and through field investigation and observation like ground diameter, DBH, height data etc. In order to the better study of bamboo forest Ecosystem, we constructed a bamboo model built software.

1) The methods and steps

(1) To organize data, To determine whether there is any abnormal data for this bamboo DBH-volume, bamboo height-volume, bamboo DBH-height, do scatter plot D^2H-volume and uses the method of three times the standard deviation and two times the standard deviation method to check abnormal data.

(2) The fitting model of the univariate and bivariate volumetric models was simulated by software, and the fitted model was used to detect the outer coincidence precision and select the optimal model.

(3) Using the software to fit the growth model, the fitted model is tested for the accuracy of the outer fit and the optimal model is selected.

(4) Using the software to fit the weight model, fitting the model to the external precision test, select the best model.

(5) Using the software to fit the gray model, the fitting model is used to detect the outer precision and select the optimal model.

2) Accuracy validation method

After the establishment of model self-checking response, the applicability of the model should be checked. Based on the precision we should determine whether the model can be used in practice. To do this, first, we need to establish a system that can correctly describe the evaluation criteria characteristics.

In statistical literature common statistical indicators and criteria for the assessment of the regression model are: the residual distribution, the sum of squared residuals Q, parameters change coefficient, interpretability, residual standard deviation, the multiple correlation coefficient R and S model extrapolation performance etc. where the parameters of stability and residual distribution randomness is determined whether the model is sound and whether it has the main basis for a comprehensive suit properties. In addition, several indicators in the inspection of forestry data Table model often considered like; total relative error of RS, the average relative error E, the mean relative absolute error RMA and forecast accuracy P.

The precision of the model test, according to the requirements of the forestry survey technology in forestry meter time should also be separately collected another set of test samples for the applicability of the model test. In addition as a general model, some of the independent variables in the production value will be beyond the scope of application of modeling samples, if the model misconduct i. e. poor extrapolation this often leads to absurd results forecast. Therefore the model should be for behavioral analysis and ultimately gives comprehensive evaluation findings. Among them, the common model of precision test and evaluation index is calculated as follows:

(1) Therelevant index (R^2). With the increasing of the number of independent variables in model the value R^2 becomes smaller. At the same time, R^2 rarely becomes the basis of choosing optimal equation. Selection of the optimal model should be based on: (i) variables as little as possible; (ii) R^2 values equal to or greater than R^2 max the maximum value (R^2). If the variable contained in the largest model also present in other models then it can be used as R^2 p-value corresponds to the value plotted. This typical figure reflects the fact that with the increase of p value, R^2 value will decrease and vice versa so as to close to the R^2 max asymptote. However, there is a point, like at starting point the value of R^2 decreases sharply, with the point p values of the corresponding model is often as the final model.

$$R^2 = 1 - \frac{SSE}{SST} \tag{3-31}$$

(2) Root-mean-square error. Root mean square error which is also called the standard error ($RMSE$). It is the sum of the squares of the observed value and true value and square root of the ratio of n observation.

(3) SSE adds up after each residual square and known as the sum of squared residuals. It said the effect of the random error. The smaller residual sum of squares the better will be fitting of the model.

$$\sum_{i=1}^{n} (y'_i - \bar{y}')^2 \tag{3-32}$$

3) The software control step for volume model (for example)

(1) To run the software, enter the main interface of the software, as shown in Figure 3-18.

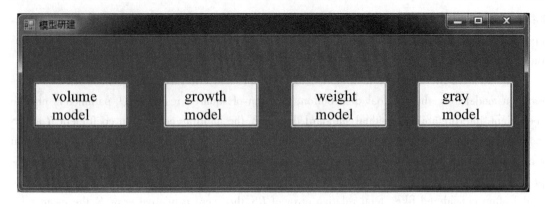

Figure 3-18   Software main interface

(2) Click the 'volume model' button; enter the unitary volume, binary volume model interface, as shown in Figure 3-19.

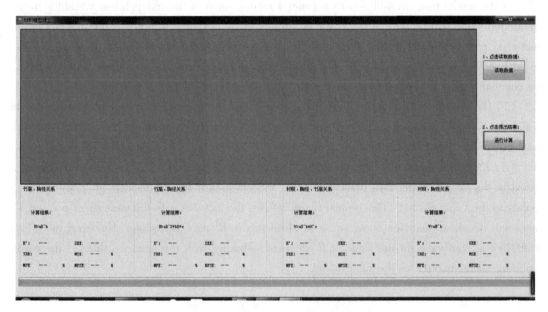

Figure 3-19   Volume model interface

(3) To import data. click the import button opens in the pop-up window to open the EXCEL data and import as shown in Figure 3-20.

EXCEL data stored in the data format as shown below (the first column in the table is of height, the second column is of diameter at breast height, third column as constants and the fourth volume) (Figure 3-21).

(4) Click the 'calculation' button, and the fitting of unitary and the binary volume model equation, as shown in Figure 3-22.

CHAPTER 3   Forest Manager's Basic Numerical Tables and Model

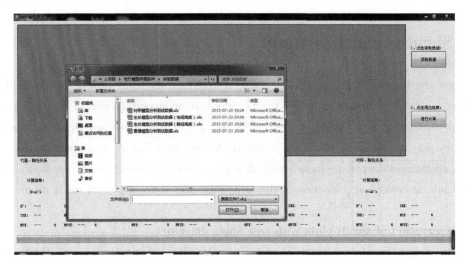

Figure 3-20   Import data

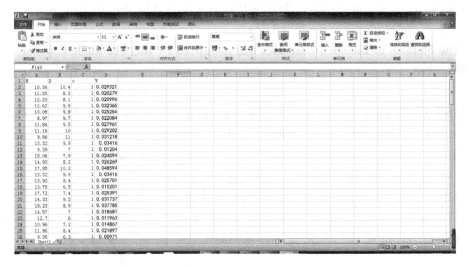

Figure 3-21   EXCEL spreadsheet

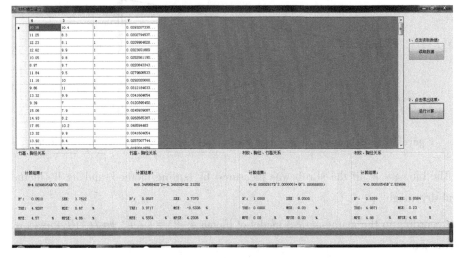

Figure 3-22   Fitting model

3.4 Shrubs model to study

Shrubs are those that have no obvious trunk, if have then the trunk is very short or in the form of clusters usually 3~6 meters of woody plants. Shrub is an important part of the ecological system, widely distributed in all parts of the world. Due to its many-branched stems and near-earth canopy they play an important role as a strong windbreak, sand-fixation and maintain the stability of soil. Underground deep roots have strong drought resistance ability, as well as shrub, plays an important role in ecological protection at the same time also is an important index of environment quality.

1) Observation of shrub plots with tree angle detector

The tree angle detector is composed of transparent ruler and standard string. The transparent ruler is 15.81 cm in length, 2 cm in width and the standard string 25 cm in length. It is connected with the geometric center of the ruler. The angle of view formed by the criterion point and the two edges of the transparent ruler is $\theta = 36°52'11''$.

2) The biomass of the shrub samples was measured by the tree angle detector

The method of measuring the biomass of the shrub plots using the tree angle instrument is as follows:

(1) Set the tree angle detector to place the rope end in front of the eye and the other hand and measure the relationship between the sight line and the target tree canopy. If images intercross it is denoted as '1', if images touch tangent '0.5' and if images are apart from each other is denoted as '0'. Among them, the intercross and tangent shrubs are counted.

(2) Measuring around in situ, count the number of shrub n ($n \leq 10$).

(3) The horizontal and vertical monolithic photogrammetry of each shrub was carried out, and the diameter d_i, crown D_i and height H_i were measured. The principle is shown in Figure 3-23.

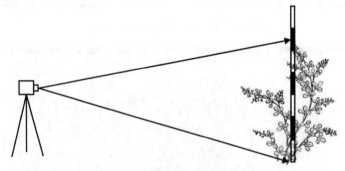

Figure 3-23 Schematic diagram of shrub level horizontal monocle photography

(4) The biomass W_i of the shrub was measured by cutting method, and its W_i included fresh weight and dry weight.

(5) According to the mathematical model of biomass: $W_i = (a_1 d_i^{b_1} + a_2 D_i^{b_2}) H_i^c$ the a_1, a_2, b_1, b_2, c are determined.

(6) The total biomass of shrub plots was calculated by mathematical model $W = \sum_{n=1}^{n} W_i$.

3) To observe the canopy density of shrubs bytree angle detector

The method of observing the canopy density of shrubs using the angle measuring device is as follows:

(1) Set the tree angle detector.

(2) Measuring the crown of the bush by tree angle detector, one hand tighten the end of the standard string placed in front of one hand around the ruler. Counting in this way: the phase angle with the tangent or tangent, the count is '1'; from the angle of view, the count is '0', the total count of Z.

(3) By mathematical model $C = K \times Z$ can be used to calculate the canopy closure value C, Where K is the canopy closure coefficient, its value of constant 1000.

3.5 Chinese dominant tree growth model

3.5.1 Data Sources

The data are mainly from the national forest inventory (NFI), which mainly includes the data of more than 7000 fixed sample plots and the data of 415000 trees in three periods. The nature of field survey is to reset the national and local sample plots and measure the sample trees. The sample plot is square with a side length of 28.28 m and an area of 0.8 hm^2. The four corners of the square are respectively called southwest, northwest, northeast and southeast corners. The specific data acquisition process is as follows:

(1) Use derived points and GPS navigation technology to find the southwest corner piles of the sample plot. The error of the position of the mark pile on the topographic map should be no more than 1mm, the measurement distance error from the derived point to the sample plot should be less than 1%, and the GPS coordinate error should not exceed 10~15 m.

(2) After finding the southwest corner pile of the sample plot, the resetting measurement of the sample plot boundary were carried out. Set up the compass at the southwest corner. Compass is set up at four corners of square plot in the order of southwest-northwest-northeast-southeast-southwest clockwise. In turn with magnetic azimuth angle 0° and 90°, 180°, 270°, and with the horizontal distance of 28.28 m, find other three corner pile, and complete perimeter measurement of the sample plot. The resetting requirement of the sample plot should reach more than 98%.

(3) After the perimeter is reset, the sample trees within its range shall be reset. The sample tree resetting is divided into two aspects: the sample tree number restting and the sample tree position restting. The resetting rate of sample trees should be more than 95%.

(4) The main contents of the in-sample survey include: each tree measurement, average tree height, each tree mark (including aluminum plate number of each tree nail, red paint mark, etc.), determination and measurement of boundary wood, etc. No error is allowed for the number

of trees with the ruler greater than 8cm, and no more than 3 trees with the ruler less than 8 cm. For trees with a diameter less than 20 cm, the measurement error should be less than 0.3 cm; for trees with a diameter greater than 20 cm, the measurement error should be less than 1.5%. For tree height measurement, if the tree height is less than 10 m, the measurement error of tree height should be less than 3%; if the tree height is >10 m, the measurement error of tree height should be less than 5%. In addition, factors such as land type, origin, forest species and high-quality tree species should not be wrong. The measurement error of canopy density, shrub coverage, vegetation coverage and total vegetation coverage should be less than 10%.

3.5.2 Growth model for dominant trees in China

In order to realize the management and sustainable development of China's forest resources, it is necessary to grasp the status of forest resources timely and accurately, especially the dynamic change of resource data and the relationship between the growth of trees and the site environment. Site environment refers to the synthesis of various environmental factors that have an impact on the growth of trees, including topography, temperature, rainfall, soil, tree density and other factors. The site environment is directly related to all aspects of forest management, such as production efficiency, economic benefits, harvesting, direction and speed of forest cultivation. Chinese dominant tree growth model is of great practical significance for accurate forest management and planning management. The quantitative management of forest resources and the prediction of forest growth can better realize the management of tending, spacing cutting, rotation cutting, replanting and transplanting under different stand conditions, and can achieve the sustainable development of forest resources while obtaining higher economic value and benefits. Based on the specific situation of forest resources in China, this study divided the complex information such as site environment and geographical location into growth pattern (site) index and growth structure (site class) index through the survey data of more than 7000 sample plots in 3 periods. The DBH growth model of the dominant trees was developed by combining 57 dominant tree species groups in China, as shown in formula (3-33).

$$\Delta Y_{t+k}^{(j)} = A_j \cdot e^{\frac{b_j}{Y_t^{(j)}} + X_N^{(m)}} \cdot X_P^{(i)} \cdot X_S^{(k)} \quad \text{or} \quad \ln \cdot Y_{t+k}^{(j)} = \ln A_j + \frac{b_j}{Y_t^{(j)}} + X_N^{(m)} + X_P^{(i)} + X_S^{(k)}$$

(3-33)

In formula 1, Y_{tj} is the dominant tree species in China, $Y_t Y_t$ is the investigated tree information, and $Y_{(t+k)} \Delta Y_{(t+k)}$ is the predicted growth amount k years later. $Y_t A_j$ is the growth speed coefficient of tree species j, $Y_t b_j$ is the growth acceleration coefficient of tree species j; $Y_t X_P^{(i)}$ is the growth pattern (site) index of the type i, and $Y_t X_S^{(k)}$ is the growth structure (site class) index of the type k. $X_N^{(m)}$ is the tree density index corresponding to sample plot m.

In addition, data standardization (normalization) is a basic work of data mining. Different influencing factors often have different dimensions and dimensional units, which will affect the results of data analysis. In order to eliminate the dimensional influence between influencing factors, data needs to be normalized.

Tree density is the number of trees per unit area, which can directly reflect the average forest area occupied by each tree in the stand. Due to the limitation of stand growth space, there is a competitive effect on the growth of trees. Therefore, stand density index has a certain impact on the growth of trees. The influence of stand density on the DBH growth of trees is summarized as $X_N^{(m)}$ in this study. And the density index of the maximum and minimum plant numbers in more than 7000 sample plots was taken as the basis for normalization, as shown in formula (3-34).

$$X_N = \frac{N - N_{min}}{N - N_{max}} \tag{3-34}$$

3.5.3 Growth pattern (site) index model of terrestrial arbor in China

Because the heat provided to the earth by solar radiation varies regularly from south to north, different climatic zones are formed. Accordingly, vegetation also forms zonal distribution, and tropical rainforest, subtropical evergreen broad-leaved forest, temperate deciduous broad-leaved forest, boreal coniferous forest, boreal tundra and polar cold desert appear successively from south to north. This kind of regularly replaced vegetation distribution along the latitude direction is called the latitude zonality of vegetation distribution. As a result of the combined action of sea and land distribution, atmospheric circulation and topography, precipitation gradually decreases from the coast to the inland. Therefore, the distribution of vegetation varies significantly with the different water conditions in the same heat belt. In the temperate regions of China, from the coast to the west to the interior, there is a change of 'forest-grassland-desert' vegetation type. This change of vegetation distribution from coastal to inland, with water as the dominant factor, is called longitude zonality. From the foothill to the top of the mountain, the annual average temperature gradually decreases with the elevation, and the growing season gradually shortens. Within a certain range, precipitation also gradually increases, wind speed increases, and solar radiation increases. Under the action of comprehensive factors, vegetation also changes with the elevation rise, which is usually manifested as a strip replacement. This vegetation zone is roughly parallel to the contour line of the hillside and has a certain vertical thickness, which is called the vertical zonality of vegetation. To sum up, the influence of regional changes on the growth process of forest trees is divided into longitude (L), latitude (B), altitude (H), temperature (T) and rainfall (R), and the above factors are summarized as the growth pattern (site) index of terrestrial arbor in China (formula 3-35).

$$X_P^{(i)} = e^{\lambda_L^{(i)} \cdot X_L^{(i)} + \lambda_B^{(i)} \cdot X_B^{(i)} + \lambda_H^{(i)} \cdot X_H^{(i)} + \lambda_T^{(i)} \cdot X_T^{(i)} + \lambda_R^{(i)} \cdot X_R^{(i)}} \tag{3-35}$$

In the formula, λ_L is the influence coefficient of growth longitude; λ_B is the influence coefficient of growth latitude; λ_H is the influence coefficient of growth altitude; λ_T is the influence coefficient of growth temperature; λ_R is the influence coefficient of growth rainfall.

In this study, the east-west and north-south boundaries of China are taken as the basis for the normalization of longitude and latitude, as shown in formula 3-36 and 3-37.

$$X_B = \frac{B - B_{min}}{B_{max} - B_{min}} \tag{3-36}$$

$$X_L = \frac{L - L_{min}}{L_{max} - L_{min}} \tag{3-37}$$

The highest and lowest elevations are taken as the basis for elevation normalization, as shown in formula 3-38.

$$X_H = \frac{H - H_{min}}{H_{max} - H_{min}} \tag{3-38}$$

The maximum annual average temperature and the minimum annual average temperature are taken as the basis for temperature normalization, as shown in formula 3-39.

$$X_T = \frac{T - T_{min}}{T_{max} - T_{min}} \tag{3-39}$$

The highest and lowest annual rainfall are taken as the basis for rainfall normalization, as shown in formula 3-40.

$$X_R = \frac{R - R_{min}}{R_{max} - R_{min}} \tag{3-40}$$

3.5.4 Growth structure(site class) index model of terrestrial arbor in China

Topography is the general term for various forms of the earth's surface. Topography can change climate and soil conditions, so topography has a great impact on the growth and distribution of trees. Slope position has a great influence on the growth of trees, and the general trend is that the upper slope position is greater than the middle slope position and the middle slope position is greater than the lower slope position. Generally, the soil conditions in the lower slope are greater than those in the middle slope and the upper slope in terms of humus content, soil layer thickness and water content, which can be reflected in the growth of trees. Slope direction also has an effect on forest growth. Shady slope has mostly mesophytic plants with more shade tolerant. The soil of shady slope has short and low intensity sunshine condition with huge humidity condition. On the sunny slope, the plants are mostly heliophytes, which show xerophytic characteristics. The sunny slope can get longer and stronger sunshine conditions, and because of higher temperature and evaporation, the soil is dry. Gradient also has a certain impact on forest growth. Most of the rainfall in the sample plots with high gradient is lost, and mud, sand and gravel are mostly deposited in the foothills. On steep slopes, due to erosion and denudation, the exposed rock surface is not suitable for the growth of trees and ferns grow more. In summary, the influence of topographic changes on the growth process of forest trees are divided into gradient(α), slope direction(β), slope position(γ) and soil thickness (h), and the above factors are summarized as the growth structure (site class) index of terrestrial arbor in China (formula 3-41).

$$X_s^{(k)} = e^{\lambda_\alpha^{(k)} \cdot X_\alpha^{(k)} + \lambda_\beta^{(k)} \cdot X_\beta^{(k)} + \lambda_\gamma^{(k)} \cdot X_\gamma^{(k)} + \lambda_h^{(k)} \cdot X_h^{(k)}} \tag{3-41}$$

In the formula, λ_α is the influence coefficient of tree growth gradient; λ_β is the influence coefficient of tree growth slope direction; λ_γ is the influence coefficient of tree growth slope position;

λ_h is the influence coefficient of tree growth soil thickness.

The gradient of the sample plot data in our study are between 0° to 60°, slope direction are divided into 0°, 45°, 90°, ⋯, 345°. Gradient and slope direction are normalized respectively, as shown in formula (3-42) and (3-43).

$$X_\alpha = \sin\alpha \tag{3-42}$$

$$X_\beta = \frac{\cos\beta + 1}{2} \tag{3-43}$$

Soil layer thickness refers to the soil thickness of plants, which is an important basic characteristic of soil and can directly reflect the degree of soil development. It is closely related to soil fertility and an important indicator for identifying soil fertility in the field. It is not only the supplementary source of soil nutrients, but also the repository of soil mineral elements, and the main index to judge the degree of soil erosion. Soil thickness is an important characterization of soil quality and an important material basis for plant growth. According to the sample plot data in this study, and the soil layer thickness is normalized, as shown in formula 3-44.

$$X_h = \frac{h - h_{\min}}{h_{\max} - h_{\min}} \tag{3-44}$$

References

方精云,郭兆迪,朴世龙,等,2007. 1981—2000年中国陆地植被碳汇的估算[J]. 中国科学(D辑:地球科学)(06):804-812.

国家林业局,2014. 国家森林资源连续清查的技术规定[M]. 北京:中国林业出版社.

国家林业局,2014. 中国森林资源报告(2019—2013)[M]. 北京:中国林业出版社.

孟宪宇,1996. 测树学[M]. 北京:中国林业出版社.

徐冰,郭兆迪,朴世龙,等,2010. 2000—2050年中国森林生物量碳库:基于生物量密度与林龄关系的预测[J]. 中国科学:生命科学,40(07):587-594.

赵敏,周广胜,2003. 中国森林植被碳储量的估算及其对气候变化的响应[C]//:中国植物学会. 中国植物学会七十周年年会论文摘要汇编(1933—2003). 北京:高等教育出版社.

Adame P, Hynynen J, Canellas I, et al., 2008. Individual-tree diameter growth model for rebollo oak (*Quercus pyrenaica* Willd.) coppices [J]. Forest Ecology and Management, 255(3-4): 1011-1022.

Adams H R, Barnard H R, Loomis A K, 2014. Topography alters tree growth climate relationships in a semiarid forested catchment [J]. Ecosphere, 5(11): 1-16.

Brown S, 2002. Measuring carbon in forests: current status and future challenges [J]. Environmental Pollution, 116(3): 363-372.

Fang J, Chen A, Peng C, et al., 2001. Changes in forest biomass carbon storage in China between 1949 and 1998 [J]. Science, 292(5525): 2320-2322.

Ford K R, Breckheimer I K, Franklin J F, et al., 2017. Competition alters tree growth responses to climate at individual and stand scales [J]. Canadian Journal of Forest Research, 47(1): 53-62.

Gómez-Aparicio L, García-Valdés R, Ruíz-Benito P, et al., 2011. Disentangling the relative importance of climate, size and competition on tree growth in Iberian forests: implications for forest management under global change [J]. Global Change Biology, 17(7): 2400-2414.

Han Y, Wu B, Wang K, et al., 2016. Individual-tree form growth models of visualization simulation for managed

Larix principis-rupprechtii plantation [J]. Computers and Electronics in Agriculture, 123: 341-350.

Huang S S, Yang Y Y, Wang Y I, et al., 2003. A critical look at procedures for validating growth and yield models [M]//: Modelling Forest Systems. Wallingford, UK: CABI Publishing: 271-292.

Jiang X, Huang J G, Cheng J, et al., 2018. Interspecific variation in growth responses to tree size competition and climate of western Canadian boreal mixed forests [J]. Science of the Total Environment(631): 1070-1078.

Laubhann D, Sterba H, Reinds G J, et al., 2009. The impact of atmospheric deposition and climate on forest growth in European monitoring plots: An individual tree growth model [J]. Forest Ecology and Management, 258(8): 1751-1761.

Le Roux X, Lacointe A, Escobar-Gutiérrez A, et al., 2001. Carbon-based models of individual tree growth: A critical appraisal [J]. Annals of Forest Research, 58(5): 469-506.

Li B, Fonseca F, 2006. TDD: A comprehensive model for qualitative spatial similarity assessment [J]. Spatial Cognition Computation, 6(1): 31-62.

Liu J, Yunhong T, Slik J W F, 2014. Topography related habitat associations of tree species traits, composition and diversity in a Chinese tropical forest [J]. Forest Ecology and Management, 330: 75-81.

Moreno P C, Palmas S, Escobedo F J, et al., 2017. Individual-tree diameter growth models for mixed Nothofagus second growth forests in southern Chile [J]. Forests, 8(12): 506.

Piao S, Fang J, Zhu B, et al., 2005. Forest biomass carbon stocks in China over the past 2 decades: Estimation based on integrated inventory and satellite data [J]. Journal of Geophysical Research: Biogeosciences, 2005, 110(G1).

Qiu Z X, Feng Z K, Song Y N, et al., Carbon sequestration potential of forest vegetation in China from 2003 to 2050 predicting forest vegetation growth based on climate and the environment-science direct [J]. Journal of Cleaner Production, 252: 119715.

Ryan M G, 2010. Temperature and tree growth [J]. Tree Physiology, 30(6): 667-668.

Scholten T, Goebes P, Kühn P, et al., 2017. On the combined effect of soil fertility and topography on tree growth in subtropical forest ecosystems a study from SE China [J]. Journal of Plant Ecology, 10(1): 111-127.

Scolforo H F, Scolforo J R S, Thiersch C R, et al., 2017. A new model of tropical tree diameter growth rate and its application to identify fast-growing native tree species [J]. Forest Ecology and Management, 400: 578-586.

Sterba H, Blab A, Katzensteiner K, 2002. Adapting an individual tree growth model for Norway spruce (*Picea abies* L. Karst.) in pure and mixed species stands [J]. Forest Ecology and Management, 159(1-2): 101-110.

Thurnher C, Klopf M, Hasenauer H, 2017. MOSES-A tree growth simulator for modelling stand response in central Europe [J]. Ecological Modelling, 352: 58-76.

Toledo M, Poorter L, Peña-Claros M, et al., 2011. Climate is a stronger driver of tree and forest growth rates than soil and disturbance [J]. Journal of Ecology, 99(1): 254-264.

Troll C, 1973. The upper timberlines in different climatic zones [J]. Arctic and Alpine Research, 5: A3-A18.

Webster C R, Lorimer C G, 2005. Minimum opening sizes for canopy recruitment of midtolerant tree species: A retrospective approach [J]. Ecological Applications, 15(4): 1245-1262.

Yang Y, Watanabe M, Li F, et al., 2006. Factors affecting forest growth and possible effects of climate change in the Taihang Mountains, northern China [J]. Forestry: An International Journal of Forest Research, 79(1): 135-147.

Zeng W S, Du H R, Lei X D, et al., 2017. Individual tree biomass equations and growth models sensitive to climate variables for *Larix* spp. in China [J]. European Journal of Forest Research, 136(2): 233-249.

Zeng W S, Tomppo E, Healey S P, 2015. The national forest inventory in China: history-results-international context [J]. Forest Ecosystems, 2(1): 1-16.

Zhang B, Lu X, Jiang J, et al. , 2017. Similarity of plant functional traits and aggregation pattern in a subtropical forest [J]. Ecology and Evolution, 7(12): 4086-4098.

Zhang L J, Peng C H, Dang Q L, 2004. Individual-tree basal area growth models for jack pine and black spruce in northern Ontario [J]. Forestry Chronicle, 80(3): 366-374.

CHAPTER 4 Forest Division

4.1 Forests division

4.1.1 The purpose and significance of forest division

The forest itself contains a huge amount of energy, it can provide support to human survival. To achieve the purpose of continue life, we must have to do research on forest resource quantity, quality, forest land types distribution, natural history, socioeconomic status and prior business. On this basis to develop a scientific and rational forest management program the first step is to improve the forest division.

Forest division is the geographical division of forest region according to the differences of the natural geographical conditions, forest resources and social and economic conditions in the region aiming at the characteristics of forestry production. Forest division in forestry can provide convenience to things like forest management and development of forest resources. Forest resources managing and utilizing activities such as the administration of all kinds of business design, afforestation, cutting and other business activities. Forest divisions are implemented in the grass roots under the principle of forestry division. It is the first step on the sub-compartment forest survey, among them the forest compartment is the smallest division unit that implemented in resources investigation. The sub-compartment is the smallest unit in forest resources calculation. Statistic of afforestation and cutting plays an important role in guiding forestry management activity recognition.

4.1.2 The forest division system of China

China has mostly state-owned forest region, the forest is commonly divided into 3 levels. Several forest plantations can be divided under the forestry administration into the forest compartments. A forest compartment can be divided into several sub-compartments. State-owned forest plantations can be divided under central station (i. e. battalion forest area or areas) and then divided into compartments, finally divided into sub-compartment. County (city) forestry bureau administration division system also commonly divided into three levels namely the county (city)-township (forestry)-village (compartment) (Table 4-1). Compartment division is the core of the forest division.

1) Forest plantation

The forest plantation is a subordinate unit of the Forestry Bureau, it is also the basic forest production unit and the basic unit of the forest management plan for management of forest re-

Table 4-1 Forest management division system in China

Forest types	Zoning system
State-owned Forestry Bureau	Forestry administration-forest, plantation-forest compartment-sub-compartment
State-owned forest plantations	Total plantation (forest plantation)- parvial field (manage area or work area)-forest compartment-sub-compartment
Collective forest area	The county (City)-town (forest plantation)-village (forest compartment)-sub-compartment

sources. In general, we use the ridge, rivers, roads or artificial marks or other product as a realm to divide forest plantation. The administration of the forest bureau in our country generally has 10 000 ~20 000 hm^2 management area. Independent state-owned forest plantation in our country has generally 10 000~30 000 hm^2 management area. In the rare forest region, the operation area of the state-owned forest plantation only 1000~2000 hm^2. Forest area of private forest plantations ranging from a few hectares to thousands of hectares.

2) Forest compartment

Forest compartment is the division for the convenience of forest resource statistical calculation and management in forest plantations. A forest plantation could be divided into many forest basic unit areas depending on size. Forest compartment is the zoning of the permanent operational unit within forest plantation for forest management, its main function is to facilitate area measurement, positioning, sorting out forest resources, forest fire prevention, and management of planning etc. Forest compartment number listed with Arabic numerals from west to east, from north to south.

3) Sub-compartment

Sub-compartment is the most basic business unit inside the forest and also the most basic unit in the inventory of forest resources and statistical calculation. The division of sub-compartment is decided on the basis of many characters like land category, forest and species category, age, density, site type or forest type, site class or site index, forest ownership and another survey factors. The average size of sub-compartment is decided according to management level commonly 3~20 hm^2. The minimum area can reflect on the basic figure as the standard and small number which is the smallest unit of management is also an auxiliary unit of sub-compartment. The small number size is decided according to the production operation skills, labor, logging, logging mode of the production process, process design and other factors. Forest plantation has a production plan, quota management, equipped with labor organization in the unit of a small number, its size is commonly a labor organization in the area of a certain time to finish.

4.1.3 Principles of forest division in our country

In the forest zone divisions shall be based on the scope of the actual situation, the future operation management and also the need for the development, utilization, and resources inventory work etc.

Division in forestry bureau should consider the following factors like type of forest resource situation, terrain, current administrative divisions and scope of the administration of the division. The size of division in the forest should consider the convenience of production, life and the traffic situation generally. When boundary line is determined it should not change arbitrarily the administration. It should not be too large and the area should be well-shaped. At the same time, the administration site should be planned within the scope of their jurisdiction.

Forest compartment districts should facilitate the forest management and the division of forest compartment should facilitate the measurement, inventory and statistics of forest resources, forest fire prevention, measures and forest management in order to utilize a variety of management of forest resources. Division of forest compartment is facilitated for long-term business activities because of its fixed geographical location and area, therefore a reasonable forest compartment division is an important content of forest resource management.

The forestry bureau, state-owned forest plantations and collective forest area with higher management level, should set up different sign on its boundary line. For natural which boundary is not obvious or forest compartment division of artificial, it should be cut the open field or set up a clear sign, and embedding forest compartment signs on the intersection of forest compartment line.

4.1.4 Method of forest division in our country

Our country has specific and detailed methods and rules in the forest division based on mountains and rivers, it requires being well-shaped at the same time. There are more detailed rules and methods in forest compartment and sub-compartment dividing.

1) Forest compartment division

The forest compartment division methods: Are divided into the following different methods and artificial division method, natural division method, compound division method.

(1) Artificial division method. Artificial division method is in the form of the square or rectangular area, the shape of the forest compartment is a neat and regular graphics. Forest compartment line is needed to cut into straight lines or broken line. The advantages of this method are firstly it simple in design, forest compartment area is of the same size, forest compartment line to easy to identify. The drawback is that the work of cutting down for forest compartmental line on volatile terrain needs heavy workload. This method is fit to be used in the flat areas and foothills of forests and part of the artificial forest(Figure 4-1).

(2) Natural division method. Natural division method is based on the natural boundaries and permanent marks inside the forest. Such as rivers, valleys, ridges, watershed and roads as a forest compartment line. Thus forest compartment areas are in the different size and shapes that are irregular. Forest compartment is usually a folder ditch in the middle of two slopes, its advantage is to keep the natural landscape and also has positive significance for sheltering forest and forests with special uses, at the same time forest compartment has a special role for a nature reserve. Natural regionalization method is suitable for the artificial forest(Figure 4-2).

(3) Compound division method. Compound division method is based on natural divisions plus artificial divisions. In compound division method area size is not consistent but too small and too large size areas are avoided, in this respect, it has more advantage than the natural zoning method. Compound division method is the main division method in mountainous areas of our country.

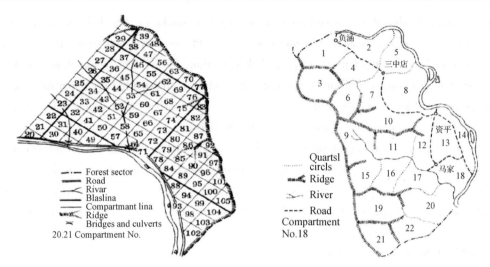

Figure 4-1　Artificial division method　　　Figure 4-2　Natural division method

Compound division method overcomes the disadvantages of the above two methods, but in the implementation, technical requirements for more complicated than the artificial division method. In regionalization of forest compartment sometimes appears the situation in which line is not easy to implement correctly (Figure 4-3).

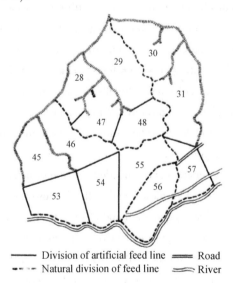

Figure 4-3　Compound division method

The size offorest compartment area should be decided according to the business purpose, management level, economic conditions and natural historical conditions. The area with better economic conditions in the south forest region should be less than 50 hm^2. North forest region compartment forest compartment area generally ranges 100~200 hm^2. In scarce forest region, nature reserve, northeast and Inner Mongolia state-owned forest region, southwest mountain area, the ecological public welfare forest inhabited areas as well as not development at the recent forest compartment area according to the need to relax the standard. In the same forest, the area of forest compartment amplitude should not exceed 50% of the required standards. For division forest compartment was too large should be prevented to inconvenience the long-term operation. The size can be less than 50 hm^2 for forest compartment area of short-rotation, forests with special usage. If it has a special scenery, tourism, natural landscape and nature forest, recreation forest compartment size and shape should consider the needs of the forest landscape and tourism as far as possible to keep the principle of natural appearance of forest compartment division.

Forest compartment zoning design should be done according to the design of forest compartment line by the use of the topographic map, aerial photographs or measurement results i. e. to cut the open forest compartment line and hanging tree in situ, also can use with color paint in forest compartment allocated tagline on both sides of the tree and intersect the forest compartment line in accordance with the provisions.

2) Sub-compartment division

Sub-compartment is the basic division unit marking on the map accurately, sub-compartment is the basic unit of the forest resources investigation, statistics and management. In order to find out the forest resources and carry out various business activities, it is necessary to divide forest compartment into different sub-compartments by certain conditions. The different land types can be divided into different sub-compartments. Stand is the forest village being divided according to the approximation of the biological characteristics and the sub-compartment is a divided area in forest compartment based on certain conditions from a management point of view. Thus, stand is the basis of sub-compartment division. Usually, a sub-compartment is a stand or may also include a few stands. In the area with good operating condition forest management highly strength may be a stand is a sub-compartment. Division of sub-compartments is basically on the basis of the following several aspects: ownership, land classification, forest category and tree species or species groups, age class or age, crown density, type of forest types and site conditions, position or status index, stand origin, grade level, the rate of grade, etc.

(1) The ownership. The ownership can be divided into the ownership of land and trees ownership (rights). Land ownership contains state-owned and collective ownership; Forest ownership contains state-owned, collective and individual, cooperation and others. In the same area of forest land, the ownership is often different such as the state, the collective, factories, mines, schools etc.

(2) Land Categories. Classification of land is mainly based on the reality of the land use ways and the covered features. Land types can be divided into different categories like land, the soil or intertidal zone above the water all the year around; Forest tree species composition, canopy density more than 0.2 or the forest crown density > 10 m. In the coniferous forest, if the accumulation of conifer is up to 65% or more, then can put this forest in the coniferous category, while in broad-leaved forest hardwood proportion is 65% or more. Mixed needle forest consists of softwood and hardwood, the proportion in the mixed forest is > 65%. Bamboo forest made of bamboo excluding DBH < 2 cm of the small miscellaneous bamboo grove. The opening in forest is made up of forest trees canopy density ranges from 0.10~0.19. Deforested land is the land that after the cutting of trees cannot meet the standard of opening blanks and not more than five years; No forest planting

area: save a number of afforestation plants or 80% after afforestation, has not been closed but hopefully forest planting area, generally refers to afforestation resent 3~5 years or less than 5 to 7 years after it was planting area; Woodland: The woodland is the land with natural regeneration, the woodland may be with medium or higher level but below the forest standard; Nursery: are the plots for forest seedling; Wasteland beyond the tree growth range: These are site conditions that are not suitable for some trees growing; Farmland: The plots with several kinds of usage for planting and grazing; Land difficult to use: The land difficult to use under the circumstances and mainly includes intertidal zone, saline-alkali land, marshland, bare land, rock land, desert, the Gobi desert, and tundra, etc.; The other Land: includes town, residential land, industry and mining land, traffic land and the land not listed; natural and artificial water inland, including rivers, lakes, reservoirs, and ponds, etc., forest stream is not of that kind(Table 4-2).

Table 4-2 Forest land classification system

SN	First level	Secondary level	Third level
1	Forestland	Tree forest	Pure forest
			Mixed forests
		Mangrove	
		Bamboo	
2	Opening		
3	Shrub	Land special provisions of the state	Shrub forest for protection
			Shrubs economic forest
		General shrubs	
4	No forest land	Non-forest planting area	
		Non-forest cultivation	
5	Viability		
6	Non felled area	Woodland deforested	
		Fire effected area	
		Another un-buck woodland	
7	Waste hills	Un-reclaimed lands suitable for afforestation	
		Abandoned land	
		Other around farmland	Sloping farmland
			The waste areas
			Other
8	Secondary forest production		

(3)Forest category. Forest category is different according to the needs of the national economy and forest benefits. Forest could be divided into ecological public welfare forest and business forest. According to regulation in 'forest law' of our country, the ecological public welfare forest includes two sorts of forest shelter forests and forests with a special use. The business forest includes timber, firewood forests and economic forests. Protective forests such as those for water supply conservation, water and soil conservation, windbreak and sand, farmland protection forest ranch, protection, belts and other shelter forest etc. Forests with special uses are divided into national defense forests, experiment forests, seed producing a forest, environmental protection forests, scenic and places of historic interest, revolutionary memorial forest, nature reserves forest, etc. Timber forest is divided into short rotation period of industrial raw material for timber, fast-growing and high-yielding program, general; firewood forests is the forest used for the production of fuelwood. The Economic forest is divided into the orchard, oil, special economic forest and other economic forest. Different effects of the forest should be divided into different sub-compartments.

(4) The advantage tree species (groups). Species composition of a different forest stand is different, its growth characteristics are different and so the economic value is also different, demand management measures are also different. Therefore, according to the different division sub-compartment from dominant tree species (or groups). When zoning sub-compartment, which the difference of dominant tree species reach 20% can be divided into different sub-compartments.

(5) The originof stand. According to forest regeneration, it is divided into the following three categories first one is a natural forest, it is a type of forest which is regenerated by natural seeding or initiation. The second one is artificial plantation, which is formed by artificial afforestation of seeding. The third one is seeding or cutting the forest, in which aerial sowing is performed. At the same time, it can be divided into two kinds the seedling forest and coppice. The different origin forest stand should be divisions for different sub-compartments.

(6) Age-class (group). The age-class group is a group having the same tree species but due to the different age level, it is divided into the different age group. In different growth stage, the management measures should be different. Generally, the trees under Ⅵ age class have one difference age level, on the age of Ⅶ which have two difference age classes, can be divided into different sub-compartments. During the investigation, it shall determine the average age of the small advantage species (group). According to the need of management, the statistic is carried out according to the age group or the age group. For the economic forest, the need for statistics and planning should be based on the growth characteristics and the growth process of division of the product for the production, first stage, rich phase and decline period of four stages of production. For the short rotation timber forest, based on the economic mature age to determine the felling age, age division.

(7) Site conditions. In the forest type (type of site conditions) level and the position (or site index) survey area is decided according to the forest (or type of site conditions) and position (site index) of different districts after we divide the sub-compartment. If the survey is not done the division can be directly done according to the slope, slope direction, slope position and topographical factors such as different districts sub-compartments. The slope is divided into different slope grade standard: Ⅰ level flat slope $0°\sim5°$; Ⅱ level is gentle slope $6°\sim15°$; For slope with $16°\sim25°$ Ⅲ level; For steep $26°\sim35°$ Ⅳ level; Ⅴ level in steep slope $36°\sim45°$; Ⅵ level risk slope is more than $46°$. Gradient magnitude difference level is divided into different sub-compartments. Slope direction is divided to the east, south, west, north, northeast, southeast, northwest, southwest and no

aspect of the slope above nine directions. Slope position, slope on a ridge, down in the valley the ground six slope positions. The level of soil thickness, based on the determination of the thickness of the +B layer in soil A layer. Landform type: Landform type is divided into very high mountain, mountain at an altitude of more than 5000 m; High mountains are those having elevation of 3500~5000 m; middle mountain, the elevation of 1000~3499 m mountain; Low mountain, mountain at an altitude of less than 1000 m; Hills, no obvious mountains, slope is moderate and the relative elevation difference is less than 100 m; Plain, flat open, ups and downs is very small.

(8) The rate of grade. According to timber stands near to over mature, economic timber wood yield level (relevant level such as Table 4-3) is divided into different sub-compartments.

Table 4-3 Separate material rate level table

The rate of material	Level stand out material rate/%			Timber tree ratio/%		
	Coniferous	Mixed needle	Broad-leaved forest	Coniferous	Mixed needle broad-leaved forest	Broad-leaved forest
1	≥70	≥60	≥50	≥90	≥80	≥70
2	50~69	40~59	30~49	70~89	60~79	45~69
3	<50	<40	<30	<70	<60	<45

Sub-compartment divisions often use three methods drawn with the aerial (satellite) photo, drawn with the topographic map, and actual measurement.

(i) Interpretation of an aerial photograph (satellite pictures) at a certain scale can truly reflect the characteristics of the land form and stand at a time. Therefore, in a region with aerial pictures (satellite pictures) one should try to use the recent aero-photograph (satellite pictures) for forest districts study and investigation. This not only can improve the survey accuracy and efficiency but also can reduce the cost. Using aerial photograph (satellite pictures) draw the sub-compartment, first, using stereoscope on aerial photograph interpretation (satellite pictures) indoors. According to the image (satellite pictures) and other interpretation factors such as color, canopy size, shape and density etc., the different characteristics decides the difference between land classes and outline the subcompartments line with drawing pen. On the basis of the contour line of sub-compartment which is drawn then compare with field observation in situ to, that is the situ annotation. If there is some inconsistency with the actual situation modify the sub-compartment contour or merge in a small field.

(ii) Draw with the topographic map. In areas having no aerial photograph (satellite pictures), we can use the existing topographic map or floor planto compare the requirement of accuracy in the investigation at an appropriate scale. By adopting the method of draw visual picture in situ, we draw sub-compartment contour on the topographic map or along the border line, forest compartment line or along the forest trail and check in forest investigation into the stand to the fixed a sub-compartment outline.

(iii) The measurement method. The measured methodis based on the survey line. In-situ direct measurement of sub-compartment boundaries, measurement usually on the basis of the control survey, using compass traverse survey or detail survey, put the sub-compartments outline into surface measurement closed error should not bigger than 1/100. This method accuracy is higher but required more manpower and material resources and time. In the absence of survey mapping data and operating high strength or have special requirements for the application of this method. After the map was prepared according to the requirements on sub-compartment calculate number and area. A small number in forest compartment use the Arabic numeral note, its writing method and forest compartment number is the same. The area is at the end of the various divisions, uses the international framing theory area as control step by step to calculate. The area of the division's units at all levels is in ha.

4.1.5 An example analysis of forest division—a case study of the experimental forest plantation in Inner Mongolia

Wangyedian survey experimental forest plantation of Inner Mongolia as an example the forest area is analyzed and the significance of forest districts to make paper.

Wangyedian survey experimental forest plantation of Inner Mongolia lies in Chifeng City which is located in the southwest of Harbin, located at longitude 118°09′ ~ 118°30′ East and latitude of 41°21′ ~ 41°39′ North, business area of 24 668.2 hm^2. Forest land area of 24 668.2 hm^2 is accounting for 100% of the total area of the land. Forest plantation forestry total volume of 1 255 756 m^3(cubic meter) of which the forestland volume of 1 253 723 m^3(cubic meters) accounting for 99.83% of the total volume of standing timber; opening volume of 333 m^3(cubic meters) accounting for 0.03%; bulk raw wood volume of 1700 m^3 cubic meters, accounting for 0.14%. Public welfare forest accounts for a big forest area and stocking volume proportion. Rich forest land tree species having a total of 12 advantageous tree species i.e. larch, spruce, mongolica, Korean pine, *Pinus tabuliformis*, elm, oak, birch, aspen, poplar trees, walnut and Chinese catalpa. Shrubland is about 1057 hm^2 main tree species of *Armeniaca siberia* and hazelnut. The forest around an area of 954 hm^2 accounting for 3.87% of the whole land area. Mainly distributed in the mountain slope, around the poor site quality, also has a small number of meadowsweet growth such as low vegetation coverage.

The forest plantation by reasonable management, protection can help in the improvement of the ecological environment, cultivate reserve of forest resources. Protection and development of forests can also improve the quality of public welfare forest and analyze, increase forest resources, strengthen forest ecological system, realize the sustainable utilization of forest resources. According to the forestry development goals, forest districts should be reasonable and effectively management to grow a natural forest.

1) The forest compartment division of Inner Mongolia Wangyedian experimental forest plantation

Wangyedian experimental forest plantation is located on the northern slope of Yanshan Mountains (near the branch of Qilaotu Mountain), is one of an important part of Mao Jingda dam in the secondary forest, the main landform for middle mountains and low mountains are undulating terrain. Mountains are inter-connected, high-lying southwest as compared to north-eastern side with an average elevation of 800~1890 m. Complex operating area of terrain is suitable for natural division method by using valley, watershed and river, to divide the compartment. But at the same time, considering the characteristics of forest resources and forest resources management level the natural division method based on the integrated division method was applied to divide the compartment (Figure 4-4).

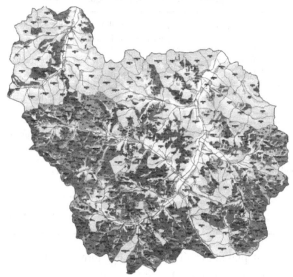

Figure 4-4 Inner Mongolia Wangyedian experimental forest plantation

Wangyedian survey experimental forest plantation in accordance with the method of comprehensive regionalization, considering the natural environment conditions, social and economic development requirements. Whole forest land divided into 2054 sub-compartments, including 1770 forestland sub-compartment, 12 opening sub-compartment, 114 scrubland sub-compartment, 44 no forest planting area sub-compartment, 802 no tree forest sub-compartment, 2 nursery land sub-compartment, 112 the sub-compartment of barren mountains and wasteland suitable for afforestation.

2) The classification of the Wangyedian survey experimental forest plantation of Inner Mongolia region

In order to achieve the prosperous forest survey industry ecological benefits, social benefits and economic benefits, in accordance with the principle of the forest classified division, the forest can be divided into ecological public welfare forest and analyze.

Ecological public welfare forest helps in the protection and improvement of the human living environment, maintaining ecological balance and save the resources, scientific experiments, forest tourism, homeland security and other needs for the main business purpose of forests. According to the forest resources planning and design investigation main technical regulation rules, combined with the actual situation of the region for forest category divisions. Shelter forest, including those for water supply conservation, windbreak, and sand; the forests with special uses include seed and forests.

In timber production, dry fields and other industrial raw materials for the main business purpose of forestland. The relevant provisions and the actual situation of the region we discuss all the region can be divided into timber forest. Timber forest is divided into the fast-growing and high-yielding timber forest(Table 4-4).

Table 4-4 The classification of the Wangyedian survey experimental forest plantation of Inner Mongolia region

The ecological public welfare forest	Shelter forest	Windbreak	Forest plantations in cropland, roadside, residential areas, which have the main function as windbreak and sand storm break
		Water conservation forests	Forests within the first ridge of the banks of the Xibo River are classified as water conservation forests
	Special use	Seed Producing forest	To foster good seed as the main goal of forests, trees, including seed, seed orchard, etc
	Shelter forest	Forests	Provide teaching and the scientific experiment place as the main purpose of forests, trees
Forest	Fast-growing and high-yielding progrm		Through the use of strong seedling and the implementation of intensive management cultivation cycle shorten and get the best value
	General timber forest		Timber rproduction as the main purpose of these forests trees

Wangyedian survey forest plantation division for the forest land, open woodland, shrub, forest planting area, not felled woodlands and other classification districts with a total area of 24 668.2 hm^2, among them: the ecological public welfare forest area of 15 739.2 hm^2, accounting for 63.80% of the area of forest land classification; We discuss an area of 8929 hm^2, accounting for 36.20% of the area of the forest land classification. Among them: focus on public welfare forest area of 14 394.4 hm^2, accounting for 91.46% of the public welfare forest area; Local public welfare forest area of 1344.8 hm^2, accounting for 8.54% of the public welfare forest area.

4.2 Forestland planning principle

4.2.1 Related concepts and division standards of China's forestland

Forestland: According to the *forest law regulations of the People's Republic of China* the forestland is included in the forest whose canopy density is more than 0.2 and also other than that bamboo, shrub, open woodland, deforested land, burned area, nursery and those land suitable for afforestation and which is planted by the government or county level people's.

(1)Forestland. Forestland is continuous area of more than 0.067 hm^2, canopy density of 0.2 or more, Forestland also including arbor and bamboo forest land.

(2)Shrub forest. Consists of shrub species or some species due to bad habitat dwarfing into shrub tree species. Shrub land specifically provides a certain economic value. In order to achieve economic efficiency for the purpose of operating shrubs, or distributed in arid, semi-arid areas and arbor growth limits for protection purposes, and cover degree of greater than 30% of the shrub land.

(3) Sparse forestland. Sparse forestland composed of tree species, continuous area of more than $0.067 \, hm^2$ and having canopy density between $0.10 \sim 0.19$ of the forest.

(4) Deforested area. Deforested land is the land that after the cutting of trees cannot meet the standard of opening blanks and not more than five years.

(5) Burned area. An are burned by fire, after fire stumpage is not up to the woodland standards and yet manual update or natural regeneration is not up to the natural seedling 3000 plants/hm^2 or saplings 500 plants/hm^2 woodland.

(6) No forest land. Forestor enclosure does not reach years, in the year's afforestation, the survival rate reach 85% or save year survival rate of afforestation is 70% (The average annual rate of afforestation survival rate was 70% and the preservation rate was 65% in the area of 400 mm in the following year). After the aerial seeding afforestation the remained seedlings 3000 plants/hm^2 or sand zone to seedling 2500 plants/hm^2, fencing natural seedling 3000 plants/hm^2 or saplings 500 above plants/hm^2, evenly distributed, no forest yet have hope to forest woodland.

(7) Fixed nursery trees and flower nursery land.

(8) Suitable for the forest. The people's governments at the county level is now planning for suitable forest land, including the barren hills wasteland, sandy area, and other forests.

In addition to the forest belonging to our land also includes:

(1) Farmland. For a variety of crops and grazing.

(2) Difficult land. Under current conditions which are difficult to use, mainly including; beaches, saline land, swamp, bare land, rock, desert, Gobi, tundra.

(3) Other land. Including urban residential and industrial land, traffic land, and not included in the category of land.

4.2.2 The concept and importance of forestland planning

4.2.2.1 The concept of forestland planning

Forestland planning in the certain region of space is for the protection of the sustainable development of ecology, economy and society premise. According to regional resource and environment carrying capacity, the existing potential for development overall planning the future population distribution, economic layout, and land use framework are the kind of features that differentiate land into woodland, farmland, difficult land, rangelands lands and others.

For forestlandplanning, one should keep the following points in mind.

(1) Guarantee regional ecological safety.

(2) Guarantee that regional ecological environment is best.

(3) Under the premise of ecological security, ensure maximum social and economic benefits of the regional forest.

Forest planning includes various types of land planning in a certain area, time frame generally involves several decades or longer. The allocation of land resources in the region, subjected to regional economic, social, ecological and environmental conditions and other constraints. After the completion of the forestland planning, we can carry out forest species planting in the region according to the biodiversity protection demand. The other things should also keep in mind while planting tree species are ecosystem productivity maintenance, soil and water resources protection, maintaining the contribution of forest on the carbon cycle, economic benefits to meet the regional social need. There are three kinds of forest planning depending on the duration i. e. short term 5 ~ 10 years, mid-term 10~20 years and long term for several decades. The tree species planting should be according to suitable tree principle so that it should not disturb the ecological balance(Figure 4-5).

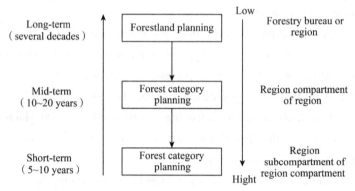

Figure 4-5 Timeframe and information characteristics difference in three kinds of forest planning process

4.2.2.2 The importance of forestland planning

Forestland is an important natural and strategic resource of our country and it is the foundation for the survival and development of the forest. It has the core position in the protection of wood and forest products supply and maintain land ecological security and play special roles in the response to global climate change. The State Council explicitly required 'woodland and farmland on the equally important position, attaches great importance to the protection of forest'. Speaking from the national level the scientific and rational land planning is the basic requirements for the protection of land ecological safety and the construction of ecological civilization.

People's demand for green space and ecological culture is growing rapidly. But at present the deterioration of the ecological systems has not been fundamentally curbed, forest degradation, soil erosion, and desertification are serious, frequent natural disasters, disaster prevention, and mitigation task is very heavy. The urgent need is to coordinate arrangements for the construction of economic and social development and ecological land demand, to ensure the sustainable development of ecological, economic and social coordination. The objective requirement is strengthening energy-saving emission reduction, improving the forestry's ability to combat climate change. The forest is the largest reservoir of terrestrial carbon-absorbing carbon and the most economical, has the advantage in this indirect reduction in response to climate change. The requirement to meet the market demand, enhance the timber and forest products supply capacity. It is the situation requirement of adapting to the regional land and forest land use pattern to arrange good timber and forest production improve the level of forest management, meet the needs of economic and social development of timber and forest products to the maximum. Woodland plays the carrier of consolidate and expand green space, improving the ecological environment, protecting the ecological security. The need is the timely adjustment of adaptation in regional development of land use pattern of forestland protection and utilization structure. The implementation of zoning strategy adapts the regional development strategy is a new period of China.

4.2.2.3 Forestland planning points

Forestland planning should fulfill the following points.

(1) Ecological benefits as main target and social, economic benefits as multiple targets;

(2) Statistical analysis of the characteristicsof the existing land and topography, soil, vegetation and meteorological;

(3) The analysis of the regional environment;

(4) The optimization models of vegetation construction, in particular, considering the difficult using the land to be green, cost benefit ratio of forest land planning and difficult use land should balance each other;

(5) Making full use of rural courtyards, open green areas and city building exterior greening and roof greening;

(6) Using the RS/GIS as the main tool for data acquisition, data management, data analysis, data output.

4.2.3 Forestland planning principle

4.2.3.1 The theory of linear programming

The mathematic model of linear programming theory has already been used to solve many problems in various fields. Multi-linear programming is the most widely used system analysis method, it studies under the limited resource constraints i. e. how to properly use and allocate the resources effectively. Mathematical language refers to seek mathematical theory and method of a linear function (also called objective function) extremism problems by the multiple constraints of linear equality and inequality. A French mathematician named Fourier in nineteenth century firstly proposed a linear programming problem. The real application by Dantzig (1947) proposed to solve linear programming problems by the simple method and provided a general method for solving the problem. With the popularity and the rapid development of computer technology and computational methods linear programming method has been rapid and have extensive application in the field of industry and trade of agricultural production, resource allocation, transportation, military and economic and also achieved significant results.

1) Multivariate linear programming model

The best expression of the optimal state of the objective function namely the objective function can reach the maximum or minimum value:

$$\max(\text{or min})Z = c_1x_1 + c_2x_2 + \cdots + c_nx_n \tag{4-1}$$

The constraints can take equality or inequality, That is:

$$\begin{cases} a_{11}x_1 + a_{12}x_2 + \cdots + a_{1n}x_n \leqslant b_1(\text{or} \geqslant b_1) \\ a_{21}x_1 + a_{22}x_2 + \cdots + a_{2n}x_n \leqslant b_2(\text{or} \geqslant b_2) \\ \cdots \\ a_{m1}x_1 + a_{m2}x_2 + \cdots + a_{mn}x_n = b_n(\text{or} \geqslant b_m) \\ x_1, x_2, \cdots, x_n \geqslant 0 \end{cases} \quad (4\text{-}2)$$

In the formula, Z—The value of the objective function;

n—The number of decision variables also known as the dimension of linear programming;

$c_1, c_2, \cdots, c_n$—Known as the coefficient of price; m is the order of the linear programming;

$a_{11}, a_{12}, +\cdots+, a_{mn}$ (or overview written as a_{ij}, $i = 1, 2, \cdots, m; j = 1, 2, \cdots, n$)—Known as system structure namely for the coefficients of the decision variables in the constraint conditions;

$b_i(i=1, 2, \cdots, m)$—Constant;

all above are any real and b_i is non-negative.

(2) The standard linear programming model

The constraints of the mathematical model are in all equations. In addition to the requirements of the constraint conditions of mathematical model for all equations, with a coefficient of +1 independent decision variables and the independent decision variables can only exist in the given constraints, other constraints do not exist in also it must be zero in the objective function coefficients.

The expression of the standard model:

$$\max(\text{or min})Z = c_1x_1 + c_2x_2 + \cdots + c_nx_n \quad (4\text{-}3)$$

Constraint condition:

$$\begin{cases} a_{11}x_1 + a_{12}x_2 + \cdots + a_{1n}x_n = b_1 \\ a_{21}x_1 + a_{22}x_2 + \cdots + a_{2n}x_n = b_2 \\ \cdots \\ a_{m1}x_1 + a_{m2}x_2 + \cdots + a_{mn}x_n = b_m \\ x_1, x_2, \cdots, x_n \geqslant 0 \end{cases} \quad (4\text{-}4)$$

4.2.3.2 The principle of linear programming forest

1) Mathematics model setup

Objective function:

$$\max Y = C(\otimes)X \quad (4\text{-}5)$$

Constraint condition:

$$A(\otimes)X \leqslant B(\otimes), X \geqslant 0 \quad (4\text{-}6)$$

In the formula: Y—Value (RMB);

C—Efficiency coefficient (RMB/hm^2);

X—Various types of land use (hm^2);

Among them: $C(\otimes) = [C_1, C_2, \cdots, C_n]$ is efficiency coefficient vector; $X = [X_1, X_2, \cdots, X_n]^T$ is the decision variable vector; $A(\otimes) = (a_{ij})_{m \times n}$ is technical coefficient matrix; $B(\otimes) = [b_1, b_2, \cdots, b_m]^T$ is the constraint coefficient vector.

2) The decision variable settings

The decision variable is needed to meet the following three basic criteria:

(1) Land use types should be complying with national standards of ' land use classification' and ' national forestry bureau of forestry land classification standards';

(2) Each variable is independent, comprehensive and not repeated;

(3) Useful data of each variable need the accuracy and availability which can be a reference to authority statistics value.

The present situation of land use as the basis starting from the protection of forest, economic development, ecological optimization requirement, we have set the following variables: X_1 (agricultural land), X_2 (the other land, including urban, residential and industrial land, traffic land, water and land), X_3 (coniferous forest), X_4 (broad-leaved forest), X_5 (conifer), X_6 (bamboo), X_7 (sparse forest), X_8 (range of tree growth within the shrub forest land), X_9 (tree growth outside the shrubland), X_{10} (slash), X_{11} (burned area), X_{12} (immature forest land), X_{13} (natural regeneration forest), X_{14} (pre-plantation), X_{15} (nursery), X_{16} (waste land), X_{17} (within the range of tree growth of wasteland), X_{18} (outside the range of tree growth of wasteland) and X_{19} (difficult to use land), show in Table 4-5.

Table 4-5 Variable table of woodland planning

Number	Variable name	Land type
1	X_1	Farmland
2	X_2	Other land, including urban, residential and industrial land, traffic land and water land
3	X_3	Coniferous forest
4	X_4	Broad-leaved forest
5	X_5	Coniferous forest
6	X_6	Bamboo forest
7	X_7	Sparse forest
8	X_8	Withinthe range of tree growth of shrub forest
9	X_9	Outsidethe range of tree growth of shrub forest
10	X_{10}	Cutting land
11	X_{11}	Burned land
12	X_{12}	Immature forest land
13	X_{13}	Natural regeneration of forest land
14	X_{14}	Reserve land
15	X_{15}	Nursery land
16	X_{16}	Waste land
17	X_{17}	Withinthe range of tree growth of waste land
18	X_{18}	Outsidethe range of tree growth of waste land
19	X_{19}	Difficult using land

3) The establishment of constraints

(1) The total land area constraint. Land area and total area should be equal to the area of S. That is:

$$X_1 + X_2 + X_3 + \cdots + X_{19} = S \tag{4-7}$$

(2) The total population constraint. Regional land population's carrying capacity is not less than the planning period of population prediction (P_8) that is:

$$M_1 \sum X_i + M_2 \sum X_j \leq P_8 \tag{4-8}$$

Where M_1 refers to the average population density (P_{da}), M_2 is the per capita urban land (P_{dc}).

$$P_{da}(X_1) + P_{dc}(X_2) \leq P_8 \tag{4-9}$$

(3) The constraint of total cultivated land. According to the area of cultivated land [(that is the basic farmland protection area (S_a)], that is:

$$X_1 \leq S_a$$

(4) Ecological protection requirement constraint:

(i) All kinds of ecological land constraints. Setting according to regional land and using planning and sample plot, especially the forest coverage rate index in the use of planning period (F_{cr}).

$$\frac{(X_3 + X_4 + X_5 + X_6)}{\sum_{i=1}^{19}} \geq F_{cr} \tag{4-10}$$

(ii) Green equivalent constraint. Comprehensive regional development planning and the minimum target of regional forest coverage rate (F_{cr}), Optimization of forest ecosystem then regional woodland area planting should be:

$$S_{forest} = F_{cr}\% \times \sum_{i=3}^{19} X_i \tag{4-11}$$

The actual area of woodland green equivalent is:

$$X_{forest} = \frac{X_4}{S_{forest}} \times 100\% \tag{4-12}$$

Another type of regional land can also produce green equivalent the formula of all green equivalent of regional land is:

$$X_{\text{Green Biomass}} = X_{forest} + QX_1 \times 0.5 \tag{4-13}$$

(iii) Urban land constraints. According to the prediction population in regional planning period (P_c) and the standard of urban land (M_L), the urban land demand value is like following:

$$X_2 \leq P_c X M_L \tag{4-14}$$

(iv) Policy control constraints. According to the regional government issued in the protection forest area, land control indicators of basic farmland and control index of construction ($S_{building}$), constraint condition is as follows:

$$X_2 \leq S_{building}$$

(v) Model constraint: $X_i \geq 0, i = 1, 2, \cdots, 19$

4) Determination of efficiency coefficient

(1) Determine the relative weight of all kinds of land use benefit coefficient. Determine weight set of all types of weight coefficient, $W_i(i = 1, 2, \cdots, 19)$, among them these land (X_{10}, X_{11}, X_{12}, X_{13}, X_{14}, X_{15}, X_{16}, X_{17}, X_{18}, X_{19}) land use efficiency is 0.

(2) Determine the profit coefficient C. All profit coefficient C of all land constitute value vector $C_i(i = 1, 2, \cdots, 19)$. C is according to per unit area, namely output coefficient per hectare multiplied the relative weight index. Assume output coefficient per unit area as B_i yuan (RMB), then $C_i = W_i \times B_i$.

(3) Objective function. In order to meet the above constraints, to ensure maximum benefit of regional forest construction, the objective function is as follows:

$$Y = \sum_{i=1}^{9} C_i X_i + X_{10} + X_{11} + X_{12} + X_{13} + X_{14} + X_{15} + X_{16} + X_{17} + X_{18} + X_{19} \quad (4\text{-}15)$$

4.2.4 Analysis of forestland planning, taking Beijing as an example

Beijing is the capital of our country and has a considerable influence on the world. Urban green development of Beijing is the main aspects and showed slow improvement to stable fast improvement, green belt, the pattern of ecotypic greenery structure, but forestry development and ecological greening have many shortcomings in Beijing, overall the development is faster and stable but the distribution is imbalanced. Greenery in suburban counties distributed more, while central urban area and urban expansion area it is distributed less. The situation of forestland in areas cannot meet the theoretical needs area, the rationality of structure layout needs improving. Therefore, the reasonable design scheme of Beijing area about forestland planning plays an important role in the development of Beijing ecological civilization construction and social economic development and at the same time provides a good example for other city's forestland construction, and plays an important role in promoting China's overall development and construction of city forestry.

In 2014 the total forest land area in Beijing is 10 345.58 km^2, accounted for 63.03% of the total area of land in Beijing area. Among them the forest area is 5995.08 km^2, woodland is 46.76 km^2, Shrub is 3191.98 km^2 and non-restocking forestland is 327.35 km^2, nursery area is 239.08 km^2, non-forest land is 37.37 km^2, a suitable area for forest is 503.46 km^2, assisted production forestland is 4.50 km^2. Assisted production occupies less land and it's difficult to interpret at a resolution of 2 meters high remote sensing images this research mainly carries out field trip according to the second data combined with literature obtained. Central city forest area is 124.15 km^2, accounting for 1.2% of the forest area in Beijing zone. Forest land area in urban expansion area is 434.15 km^2, accounting for 4.2% of the Beijing forest area, suburban woodland area is 9816.33 km^2, accounting for 94.88% of forest area in Beijing.

The following table shows forestland distribution in Beijing (Table 4-6):

Table 4-6 Statistics table of forestland area in Beijing ($km^2/\%$)

Type		Area	Proportion of Forest land	Proportion of Land
Forest land	Sorest land	5995.1	57.95	36.53
	Sparse forest	46.76	0.45	0.28
	Shrub forest	3192	30.85	19.45
	Afforestation	327.35	3.16	1.95
	Nursery	239.08	2.32	1.5
No forest land	No tree forest	37.37	0.36	0.2
	Suitable for forest	503.46	4.87	3.1
	Auxiliary forest	4.5	0.04	0.02
Non-forest land		6067.7		36.97

As seen from the above data there are all forest types in the Beijing area, most of them are forest land, accounting for 57.95% of total forest area, which consists mainly of coniferous forest and mixed forest. Secondly, the shrub land occupancy area is large accounting for 30.85% of the total area of the study area of woodland. The two kinds of woodland have very high ecological benefits are mainly distributed in the mountains and cities on both sides of the road, which are closely related to the ecological benefit being maximization, the benefits of forests being maximization, and improving the mission of the Beijing area ecological environment and air quality.

In the slash afforestation cost at least 9963.1 RMB/hm^2 estimated; the deforested soil texture and environment has been relatively mature; farmland need for the government to acquire subsidies, therefore, afforestation cost is high, for 99 654.7 RMB/hm^2, similar to cutting reforestation cost 10 times; sand and barren wasteland afforestation are difficult, the cost is 43 591.5 \$/hm^2 and 59 802.3 RMB/hm^2; the cost is 23 665.3 RMB/hm^2 in burned land.

By the optimal linear programming principle, when the expense of the quota is known, the largest area of the plantation land for afforestation (hectare):

Objective function:

$$\max S(x) = \sum V_i X_i \quad (4\text{-}16)$$

$$\max S(x) = 9963.1X_1 + 23665.3X_2 + 43591.5X_3 + 59802.3X_4 + 99654.7X_5 \quad (4\text{-}17)$$

Constraint conditions:

$$X_5 < 2333$$

$$\sum_{i=1}^{n} X_i \geq 58339$$

$$\sum_{i=1}^{n} X_i \leq 102325$$

$$\sum_{i=1}^{n} X_i V_i \geq 3480340000$$

$$X_1 \geqslant 373 \quad X_2 \leqslant 3364 \quad X_3 \geqslant 47365 \quad X_4 \leqslant 2961$$

By LINGO linear programming software, it can be referred:

$$X_i(373, 1281, 38105, 357, 10200)$$

According to the above calculation results, it can be concluded that in the planning years (2014-2020), under the constraints of forest coverage, carbon balance, capital, national farmland 'red line' and area, the area available for afforestation is 3.73 km². The burned area can be used for afforestation 12.81 km², barren hills and wasteland can be used for afforestation area 381.05 km², barren sand can be used for afforestation area of 0.357 km², farmland can be used for afforestation area of 102.00 km². In this plan, by 2020, Beijing area will increase 503.16 km² forest area, increase forest land area for the majority of barren hills and wasteland afforestation, followed by agricultural land afforestation, for cutting the plot will be all planning for afforestation.

For forest planning in Beijing area, this study uses the remote sensing data from GF-1 in 2014 and the survey data of forest sub-compartment division in 2009. After the existing woodland distribution analysis, the theory of the demand for forest land analysis determines the selection and area of afforestation. Finally, choose the net present value of forest carbon sequestration benefits as the standard, and get Beijing 2014-2020 forest planning and results as follows:

(1) In order to obtain the largest forest carbon sequestration benefits, by establishing the linear programming model of Beijing forest distribution in 2014-2020 to carry on forest planning. The forest planning model is as follows:

Objective function:

$$\max Y = 1459.38X_1 + 1108.83X_2 + 862.22X_3 + 652.74X_4 \tag{4-18}$$

Constraint conditions:

Forest coverage rate constraint:

$$\frac{\sum_{i=1}^{5} = X_i}{16413.14} \geqslant 30\%$$

Carbon balance constraint theory:

$$\sum_{i=1}^{n} X_i \geqslant 500$$

Beijing farmland 'red line' policy constraints

$$X_5 = 23333$$

The requirements about the demand of increasing construction area to forest area:

$$41857 \leqslant \sum_{i=1}^{n} X_i \geqslant 102325$$

Afforestation cost constraints:

$$\sum_{i=1}^{n} X_i V_i \geqslant 3480340000$$

(2) Forest area will going to increase 503.16 km² in Beijing by 2020. The increase of forest land is mainly by reforestation in areas were deforested or areas burned by fire and afforestation of

agricultural lands in several ways. The plantation afforestation area is listed in the following(Table 4-7):

Table 4-7 The statistics table of afforestation area (km^2)

Afforestation type	No tree forest		Suitable for forest		Farmland
	Stump land	Burned area	Wasteland	Sandy land	
Existing area	3.73	33.64	473.85	29.61	2629.09
Planning afforestation area	3.73	12.81	381.05	3.57	102.00

(3) Based on high resolution remote sensing images called GF-1 and the survey data of the second type of forest resources in Beijing area, optimal linear programming model of Beijing forest area development are made to calculate plantation species and afforestation area from 2014-2020 in Beijing area, the following table shows the result(Table 4-8):

Table 4-8 Afforestation area distribution map in 2014-2020 about Beijing area(km^2)

Land type	Coniferous forest	Mixed forest	Broad-leaved forest	shrub forest	Total
Stump land	1.33	1.01	0.79	0.60	3.73
Burned area	4.58	3.48	2.70	2.05	12.81
Wasteland	136.11	103.57	80.40	60.97	381.05
Sand land	2.14	0	1.43	0	3.57
Farmland	36.43	27.73	21.52	16.32	102.00
Total	180.59	135.79	106.84	79.94	503.16

There are deforested land, burned area, barren hills and wasteland, sand and farmland for afforestation in Beijing area in 2014-2020. In these plantations, in order to obtain maximum total benefit in the forest as the ultimate goal, and forest solid carbon net present value of benefits as the constraints of the planning, we planned coniferous forest as the largest afforestation area about 180.59 km^2, secondly for mixed forest about 135.78 km^2, broadleaf forest, and shrub forest are 106.84 km^2 and 79.94 km^2.

Coniferous forest, mixed forest, and broad-leaved forest belong to the forest in the forest division. Calculated by the above analysis, we can get the land planning (2014-2020) in Beijing area as following(Table 4-9):

Table 4-9 Forest land use planning scheme (2014-2020) in Beijing area(km^2)

Forest land	Existing area	Planning area	Increase or decrease (+/-)
Forest land	5995.08	6418.29	423.21
Shrub forest	3191.98	3271.92	79.94
Sparse forest	46.76	46.76	0
Nursery	239.08	239.08	0
Auxiliary forest	4.50	4.5	0
Afforestation	327.35	327.35	0
No tree forest	37.37	20.83	-16.54
Suitable for forest	503.46	118.84	-384.62

4.3 Planning principles of forest species

4.3.1 Division of forest types

4.3.1.1 Forest category

China classifies forest into the ecological public welfare forest and commercial forest according to its dominant feature.

(1) Ecological Forest. Including shelterbelts and special purpose forests. The main functions of ecological forests are as follows to protect and improve the human environment, maintaining ecological balance, preservation of species resources, science experiments, forest tourism, homeland security and other needs, involving woodland, shrubland, and other wooded land.

Ecological Forest divided into national public forest and local public forest. National public forests are the public forest that in accordance with relevant state regulations delineated by the local people's government and the verification by the competent department of forestry under the State Council. The local public forest is the delineated public forest that delineated by the local People's Government, in accordance with the relevant provisions of national and local and verification identified by forestry administrations of the same level.

The ecological public welfare forest is divided into special, key and general 3 grades according to the protection level. According to: *Ecological Forest Construction Planning Design Principles* (GB/T 18337.2—2001) and provision of the State Council Forestry Department, national public welfare forest in accordance with the ecological niche differences are generally divided into special and key ecological public welfare forest, the local public welfare forest in accordance with the ecological location difference is generally divided into the key and general ecological public welfare forest

(2) Commodity Forest. This type of forest includes woodland, shrubland, and other wooded land, with the main purpose of production of wood, bamboo, firewood, dried and fresh fruits and other industrial raw materials, including timber, firewood and forest.

4.3.1.2 Forest species division

Depending on the business objectives forest, woodland and shrubland is divided into 5 Forest species, 23 subspecies, see Table 4-10.

1) Protection forest

Woodland, sparse forest and shrubland on the main purpose of ecological protection.

(1) Water Conservation Forest. Water conservation forests are those which are used to conserve water, to improve the hydrological conditions and to regulate region's water cycle. The other main function of these forests is to prevent the rivers, lakes, and reservoirs from silting and the protection of drinking water sources for the primary purpose of woodland, sparse forest, and shrubland. Who have one of the following conditions may be classified in water conservation forest:

(i) In the process of 500 km above the river's birth place of the catchment area, mainstream in the ridge of two sides of the mountain natural terrain of first level and second level tributary.

(ⅱ) The river following processes in the 500 km, but the region concentrated rainfall, have an important impact on the downstream industry and agriculture, its source of the river, mainstream and the catchment area of cross-strait mountain natural terrain in the first layer within the ridge.

Table 4-10 Forest species classification system

Forest category	Forest species	Subspecies
Ecological forest	Shelterbelt	Water conservation forest
		Soil and water conservation forest
		Sand-fixing forest
		Pasture farmland shelterbelts
		Waterfront protection forest
		Road protection forest
		Others
	Special-use forests	Defense forest
		Experimental forest
		Seed production forests
		Forest protection
		Scenic forests
		Monuments and revolution memorial forest
		Forest nature reserve
Commercial forest	Timber	Short rotation timber
		Fast growing timber
		General wood
	Firewood forest	Firewood forest
	Economic forest	Orchards
		Edible raw material forest
		Forest chemical raw material forest
		Medicinal forest
		Others

(ⅲ) Large and medium-sized reservoirs and lakes around the mountain natural terrain within the first layer of the mountain ridge or within the level of 1000 m, small reservoirs and lakes around the natural terrain within the first layer of the mountain ridge or within the level 250 m.

(ⅳ) The snow line within 500 m and glacier peripherals within 2 km.

(ⅴ) Protect the urban drinking water sources of the forest, woodland and shrubland.

(2) Soil and water conservation forest. The main purpose of soil and water conservation forest is to slow surface runoff, reduce erosion, prevent soil erosion, maintain and restore land fertility for the primary purpose of woodland, open woodland, and shrubland. Which have one of the following conditions may be classified as soil and water conservation forest:

(ⅰ) Northeastand southwest region (including eastern Inner Mongolia) where the slope is

25°, northwest and other regions where the slope is above 35°, the Southeast region where the slope is above 45°, if these areas become open spaces due to deforestation then it will cause serious soil erosion.

(ii) Barren soil, bare rock, post-harvesting difficult to update or difficult to restore the ecological environment.

(iii) Serious soil erosion in the Loess Plateau hilly region due to, gully erosion, rocky mountain gully, geological structure prone to osteoporosis and other debris flow area.

(iv) Woodland, openwoodland and shrubland within the ridge watershed 300 m range.

(3) Sand-fixing forest. The sand-fixing forest is the forests used to reduce wind speed and prevent or mitigate erosion and fixed sand. Protecting arable land, orchards, crops, pastures from sandstorms.

Forests which have one of the following conditions may be classified as sand-fixing forest:

(i) Strength eroded areas, common flow, semi-flow sand (hill, ridge) or erosion monadnock area.

(ii) Region within 250 m from the sand and desert outside 100 m from the oasis.

(iii) Coast matrix type sandy, muddy areas along the prevailing wind direction from landing within a fixed range of 1000 m of coastline, within 200 m range of other directions.

(iv) Coral Island evergreen.

(v) Other sand-infested areas.

(4) Pasture farmland shelterbelts. To protect farmland, pasture natural disaster reduction, improve the natural environment, safeguard agricultural production conditions the primary purpose of woodland, open woodland, and shrubland. Which have one of the following conditions may be classified as cropland pasture protection forest:

(i) In the farmland, pasture outside the realm 100 m range, with sandy areas within the range of 250~500 m border.

(ii) To prevent mitigation, natural disasters in forest fields, pastures, terraces, hills and other places set shelterbelts, woodlots.

(5) Bank protection forest. Bank protection forest is to avoid riparian, lakeshore/coastal erosion or collapse. Plantation grew at the banks of river, lakes, etc.

(i) In the range of both sides of main rivers 200 m and major tributaries 50 m, including the bed of wild geese forest.

(ii) In the both sides of bank with 10 m range.

(iii) Forest, woodland and shrubland in the 500 m range of mangrove or coastal area.

(6) Road protection forests. To protect the railways and highways from wind, sand, water, snow for the primary purpose woodland, open woodland, and shrubland. Which have one of the following conditions may be classified as road protection forests:

(i) Forest areas, mountain national highway trunk railway embankment and both sides (outside of the line of fire, the same below) slopes or flat area within 200 m, non-forest, hilly, flat and sandy area within each 50 m.

(ii) Forest, mountains, sandy areas of the province, county feeder roads and railway embankment on bothsides within 50 m other regions within a 10 m range.

(7) Other protection forests. The forest grew as to protect from fire, snow, fog, smoke as the main purpose is called protection forest.

2) Special purpose forests

The main business purpose of the forest is to save the species resources, protect the ecological environment, defense, forest tourism and scientific experiments as the open woodland and shrubland.

(1) Defense Forest. Defense forest is used to shield military installations and military barriers for the primary purpose of woodland, sparse forest, and shrubland. Who have one of the following conditions may be classified as national defense forests.

(i) Woodland, sparse forest and shrubland in the border areas and its width by the provincial demarcation in accordance with relevant requirements.

(ii) Approved by the forestry department in charge, of military installations around the woodland, sparse forest, and shrubland.

(2) Experimental forest. Experimental forest is that forest which provides a place for teaching or scientific experiments for the primary purpose of woodland, sparse forest and shrubland, including scientific research and teaching forestry practices, forest science education, location observation forests.

(3) Seed production forest. Seed production forest is used for the production of seeds. The primary purpose of woodland, sparse forest, and shrubland, including seed stands, seed orchard, progeny test forest, scion, mining roots garden, arboretum, germplasm, and gene save forests.

(4) Environmental protection forests. To purify the air and prevent pollution, reduce noise, improve the environment for the main purpose of distribution in urban and suburban, in the industrial and mining enterprises, the residential and green area of village forest land, sparse forest, and shrubland.

(5) Scenic forests. To meet human ecological needs, beautify the environment as the main purpose, located in scenic spots, forest parks, resorts, ski resorts, hunting grounds, city parks, woodland countryside parks and sightseeing places within the sparse forest and shrubland.

(6) Monuments and revolutionary memorial forest. Located revolutionary monuments and memorial places (including natural and cultural heritage sites, historical and revolutionary sites area) of woodland, sparse forest, and shrubland as well as memorial forest, forest culture, old trees were wood, etc.

(7) Forest nature reserve. Within nature reserve levels the nature reserve is to protect and restore ecosystems and typical precious, rare animal and plant resources and their habitat in original and natural condition. Preservation and reconstruction of the natural landscape is the main purpose of woodland, sparse forest, and shrubland.

3) Timber forest

Woodland and sparse forest having wood or bamboo production as the main purpose.

(1) Short rotation timber. The main purpose of this forest is to produce pulpwood and specialized industrial use of wood raw material, take intensive management measures to make a directed cultivation.

(2) Fast growing timber. Through the use of better seeds and intensive management is the arbor woodland whose forest growth indicators have reached the appropriate species plantation national or industry standards.

(3) General using timber. Other woodlands having the main purpose is to make the production of wood and bamboo.

4) Firewood forest

Woodland, open woodland and shrubland having the main purpose are to produce thermal energy fuel.

5) Economic forest

Woodland and shrubland which main operating purpose is to produce oil, fresh and dried fruit, industrial raw materials, medicines and other by-products.

(1) Fruit forest. The main purpose is to produce all kinds of fresh and dried fruit.

(2) Consumption of raw material forest. The main purpose is to produce edible oils, beverages, seasoning, and spices.

(3) Chemical raw material forest. The main purpose is to produce resins, rubber, cork, single lemon and other non-wood forest products.

(4) Medicinal forest. The main purpose is to produce herbs and medicinal raw materials.

(5) Other economic forest. The main purpose is to produce other special products.

According to the eighth national forest, inventory results the national forest area is 207690000 hectares. Compared with the seventh national forest inventory results a net increase in the forest area is 12.23 million hectares having an increased rate of 6%. In total of national forest area the arbor forest area is 164.6 million hectares and accounting for 80%; economic forest area is 20.56 million hectares, accounting for 10%; bamboo forest area is 6.01 million hectares, accounting for 3%; special provisions of national shrubbery areas are 14.38 million hectares, accounting for 7%. Inner Mongolia, Heilongjiang, Yunnan, Sichuan, Tibet, Guangxi, Hunan, Jiangxi have larger forest areas than other provinces (accounting for more than 5% of the country's total forest area). The total forest areas of the eight provinces are 108.82 million hectares, accounting for 48% of the country.

Distinguishing forest area by tree species, shelterbelt forest are 99.67 million hectares, accounting for 48%; special forest are 16.31 million hectares, accounting for 8%; timber forest is 67.24 million hectares, accounting for 33%; firewood forest is 1.77 million hectares, accounting for 1%; economic forest are 20.56 million hectares, accounting for 10%. According to different major forest uses protection forest and special forest are classified as public forest, the timber, firewood forest and economic forest are classified as commercial forest. The ratio of public forest and commercial forest area is 56 : 44.

4.3.2 Forest species planning principles

Forest species planning is a problem related to multiple relationships, multi-event, multi-objective comprehensive decision problem. How to put multiple targets set as a whole, integrated, comprehensive analysis of the relationship between the rational decision-making program planning goals. To achieve planning objectives, it is necessary to have a kind of long-term forest planning and reasonable layout to determine the direction of regional development, in order to obtain the most comprehensive benefits (social, ecological, economic) for the target. Forest species planning need fully consider the following issues:

(1) To coordinate comprehensive ecological, economic, land use and social development.

(2) Focus on people, let the people living in the best ecological environment.

(3) Improve the living green space configuration.

(4) Multi-objective the largest commercial forest economic, ecological public welfare forest ecological benefit the best taking into account the protection forest and forest viewing.

(5) UAV, the image super-station instrument as data acquisition means.

(6) Gray system thinking and modeling to solve forest species linear programming problems.

1) Mathematical model

The objective function is $Y = C(\otimes)X \to \max$

Restrictions is $\otimes (A)X \leq b, X \geq 0$

Where in $X = [X_1, X_2, \cdots, X_m]^T$; $C = [C_1, C_2, \cdots, C_n]$, $C_j = (j = 1, 2, \cdots, n)$ can be a gray number.

Matrix of constraint coefficients $\otimes (A)$ is

$$\otimes A = \begin{bmatrix} \otimes a_{11} & a_{12} & \cdots & a_{1m} \\ \vdots & \vdots & \vdots & \vdots \\ \otimes a_{n1} & a_{n2} & \cdots & \otimes a_{nm} \end{bmatrix}$$

Whitening matrix of $\otimes (A)$ is

$$(A) = \begin{bmatrix} \otimes a_{11} & a_{12} & \cdots & a_{1m} \\ \vdots & \vdots & \vdots & \vdots \\ \otimes a_{n1} & a_{n2} & \cdots & \otimes a_{nm} \end{bmatrix}$$

In objective function, X_j —Forest types(hm^2);

C_j —benefit factor ($¥ hm^2$).

In restrictions, a is constraint coefficient (Unit depends on the specific constraints); b_i is constraint constant (Unit depends on the specific constraints).

2) Variable settings

Variable settings should meet the following criteria:

(1) Forest species settings to meet the '*National Forest Inventory* (NFI) *Technical Regulations*' and try to reflect the actual situation of the region.

(2) Each variable geographically independent, not overlap and has the typical comprehensive.

(3) The effectiveness of each variable can be obtained in order to determine the effectiveness factors for each forest types.

According to China's forest types classified system standards were set at following variables (Table 4-11):

Table 4-11 Forest species divided system standard

Serial number	Variable name	Implication
1	X_1	Water conservation forest
2	X_2	Water and soil conservation forest
3	X_3	Sand-fixing forest
4	X_4	Pasture farmland shelterbelts
5	X_5	Waterfront protection forest
6	X_6	Road protection forest
7	X_7	Defense forest
8	X_8	Experimental forest
9	X_9	Seed production forests
10	X_{10}	Forest protection
11	X_{11}	Scenicand monuments revolution forests
12	X_{12}	Memorial forest
13	X_{13}	Forest nature reserve
14	X_{14}	Short rotation timber
15	X_{15}	Fast growing timber
16	X_{16}	General wood
17	X_{17}	Firewood forest
18	X_{18}	Orchards
19	X_{19}	Edible raw material forest
20	X_{20}	Forest chemical raw material forest
21	X_{21}	Medicinal Forest
22	X_{22}	Others

3) Restrictions

(1) Forest area constraint that the sum of all species Forest area forest area should be equal to the sum of the planning period(S):

$$\sum_{i=1}^{22} X_i = S \qquad (4\text{-}21)$$

(2) Environmental protection constraint namely in accordance with the requirements of ecological protection in a particular period the proportion of ecological public welfare forest and commercial forest will be adjusted to meet the actual needs of the planning period, it is assumed according to plan as ecological public welfare forest area S_1, commercial forest area S_2, then:

$$\sum_{i=1}^{13} X_i = S_1 \qquad (4\text{-}22)$$

$$\sum_{i=14}^{21} X_i = S_2 \qquad (4\text{-}23)$$

(3) Timber demand constraint in order to ensure the timber volume output to meet regional needs in planning period and must guarantee a acertain timber a acreage and to consider the pro-

portion of the three timber subspecies($a: b: c$), scilicet.

$$\sum_{i=14}^{16} X_i = S_{21} \quad (4\text{-}24)$$

$$\begin{cases} \dfrac{X_{14}}{S_{21}} = a \\[6pt] \dfrac{X_{15}}{S_{21}} = b \\[6pt] \dfrac{X_{16}}{S_{21}} = c \end{cases} \quad (4\text{-}25)$$

(4) Regional fuel energy demand for constraints. Namely, firewood forest area (S_{22}) to meet fuel energy demand:

$$X_{17} = S_{22}$$

(5) Economic and social development needs, which is to ensure the region oil, fresh and dried fruit, work materials, medicinal herbs and other forest products demand in planning period, economic forests (S_{23}) for each forest category planning to meet a certain percentage, such as ($d_1 : d_2 : d_3 : d_4 : d_5$):

$$\sum_{i=18}^{22} X_i = S_{23} \quad (4\text{-}26)$$

$$\begin{cases} \dfrac{X_{18}}{S_{23}} = d_1 \\[6pt] \dfrac{X_{19}}{S_{23}} = d_2 \\[6pt] \dfrac{X_{20}}{S_{23}} = d_3 \\[6pt] \dfrac{X_{21}}{S_{23}} = d_4 \\[6pt] \dfrac{X_{22}}{S_{23}} = d_5 \end{cases} \quad (4\text{-}27)$$

(6) Mathematical model demands constraints:

$$X_i \geq 0, \quad i = 1, 2, \cdots, 22 \ _\circ$$

4) Efficiency coefficient

(1) Determine effectiveness of forest types relative weighting factor. Determine the Forest types of weight coefficients constituting the weight set, $W_i(i = 1, 2, \cdots, 22)$, to determine the weight coefficients can be used an expert scoring method or according to the regional real economic statistics. In determining the various forest types weighting coefficients should have an overall consideration of all forest types simultaneously exert social and ecological benefits and economic benefits.

(2) To determine the effectiveness of forest species number C. The effectiveness of forest spe-

cies number C constitute value vector $C_i (i = 1, 2, \cdots, 22)$. Various forest species benefits for the output per unit area is B_i yuan, $C_i = W_i \times B_i$.

5) The objective function

$$Y = \sum_{i=1}^{22} C_i X_i \qquad (4\text{-}28)$$

4.4 Sub-compartment afforestation design

Sub-compartment is the smallest unit of afforestation division. This division can be drawn accurately by forest resources investigation, statistics which are the basic unit of the management. Sub-compartment afforestation design based on land utilization and ecological benefit maximization as the main goal, make research in the area of forest land distribution in structure arrangement is reasonable the best area.

4.4.1 Digital optimized afforestation

Digital optimized afforestation adjust the forest age classes by thinning and replanting step-by-step, use the minimum and maximum planting area as constraint conditions and evaluated the best afforestation pattern by the linear programming model. Known to the limit of stand growth at certain year as n_0 I, II, III age level corresponding to the fixed number of year is set to $n_0/3$, $2n_0/3$ and n_0, N_{I}, N_{II}, N_{III} respectively present I, II, III, the optimal growth density.

(1) At the beginning of afforestation, afforest I age class of trees as N_{I}.

(2) Tree growth when reached the age class II, felling and replanting at this time will be implemented to adjust the age-class structure and adjusted in the following manner:

$$\begin{cases} \min(x_{11} + x_{31}) \\ x_{11} + x_{21} = N \\ x_{21} \geq (N + N)/3 \\ x_{31} \geq N/3 \\ x_{21}/N_+ x_{31}/N = 1 \end{cases} \qquad (4\text{-}29)$$

x_{11}, x_{21}, x_{31} respectively represent the first adjustment of the amount of thinning, reserve capacity, replanting.

Solving the optimal solution for:

$$\begin{cases} x_{11} = N/3 \\ x_{21} = 2N/3 \\ x_{31} = N/3 \end{cases} \qquad (4\text{-}30)$$

(3) Stand growth to reach third instar level and necessary felling and replanting performed to adjust the age-class structure. Which can be adjusted in the following manner:

$$\begin{cases} \min(x_{12} + x_{22} + x_{52}) \\ x_{12} + x_{32} = N/3 \\ x_{22} + x_{42} = 2N/3 \\ x_{32} \geq N/3 \\ x_{42} \geq N/3 \\ x_{52} \geq N/3 \\ x_{32}/N + x_{42}/N + x_{52}/N = 1 \end{cases} \quad (4\text{-}31)$$

x_{12}、x_{22}、x_{32}、x_{42}、x_{52} respectively the second adjustment II age class of thinning amount, the amount of thinning III age level, the amount of retained II age, III age class reserved quantity, the amount of replanting I age level.

Solving the optimal solution for:

$$\begin{cases} x_{12} = N/3 - N/3 \\ x_{22} = 2N/3 - N/3 \\ x_{32} = N/3 \\ x_{42} = N/3 \\ x_{52} = N/3 \end{cases} \quad (4\text{-}32)$$

Adjusted to ensure that two thinning and replanting of maximum minimum workload and land utilization, optimized digital afforestation.

4.4.2 Nine grid afforestation

In this type of afforestation area is divided into nine parts, through clear cutting, thinning and transplanting methods. By adjusting the age class structure eventually, build into afforestation pattern as shown.

Age class I	Age class II	Age class III
Item no: 1	Item no: 2	Item no: 3
Age class II	Age class III	Age class I
Item no: 4	Item no: 5	Item no: 6
Age class III	Age class I	Age class II
Item no: 7	Item no: 8	Item no: 9

A broad-leaved tree species growth rule, known at the time of year of A_I optimal growth density of N_I, thus calculated, $N_I/2$, $N_I/4$ fixed number of year of the corresponding growth of A_{II}, A_{III}, three-time points corresponding age class with I, II, III. Afforestation tree species can be divided into coniferous and broad-leaved forest, the best mixing ratio of k_0: 1.

(1) During the initial afforestation, it can be divided into 3 × 3 nine grid, from left to right, from top to bottom rules numbers 1 to 9, I age level by coniferous trees N_I/k_0 strains and broadleaf N_I afforestation strains.

(2) By the time the stand growth up to the end of A_1, 2, 4, 9 blocks broad-leaved forest to implement all or all of the transplanting and replanting I age level broad-leaved forest $N_1/3$, then the remaining six plots for thinning or transplanting II age level broad-leaved forest $N_1/3$.

(3) When growth to the end of A_{II}, 1, 6, 8 blocks broad-leaved forest to implement all or all of the transplanting A total $N_1/6$, and replanting I age level broad-leaved forest, A total of $N_1/3$, 2, 4, 9 plot thinning or transplanting broad-leaved forest $N_1/6$, will block 3, 5, 7 total $N_1/12$ thinning or transplanting broad-leaved forest plants.

After two adjustment, stand around the block structure and density, to achieve the optimal land utilization, late A_{III} can cutting by using 3, 5, 7 plot III age level of broad-leaved forest and replanting I age level of broad-leaved tree species, adjust the density of other fields, in turn, order to harvest the diameter lumber and the purpose of the effectiveness of ecological protection benefits.

4.5 Small-scale precision business planning and design

4.5.1 Background

Forest resources are the basis for human survival and sustainable development. As the country with the largest plantation area in the world, the state management department attaches great importance to the development of plantation, but in practice, there are still some shortcomings in the management of the company, which are embodied in.

The plantations are mostly pure forests, the tree species structure is single, the stability of the ecosystem and the ability to resist disasters are weak, the soil fertility decline leads to the productivity level, and the forest landscape is monotonous. And lead to a reduction in biodiversity.

The traditional afforestation indicators of plantation are not clear, the management status is relatively extensive, and more attention is paid to the economic value of forest trees, and the lack of correct understanding of its ecological value leads to low overall quality of forests.

The traditional plantations are mostly forests of the same age, and most of them adopt the policy of large-scale clear cutting, which leads to low forest quality, uneven distribution and unreasonable institutions.

The important role of forest resources has become increasingly prominent, and the management and management of plantation has entered a new stage of cutting from high-quality to high-quality precision management. We use the space allocation matrix method to make a great innovation and breakthrough for the precision management planning of the forest small class, which largely solves the disadvantages and shortcomings of the traditional artificial same-age pure forest.

In order to improve the efficient and sustainable use of the growth of plantation forests, the precise management planning design features of the forest small class are: dividing the land into 4×4 total 16 plot units, according to the mix ratio of 25%, 50% or 75%. Tree species, in which each unit tree species is planted with 16 trees at equal intervals, combined with the time required for the trees to grow to different scales, the spatial allocation matrix is used to determine the harvesting and replanting conditions, and the final matrix is the first, II, III and IV paths. The proportion of trees in the order is 25%.

4.5.2 Specific implementation plan

The precise operation planning and design of the forest small class is of great significance both for the rational use of forest resources and for the efficient and sustainable management of plantations. The specific implementation process is as follows:

(1) Determine the afforestation area and a certain, generally, the ratio of tree species to other tree species is generally set at 25%, 50% and 75%. For a mixture ratio of 25%, the tree species are planted in a diagonal direction of the 4×4 plot unit; for a mixture ratio of 50% Set the odd-numbered plot unit planted in the odd-numbered rows of the tree species, and the even-numbered plot-block units in the even-numbered rows; for the 75% blending ratio, the tree species is set to be planted on other plot units except one diagonal.

(2) For each plot unit, according to the row spacing D of the d_{IV} diameter step that the tree grows to grow, the planting 4×4 totals 16 locations, planting the seedlings at t_0, and assuming the DBH The size is d_0; when t_1 arrives, all the trees reach the d_I scale, 50% of which is cut by thinning and the seedlings are replanted at the felling position; when it reaches t_2, 50% reaches d_I respectively. The diameter step, 50% reaches the d_{II} diameter step, respectively thinning the d_I diameter step and the d_{II} diameter step by 50% and replanting the seedling at the felling position; when reaching t_3, respectively, 50% reaches d_I diameter step, 25% reaches d_{II} diameter step, 25% reaches d_{III} diameter step, 50% of thinning d_I diameter step and replants seedling at felling position; reaches t_4 time, d_I, d_{II}, d_{III} and d_{IV} all reached 25%, and the purpose of adjusting the same age plantation to the mixed forest of different ages was achieved. The above description is clear through the matrix form, wherein the real area ratio matrix A_1, the harvest area ratio matrix A_2, and the planting area ratio matrix A_3 are as follows:

Reality area ratio matrix A_1: (d_{tji} represents j time, the proportion of trees in the i-th scale)

$$A_1 = (d_{tji}) = \begin{pmatrix} d_{t0i} \\ d_{t1i} \\ d_{t2i} \\ d_{t3i} \\ d_{t4i} \end{pmatrix} = \begin{pmatrix} d_{t00} & d_{t01} & d_{t02} & d_{t03} & d_{t04} \\ d_{t10} & d_{t11} & d_{t12} & d_{t13} & d_{t14} \\ d_{t20} & d_{t21} & d_{t22} & d_{t23} & d_{t24} \\ d_{t30} & d_{t31} & d_{t32} & d_{t33} & d_{t34} \\ d_{t40} & d_{t41} & d_{t42} & d_{t43} & d_{t44} \end{pmatrix} = \begin{pmatrix} 1 & 0 & 0 & 0 & 0 \\ 0.5 & 0.5 & 0 & 0 & 0 \\ 0.5 & 0.25 & 0.5 & 0 & 0 \\ 0.25 & 0.25 & 0.25 & 0.25 & 0 \\ 0 & 0.25 & 0.25 & 0.25 & 0.25 \end{pmatrix}$$

A_2: Harvest area ratio matrix A_2:

$$A_2 = (d_{tji}) = \begin{pmatrix} d_{t1i} \\ d_{t2i} \\ d_{t3i} \\ d_{t4i} \end{pmatrix} = \begin{pmatrix} d_{t11} & d_{t12} & d_{t13} & d_{t14} \\ d_{t21} & d_{t22} & d_{t23} & d_{t24} \\ d_{t31} & d_{t32} & d_{t33} & d_{t34} \\ d_{t41} & d_{t42} & d_{t43} & d_{t44} \end{pmatrix} = \begin{pmatrix} 0.5 & 0 & 0 & 0 \\ 0.25 & 0.25 & 0 & 0 \\ 0.25 & 0 & 0.25 & 0 \\ 0 & 0 & 0 & 0 \end{pmatrix}$$

Planting area ratio matrix A_3:

$$A_3 = (d_{tji}) = \begin{pmatrix} d_{t0i} \\ d_{t1i} \\ d_{t2i} \\ d_{t3i} \\ d_{t4i} \end{pmatrix} = \begin{pmatrix} d_{t00} \\ d_{t20} \\ d_{t30} \\ d_{t40} \\ d_{t40} \end{pmatrix} = \begin{pmatrix} 1 \\ 0.5 \\ 0.5 \\ 0.25 \\ 0 \end{pmatrix}$$

Among them: d_{tji} in A_1, A_2, and A_3 represents the proportion of trees in the i-th scale.

References

北京市房山区志编纂委员会,1999. 北京市房山区志[M]. 北京:北京出版社.
蔡体久,潘紫文,朱德柱,等,2002. 黑龙江省林业生态经营区系与林种规划[J]. 东北林业大学学报(05):32-35.
陈昌鹏,吴保国,2004. 林业数据仓库的设计[J]. 农业网络信息(04):30-32.
陈昌鹏,吴保国,贾永刚,等,2004. 数据挖掘技术在森林资源信息管理中的应用[J]. 河北林果研究(02):149-153.
陈建成,张青,1992. 用统计决策理论决定标准年伐量[J]. 林业资源管理(06):73-79.
陈文伟,黄金才,2002. 数据挖掘技术[M]. 北京:北京工业大学出版社.
陈友荣,2004. 森林火灾损失的分类[J]. 福建林业科技(02):65-67,74.
程锡礼,殷国新,2003. 数据挖掘:零售企业提升竞争力的利器[J]. 江苏商论(12):25-26.
迟庆云,2005. 决策树分类算法及其应用[J]. 枣庄学院学报(05):29-32.
储菊香,2000. 森林火灾损失评估系统FFIREGIS的研制与开发[J]. 林业资源管理(05):56-58.
邓尚民,韩靖,2007. Clementine在电子商务环境中的数据挖掘应用[J]. 现代图书情报技术(10):62-65..
董静曦,段永智,高常寿,等,2002. 昆明市森林分类经营区划研究与示范[J]. 西南林学院学报(01):23-28.
杜纲,王世忠,1994. 造林方案的优化及其计算机辅助系统[J]. 数理统计与管理,13(5):5.
范文杰,1987. 多目标灰色局势决策与林种规划[J]. 中南林业调查规划(04):13-17.
方晓平,刘爽,1999. 最优化理论与方法在人工造林规划中的应用[J]. 铁道科学与工程学报,17(3):34-38.
冯健文,林璇,陈启买,2006. 决策树在银行特约商户分析中的应用研究[J]. 计算机工程与设计,27(24):5.
冯仲科,姚山,刘永霞,2007. 从林业信息数据采集到森林知识获取[J]. 北京林业大学学报(S2):7.
冯仲科,余新晓,2000. "3S"技术及其应用[M]. 北京:中国林业出版社.
冯仲科,臧淑英,姚山,2006. 基于广义"3S"技术的森林资源经营管理系统建设[J]. 测绘工程,15(2):6.
高正红,沈学利,2007. Apriori算法在超市决策中的应用[J]. 长春工程学院学报(自然科学版)(01):63-66.
巩帅,2006. 交通流量数据的分类规则挖掘[J]. 计算机工程与应用(06):219-220,232.
桂现才,彭宏,王小华,2005. C4.5算法在保险客户流失分析中的应用[J]. 计算机工程与应用,41(17):4.
河北省林业厅组织编写,1990. 实用工程造林[M]. 北京:中国林业出版社.
胡波,吴保国,陆道调,2005. 基于ASP.NET的造林专家系统[J]. 林业资源管理(1):4.
胡红玲,张怀清,2006. 退耕还林造林决策支持系统开发[J]. 林业科学(S1):120-126.
胡林,2006. 基于知识发现技术的林火研究[D]. 北京:北京林业大学.
胡勇,胡玲,2006. 基于C4.5算法在水利水电建筑工程专业成绩分析中的应用[J]. 高等建筑教育(04):108-111.
黄圣乐,1989. 用线性规划单纯形法在计算机上进行控制环节辨识[J]. 同济大学学报(02):259-264.
黄月琼,周元满,刘素青,2002. 人工神经网络约束模型在造林规划上的应用研究[J]. 江西农业大学学报,24(3):3.

景利萍,胡林林,2007.白泉山生态公园的林种规划及树种选择[J].内蒙古科技与经济(21):24-25.
亢新刚,2012.森林资源管理[M].北京:中国林业出版社
国家林业局,2013,森林资源数据采集技术规范 第1部分:森林资源连续清查:LY/T 2188.1-2013[S].北京:中国标准出版社.
李春华,1993.线性规划方法在"下料"中的应用[J].运筹与管理(01):26-29.
刘于鹤,林进,2013.加强防护林经营改善生态与民生[J].林业经济(02):3-10+96.
沈国舫,关玉秀,周沛村,等,1979.影响北京市西山地区油松人工林生长的立地因子[J].北京林学院学报(00):96-104.
谭伟,冯仲科,张雁,等,2006.基于组件 GIS 的造林小班地形分析的研究——以造林小班坡向为例[J].北京林业大学学报(02):91-95.
陶凤玲,宛士春,2004.利用"准最优基"简化单纯形法求解过程[J].武汉大学学报(工学版)(01):68-71.
王霓虹,2002.基于 WEB 与"3S"技术的森林防火智能决策支持系统的研究[J].林业科学(03):114-119.
韦鸿雨,2007.方格网法在城市公共绿地建设规划中的应用[J].山西建筑(18):336-337.
杨广斌,李亦秋,安裕伦,2006.基于网格数据的贵州土壤侵蚀敏感性评价及其空间分异[J].中国岩溶(01):73-78.
姚树人,文定元,2002.森林消防管理学[M].北京:中国林业出版社.
野上启一郎,陈建成,张青,1993.模糊目标规划法在林业经营计划中的应用[J].林业资源管理(05):61-66.
张国防,1998.森林火灾直接和间接经济损失[J].森林防火(01):23-25.
郑宏,张玉红,2003.黑龙江省林火信息管理与火灾损失评估系统的设计[J].森林防火(04):18-21.
朱万昌,梁延海,张静,等,2004.黑龙江省大兴安岭森林功能研究[A].中国林学会.中国林学会 2004 年年会论文集[C].中国林学会:中国林学会,170-173.
Han J G, Ryu K H, Chi K H, et al., 2003. Statistics based predictive geo-spatial data mining: Forest fire hazardous area mapping application[C]. Asia-Pacific Web Conference. Springer-Verlag.
Isaev A S, G. N. Korovin G N, S. A. Bartalev S A, et al., 2002. Using remote sensing to assess Russian forest fire carbon emissions [J]. Climatic Change, 55(1-2): 235-249.
Pan P Q, 2010. A fast algorithm for linear programming [J]. J Comp. Math, 28(6): 837-847.
Perminov V. 2002. Numerical solution of reynolds equations for forest fire spread [J]. Lecture Notes in Computer Science, 2329(2329): 823-832.
Wang G, Jiang B, Zhu K, et al., 2010. Global convergent algorithm for the bilevel linear fractional-linear programming based on modified convex simplex method [J]. Journal of Systems Engineering and Electronics, 21(002): 239-243.

CHAPTER 5 Technical Principles of Forest Inventory

5.1 Forest survey techniques overview

National continuous forest inventory is also referred to as first class survey. The continuous inventory is a repeated investigation of forest resources in the same area at the appointed time. The method is sample investigation based on the principle of mathematical statistics. The survey object is fixed sample plots which are set up in the survey area and sometimes few temporary sample plots are added.

5.1.1 Objectives and tasks

National Forest Resource Continuous Inventory (first class inventory) is a forest resource survey method mainly using fixed sample plots for making a periodic review. The purpose is to grasp the macro situation and dynamic situation of forest resources. It takes the province (municipalities, autonomous region, hereinafter referred to as province) as a unit. It is an important part of the comprehensive monitoring system of the forest resources and ecological conditions. Continuous forest inventory results in an important basis of reflecting national and provincial forest resources and ecological conditions, formulating and adjusting forestry policy. Supervising and checking tenure target responsibility system of the growth and decline of forest resources.

National Forest Resources Continuous Inventory tasks are to regularly and accurately identify national and provincial forest resources quantity, quality and dynamics, grasp the status and tendency of forest ecological system, the comprehensive evaluate forest resources. Specific works include:

(1) To formulate the work plan, technical scheme and operation rules of the continuous inventory of forest resources.

(2) To complete sample setting, external investigation, and supplementary data collection.

(3) To complete statistics, analysis, and evaluation of forest resources and ecological conditions.

(4) To regularly provide the national and provincial continuous inventory results of forest resources.

(5) To establish the National Forest Resources Inventory database and information management system.

5.1.2 Historical review of first class inventory

From 2004 to 2008, China announced the Seventh National Forest Resource Continuous Inventory (first class inventory). As compared with five years ago, forest coverage rate changed from 18.21% to 20.36%; forest stock increased from 124.5 billion cubic meters to 137.2 billion cubic meters. Forest area increased from 175 million hectares to 195 million hectares; artificial forest preservation area was 0.62 million hectares, ranking first in the world. China at present is one of the few countries in the world that announces forest data accurately, timely and systematically and the above achievements are due to the establishment of a reliable National Forest Resources Inventory System.

After the foundation of new China, China learned class visual investigation from the Soviet Union. A large-scale forest survey work had been launched. National forest resource data was collected via smaller unit to larger one i.e. the small group- class- partition-forest farm- forestry bureau- county- province. In 1965, the first national forest resource data summary was completed. However this approach does not have complete coverage, cannot meet the needs of the development and data accuracy is not stable, the cycle is too long. In order to complete more efficient investigation of forest resources, new ways must be studied.

1) Introduction and rapid promotion of new technologies

Forest sampling survey technology based on mathematical statistics was introduced by teachers of Beijing Forestry College in 1963. Daping Forest Farm in Rucheng County and Hunan Province performed stratified sampling test in 1964 and achieved good results. Huaping County of Yunnan Province and Xinlin Forestry Bureau of Great Khingan in 1965 carried out further stratified sampling survey of technology production. The large area measured data verified the success of the technology. As a result, the Ministry decided to promote the sampling survey in large scale and in 1966 completed the task of forest sampling survey of about 2000000 hectares area. The forest sampling survey method has been widely recognized.

2) The 'Fourth Five-year Plan' the first check to complete the National Forest Inventory in required time

In order to formulate the forestry policy and all kinds of planning, the Ministry of Agriculture and Forestry in 1973 required quickly and accurately identify the status of China's forest resources in the 'fourth five-year plan' period. The Ministry of Agriculture and Forestry also attempted to take county bureau as survey unit in Jiwen Forestry Bureau, Greater Khingan Range and Huitong County, Hunan Province and apply a variety of sampling techniques by setting permanent sample plots. The project achieved success, through the method of the project, the Ministry of Agriculture and Forestry carried out forest resources inventory by taking county bureau as a unit in the country during 1973 to 1976. Survey results show that except Taiwan Province, the national forest area is 1.22 million hm^2, the forest coverage rate is 12.7% and forest stock volume is 87 billion m^3. For the first time in China, the National Forest Resources Inventory was completed by adopting a unified survey method.

3) The second check — the establishment of the National Forest Resources Continuous Inventory System

In 1977, the Ministry of Agriculture and Forestry in Jiangxi Province carry out a national continuous investigation on pilot work. The purpose was to find an important basis that can provide the government with the changing tendency of accurate forest resources quantity and quality, as well as forest resources growth and decline. Forestry survey workers and agricultural and forestry colleges and universities teachers from 27 provinces participated in the work and lay a solid foundation for the establishment of the province (autonomous regions and municipalities) national level Forest Resource Continuous Inventory System with the characteristics of fixed sample review. At the end of 1981, except for Tibet, Shanghai, Tianjin, Taiwan the system was completed in 26 provinces, autonomous regions, and municipalities. At the end of the investigation, 13. 7 million permanent sample plots and 2. 3 million temporary sample plots had been set and the 'Fifth Five-year Plan' inventory database was established. It showed that China's forest resources inventory technology developed from the static inventory in the past to a new stage of dynamic monitoring.

4) 'The Third Check'-first review of forest continuous inventory

From 1984 to 1988, China has completed the first review of the continuous inventory of forest resources in the provinces except for Tibet, Taiwan and in 1989, the Ministry of Forestry announced the results of the latest forest survey. In the inventory of national survey, 140 thousand review sample plots and 111 thousand new sample plots were set. The total number of samples reached 255 thousand.

5) The fourth check-establishment of the National Forest Resources Monitoring System

In order to grasp the forest resource status and dynamic change and to forecast the development trend of forest resources in February 1989, the Ministry of Forestry required the establishment of monitoring system of forest resources and asked the provinces, municipalities and autonomous regions to submit all the data results to the Ministry of Forestry after the completion for summary. National Forest Resource Continuous Inventory results would be released every 5 years. As a result, the forest resources monitoring work became more standardized. From 1989 to 1993, China not only completed the second review of the National Forest Resources Inventory but also established a continuous inventory system in the Tibet Autonomous Region. In the survey 227. 2 thousand ground sample plots were studied, 106. 2 thousand satellite images sample plots and aerial photographs sample plots were interpreted. Some provinces had further improved the layout and used remote sensing data to rich inventory results.

6) The national forest resources monitoring system is running normally

From 1994 to 1998, under the unified deployment of State Forestry Department, provinces had planned continuous review of forest resource work and the fifth check was carried out. 18 million ground sample plots were studied and more than 90 thousand satellite images sample plots and aerial photographs sample plots were interpreted. During the period of 1999 to 2003, the Sixth National Forest Resources Inventory was carried out. For the first time, desertification and wetland investigation were contented into the inventory system. Forest Resource Continuous Inventory adopted palm the computer (PDA) data acquisition technology and the generalized '3S' technology that is Remote Sensing (RS), Geographic Information System (GIS) and Global Positioning System (GPS). The number of the permanent sample plots reached 415 thousand. Remote sensing sample plots number reached 2. 8444 million. Except for Taiwan Province, Hong Kong SAR, and Macao SAR, ground permanent, sample plots had covered the provinces, city and autonomous region. During 2004 to 2008, the seventh inventory was carried out. The measured permanent sample plots reached 415 thousand and remote sensing sample plots number reached 2. 8444 million. The data concern forest ecology status the function and benefit and forest resources quantity, quality, structure, distribution status and dynamic. According to the results of the seventh inventory the forest ecosystem service function value was more than 10 trillion Yuan and the national forest vegetation total carbon reserves was 7. 811 billion tons. In the sixth and seven inventories although a large number of remote sensing samples have been interpreted and the remote sensing data that is mainly confined to the preparation of forest distribution map not used for forest resources data. From 2009 to 2013, China carried out the Eighth National Continuous Inventory. Compared with the Seventh Continuous Inventory the result shows that forest resources in our country have the following four characteristics.

First, the total amount of forest keeps growing. Forest area increased from original 1. 95 million hectares to 208 million hectares. Forest coverage rate increased from originally 20. 36% to 21. 63% and forest stock volume also increased from 13. 721 billion m³ to 15. 137 billion m³.

Second, forest quality is further improved. Volume per hectare increased 3. 91 m³ and reached 89. 79 m³. Annual growth per hectare increased to 4. 23 m³ and due to the significant increase in the total forest the quality of forest ecological function also strengthened.

Third, the quantity and quality of natural forests increased steadily. The natural forest area increased 2. 15 million hectares and reached 121. 84 million hectares. Volume increased 894 million m³ and reached 12. 296 billion m³.

At last artificial forest developed rapidly. The investigation showed artificial forest area increased from the nearly 61. 69 million hectares to 69. 33 million hectares and the volume increased 522 million to 2. 483 billion m³. The artificial forest area continued to maintain first in the world.

The results of the 8 inventory data and the estimated carbon contribution in China are showed as follows Table 5-1 to Table 5-3 (part of the data from Zhou Changxiang, 2012):

Table 5-1 Accumulation of forest area

Inventory interval	Forest area / $10^4 hm^2$	Forest stock / $10^4 m^3$	Forest coverage / %
No. 1 (1973–1976)	12186. 00	865579. 00	12. 7
No. 2 (1977–1981)	11527. 74	902795. 33	12. 0
No. 3(1984–1988)	12465. 28	914107. 64	12. 98
No. 4 (1989–1993)	13370. 35	1013700. 00	13. 92
No. 5(1994–1998)	15894. 09	1126659. 14	16. 55
No. 6 (1999–2003)	17490. 92	1245584. 58	18. 21
No. 7(2004–2008)	19545. 22	1372080. 36	20. 36
No. 8(2009–2013)	20800. 00	1513700. 00	21. 63

CHAPTER 5   Technical Principles of Forest Inventory

Table 5-2   The stumpage and forest vegetation carbon storage

Inventory interval	Stumpage / 10^4 m^3	Forest carbon storage / 10^4 t	CO_2/ 10^4 t
No. 1 (1973–1976)	953227	493485.62	1809628
No. 2 (1977–1981)	1026059.88	531191.2	1947896
No. 3 (1984–1988)	1057249.86	547338.25	2007108
No. 4 (1989–1993)	1178500	610109.45	2237292
No. 5 (1994–1998)	1248786.39	646496.71	2370725
No. 6 (1999–2003)	1361810	705009.04	2585292
No. 7 (2004–2008)	1491268.19	772029.54	2831058
No. 8 (2009–2013)	—	842700.00	—

Table 5-3   The stumpage and forest vegetation carbon storage

Inventory interval	Annual average forest trees net growth /10^4 m^3	Annual average cut/10^4 m^3	Average annual trees increase /10^4 m^3	Net growth of forest carbon /10^4 t	CO_2 absorption /10^4 t	Carbon removal /10^4 t	Carbon storage /10^4 t
1973–1976	22692	19653	3039	11747.65	43079.02	10174.36	1573.29
1977–1981	27532	29410	-1878	14253.32	52267.39	15225.56	-972.24
1984–1988	32946	34483	-1537	17056.14	62545.45	17851.85	-795.70
1989–1993	41912.35	32794.31	9118.04	21698.02	79567.38	16977.61	4720.41
1994–1998	45752.45	37075.18	8677.27	23686.04	86857.51	19193.82	4492.22
1999–2003	49674.63	36538.13	13136.5	25716.56	94303.47	18915.79	6800.77
2004–2008	57157.38	37911.71	19245.67	29590.38	108508.9	19626.89	9963.48

5.1.3   First class inventory content

The main object of the National Forest Resources Inventory (a kind of inventory) is to get knowledge about the forest resources and its ecological condition. Main contents include:

(1) Land utilization and cover, includes land type (land type), area and distribution of vegetation type.

(2) Forest resources include the quantity, quality, structure and distribution of forests as well as the forest area and accumulation according to the age group, forest type, and tree species. It also includes growth and consumption and its dynamic changes.

(3) Ecological status, including the forest health status and ecological function, the diversity of forest ecosystem, the area and distribution of land desertification, desertification and the type of wetland and its dynamic change.

5.1.4   First class inventory method

According to the 2003 *National Forest Resources Continuous Inventory Main Technical Regulations*, one kind of survey processis preliminary preparation → basic survey method determination → area measurement → laying fixed sample plot → fixed sample plot building sign → plots survey →

Investigation of other factors in each plot.

(1) Preparation. The provincial forestry department was establishedbet ween CFI group and office. The formation of the investigation team, formulate relevant work schemes and rules for the survey. Organize technical training and to prepare basic tables and to pographic map, a variety of survey tools, survey data, and other relevant information.

(2) Basic survey method. The principle of basic survey method should be used in order to set the fixed sample plot or the configuration of tem porary sample plots and with remote sensing technology to implement the survey method.

(3) Area measurement. Estimation of forest area by using the method of estimating the number of Specific methods, according to the system sampling formula:

$$p_i = \frac{m_i}{n} \tag{5-1}$$

$$S_{p_i} = \sqrt{\frac{p_i(I - p_i)}{n - I}} \tag{5-2}$$

In the formula, n is the total sample number; MI is typed (including land types, vegetation type, forest types and a variety of other land classification attribute); I is the number of sample plots, area as a type I to estimate the number, area of type I is to estimate the value of standard deviation.

$$\hat{A}_i = A \cdot p_i \tag{5-3}$$

In the formula, the area of type i is estimated and the A is the total area.

$$\Delta_{A_i} = A \cdot t_\alpha \cdot S_{p_i} \tag{5-4}$$

Form in, the error limit of the estimated value of the type I area is the reliability index. Area estimation interval for type I.

$$P_{A_i} = \left(I - \frac{t_a \cdot S_{p_i}}{p_i}\right) \cdot 100\% \tag{5-5}$$

The sampling precision of the estimated value of the type I area.

(4) The layout of fixed sample plots. According to the systematic sampling, the fixed plots should be laid in the inter section km grid of 1/50000, 1/100000 to pography of the country, and the GIS should be used to ensure that the samples are laid. Fixed sample shape is generally square, rectangular, circular or angle gauge sample, the fixed sample area is generally 0.0667 hm^2. The plot numbers are numbered sequentially from northwest to southeast and permanently fixed.

(5) To set up a sign for a fixed plot. Including the landmarks, the samples, among them, the sample should be included in the fixed marking starting from the southwest corner of the standard plot then northwest, northeast, southeast corner of the right angle pit or corner pile. Marking in circular plots should be at the border from the east, south, west, north to establish a pit and other markers. All the samples in the sample plots were set up in accordance with the fixed sample. The position of the sign should be set at the base of the trunk, and the paint line can be fixed to the trunk high position. When the layout sample cannot receive the GPS signal or the signal is unsta-

ble, should record the lead mark.

(6) Fixed sample survey. Fixed sample plot survey follows certain basic principles. In the retest sample and set up additional samples for accurate positioning by GPS, to sample 75 factors were investigated. These factors include: number, type, altitude, slope degree, slope, vegetation type, vegetation cover, forest type, origin, average age, average tree height, average DBH, canopy density, stand volume, growth, consumption, in addition to the cross angle forest sample plot survey and remote sensing interpretation recorded link. For example, the sampling method for estimating the volume of the stock is simple systematic sampling.

(i) Sample mean:

$$\overline{V}_i = \frac{1}{n} \sum_{j=1}^{n} V_{ij} \qquad (5\text{-}6)$$

In the formula, V_{ij} is the type I of the first j type of accumulation.

(ii) Sample variance:

$$S_{V_i}^2 = \frac{1}{n-1} \sum_{j=1}^{n} (V_{ij} - \overline{V}_i)^2 \qquad (5\text{-}7)$$

$$S_{\overline{V}_i} = \frac{S_{V_i}}{\sqrt{n}} \qquad (5\text{-}8)$$

(iii) Total estimated value:

$$\hat{V}_i = \frac{A}{a} \cdot \overline{V}_i \qquad (5\text{-}9)$$

In the formula, A is the total area, a is the sample area and the total amount of the I type are estimated.

(iv) The error limit of the total estimated value:

$$\Delta_{V_i} = \frac{A}{a} \cdot t_a \cdot S_{\overline{V}_i} \qquad (5\text{-}10)$$

Form t_a is reliability index. The estimated range of the total estimated value is $\overline{V}_i \pm \Delta_{V_i}$

(v) Sampling precision:

$$P_{V_i} = \left(1 - \frac{t_a \cdot S_{V_i}}{\overline{V}_i}\right) \cdot 100\% \qquad (5\text{-}11)$$

When the object is tree species the diameter at breast height is 5.0 cm. To determine the size of the object, the main consideration of the characteristics of forest trees, tree species need to check the size of shrubs, and trees without the need to check the size. When the ruler is used, all the steel ruler is used and the height of the 1.3 m height of the upper part of the trunk is detected and the reading is accurate to 0.1 cm. Each wooden ruler needs to be recorded in detail, including the number of samples, the types of trees, the size of the type, tree species and code, diameter at breast height, and so on. In addition, it should also be measured in each individual plant samples of the horizontal distance and azimuth (or other location measurement data) to map the distribution of samples. At the end of the study, the tree height, the degree of desertification, forest disaster investigation, and other factors were also investigated.

5.1.5 First class inventory accuracy standards

To the province as a whole, the overall sampling accuracy that is the province of the sampling precision (at 95% reliability, the same below); a province is divided into a plurality of vice general. The overall accuracy of sampling by the vice general by stratified sampling was a joint estimation.

1) The status of forest resources sampling accuracy

(1) Forest area. Where there is forest land area of the province's land area of more than 10% of the province, the accuracy of more than 95%; and the rest of the provinces is more than 90%.

(2) Artificial forest area. The forest area of 5% of the forest land area is more than and the accuracy is above 90% and the other provinces are over 85%.

(3) The stumpage. Where the stumpage more than 500 million cubic meters the precision is more than 95%, Beijing, Shanghai, Tianjin, is more than 85% and other provinces is more than 90%.

2) Stumpage dynamic accuracy

(1) Of the total amount of growth. Stumpage 500 million cubic meters, above 90%, the rest of the provinces for more than 85%.

(2) Of the total consumption. Stumpage 500 million cubic meters, above 80%, other provinces are not specified.

(3) The stumpage net increment, should make a change in the direction of judgment.

3) For a fixed sample and sample reduction rate

Fixed sample up to 98% above, the fixed sample more than 95%.

4) The allowable error to investigate

(1) Point positioning. Stake location in topographic map.

(2) Measuring the number of trees. Greater than or equal to 8 cm should gauge the number of trees does not allow a margin of error of less than 8 cm should be measuring the number of trees, allowable error is 5%, and a maximum of 3 strains.

(3) Diameter measurement. Diameter less than 20 cm of trees, measurement error is less than 0.3 cm; diameter at breast height is greater than or equal to 20 cm of the trees, the measurement error is less than 1.5%.

(4) Tree height measurements. When the tree height is less than 10 m, the measurement error is less than 3%; when the tree is higher than or equal to 10 m, the measurement error is less than 5%.

(5) Factor, origin, tree species, dominant species etc. should not be wrong.

5.2 The second class forest management survey

Forest managers survey, also known as a second class survey of forest resources planning and design. Specifically refers to the management of forest resources of enterprises, institutions or administrative units targeted for the development of forest management plans and inspection. Including the evaluation of forestry zoning effects of forest management, forest resources survey carried out dynamically. The second class survey is generally of 10 years at a high management level of regional or unit may be once every five years.

5.2.1 Task and purpose of the second class survey

1) Primary mission

In the investigation includes identifying the type, quantity and quality of forest and the natural resources, social and economic conditions. Evaluate and check on the status of forest resources management.

2) Purpose

(1) To develop and revise management plans on regions basis.

(2) Check, analyze and evaluate the effectiveness of forest management.

(3) For the development and evaluation of forestry policy, methodology, implementation and effectiveness of the regulations, and provide a revised basis for it.

(4) Provide basis for regional and national authorities.

5.2.2 Content of the second class survey

Forest manager's survey is one of the main forest management work, which mainly includes the following four aspects.

5.2.2.1 Forestry production conditions survey

1) Natural conditions survey

Natural conditions are the environmental conditions of forest resources, which play a decisive role in the itinerary, succession, growth, type structure, function, quantity and quality of forest resources. The investigation of natural conditions includes the following:

(1) Geographical conditions. Mainly geographical conditions are the location of administrative divisions, flora, terrain conditions, the mountains, and water status.

(2) Geological conditions. Including soil types, scale, distribution; a variety of soil fertility status, such as the content of organic matter content, N, P, K and other elements; soil physical properties, such as thickness, texture, and structure; soil formation conditions, such as soil rock weathering degree.

(3) Water resources. Water resources include the length of the river flow, water level, wetland status, water volume, depth and impact on forests as well as other flora and fauna; utilization of water resources, water resources and land survey of the surface area of erosion.

(4) Climate, weather conditions. Climate, weather conditions focusing on forest resources have the most direct effect, like Temperature: maximum and minimum temperature, average temperature, annual accumulated temperature, length of growing season, early frost, late frost period, etc; precipitation: mainly annual rainfall, seasonal distribution of precipitation and other forms; other

weather phenomena: time wind appeared, strength, possible effects on forest growth conditions and the like.

2) Socio-economic conditions survey

(1) The relationship between forestry and agriculture, animal husbandry, fisheries, industry and similar.

(2) The relationship between forests and regional social and environmental.

(3) Zone configuration forestry, foresttenure, and other conditions.

(4) Forest products market situation.

(5) Contribution to regional social forestry economy.

(6) Transport status.

(7) Population, labor force status.

3) Forest management history survey

We should focus on forestry management history positive and negative aspects of the experience. Forestry management history survey content includes the following aspects:

(1) History of forest management structure. Include the statusof forest management institutions establishment and other aspects of the scale.

(2) Investigation. Include the number of investigations, each time the state of technology, content, and methods of investigation.

(3) Forest management plan. Implementation of the forest management plan preparation, the management plans.

(4) Forest management situation. Harvesting is one of the most important aspects of the history of management; forest resources of other aspects; multiple resource utilization in addition to the common forest management activities outside.

(5) Environmental conditions. Includes two aspects: one is whether the environmental function and role played by changes in the intensity of forest, how the trend; on the other hand, forest resources change their environment ie soil, water, socio-demographic, changes in employment status.

(6) Forestry enterprise management. Include business organizations, the number and quality of staff level practitioners, operating profit status, productivity, market demand and other products.

5.2.2.2 Sub-compartment investigation

Sub-compartment investigation is a job relates most extensive area, the maximum amount of work in the second class survey. Sub-compartment is partitioned based on the stand and stand is basically the same forestry, biological characteristics and surrounding forest lots have obvious differences of forest area, forest resources constitute the fundamental group unit. In order to carry out the management of forest resources scientifically and reasonably, it is necessary to carry out the information of forest resources into every forest. Sub-compartment survey is to implement a variety of forest stand survey factors in each stand.

1) Content of sub-compartment survey

(1) Survey of ground. Type forest land resources planning and design is divided into two major forest and non-forest land categories. Among them, the forest land is divided into eight categories, see Table 5-4.

Table 5-4 Woodland classification system tables

Number	Level one	Level two	Level three
1	Woodland	Trees	Pure forest
			Mixed
		Mangrove forest	
		Bamboo forest	
2	Open forest land		
3	Shrubland	Special provisions of the state bushes	
		Other shrub	
4	Afforest land	Artificial afforestation into woodland	
		Natural regeneration	
5	Nursery lands		
6	No standing woodland	Cut-over land	
		Burned area	
		Other no standing woodland	
7	Suitablel and for forest	The waste mountains and land	
		The waste sands land	
		Other suitable land for forest	
8	Auxiliary production of woodland		

(2) Sub-compartment survey factors. Sub-compartment survey factor should be recorded. These factors include the following aspects i. e. location, ownership, land, topography, soil, understory vegetation, site type, rank, management measure type, forest category, layer forest, species composition, dominant species, average age, average height, average diameter, average dominant tree height, canopy density, per hectare of trees and volume per hectare. The fallen dead wood accumulation, plant diseases and insect pests, fire and other(Figure 5-1).

2) Methods of sub-compartment survey

(1) Plot survey. Plot survey is the method of making plots in any sample area. A sampling includes random sampling, mechanical sampling, stratified sampling, etc. Likewise shape of round, rectangular, square, etc. The factors of sampling error and accuracy were calculated by the method of mathematical statistics and the total corresponding value was estimated with the measured value of the sample plot. Sampling selection can be determined by factors such as reliability, observation variation coefficient, error requirement and so on.

(2) Standard investigation methods. Within the geographical scope of the investigation, the results of the standard method of investigation estimated the overall value. Standard ground gener-

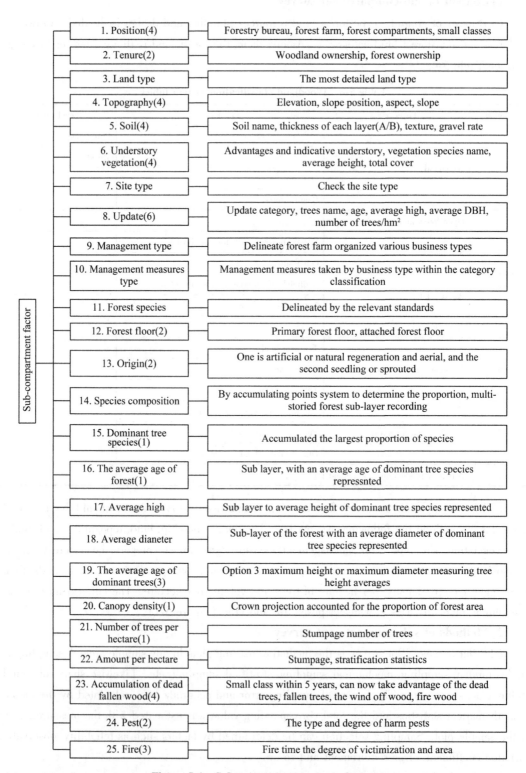

Figure 5-1 Sub-compartment survey factor

ally rectangular shape, ribbon and so on.

Implementation steps: Set the standard and then measured with the results estimated accumulation amount and other units within the sub-compartment size and the entire range of sub-compartment factor.

The key to reducing the error using a standard survey method is selected criteria as possible 'standard' that is the standard way to have the status of the overall survey average. In addition, the area of standards must be guaranteed a certain amount, if the area is too small it is difficult to ensure that the selected sample representative, will have a large deviation when estimating overall. However, the standard area of increased workload over the General Assembly to increase the investigation time and operating costs. Also, the standard layout should try to avoid systematic errors.

(3) Visual survey. In the state of forests relatively simple case, the choice of investigations by investigators with this side assessment.

(4) Angle gauge survey. Basal area per hectare were factors such investigations with high work efficiency characteristics with an angle gauge. In good condition sub-compartment perspective, investigators in the case of relevant experience foot can use this method.

Angle gauge constant selection depending on forest size. Point angle gauge layout should follow the principle of random and systematic errors to avoid errors forest edge.

Great angle gauge operations and the observer's experience, habits, visual relationships. Analysis showed that large tree measurement error is common with an angle gauge. When using an electronic angle gauge having a telescope, estimates estimation accuracy improved significantly. Research project successfully developed horizontal and vertical lines and sampling tools such as algorithms, automatic real-time recording and automatic, accurate, three-dimensional tree measurement. Verification results show that compliance with the corner of a small difference in measurement results point repeatedly.

(5) Regression estimation method. Using measured values of small groups to establish regression relationship. Calculate the size of small units per unit area and other factors the amount of value, known as the regression estimation method. There are many regression estimation methods, the most commonly used is the use of aerial photo interpretation value. The survey values are used for the establishment of small regression estimation method. The working steps of this method are as follows:

(ⅰ) Select a recent photograph of the large-scale aerial (1 : 25000 or more).

(ⅱ) The sub-compartment interpretation, aerial uppercut in the forest land plot in sub-compartments and borders.

(ⅲ) The sub-compartment remit and quadrature.

(ⅳ) The interpretation of each stock volume in sub-compartments.

(ⅴ) In accordance with the principle of randomly selected a certain number of sub-compartments (also found in sub-compartments can be extracted by mechanical sampling method) were measured.

Measured in sub-compartments which are drawn should be all-wood forest each gauging method for determining the amount accumulated over the entire range. But in reality, in the full range of sub-compartments each forest wood gauging, field work is too difficult to implement, so commonly used in small precision sampling to assess the amount of accumulation, replace the whole forest wood per gauging. A number of samples in sub-compartments with strong accuracy minimum should meet the large sample ($n>50$) the requirements of each area of the sample unit may be in the $0.01 \sim 0.02$ hm^2. Drawn in small groups, in addition to actual accumulation amount, the other factors can be found stand investigation, visually or measured with the visual part of the method of combining.

Regression equation. The interpretation and measured values of the regression equation method steps:

(i) Plotted on a scatter plot graph paper and measured values of the interpretation, analysis of the distribution trends, determine the type of regression equation then interpretation and measured values measured in sub-compartments and estimate the parameters of the regression equation. For example, the regression equation $y=A+Bx$, where A, B as a parameter.

(ii) If existing regression estimation software also can be directly interpreted and measured values input to identify the type of regression equation and parameter values.

Estimated values calculating overall stock volume per hectare, variance estimates, error limit, precision, range estimation and the like.

(6) Remote sensing measurement: Known points: ground sampling / plot survey data (DBH, tree height, crown), quantitative factors (band gray values and ratio terms) of remote sensing data and geographical environmental factors (soil, slope, aspect and elevation). Through GPS plot / sample point, inversion value: forest biomass, forest stock, forest asset value.

3) The overall volume by sampling control

(1) General settings. State Forestry Bureau, forest farm as a unit is to set the overall the forest.

Key forestry county as a unit is to set the overall.

(i) General forest counties or Shaolin County to the city (region) set the overall.

(ii) In the second category the survey, regardless of the overall size, must ensure that the investigation method and investigation of the incident in the overall context of consistency, to ensure accuracy.

(2) Accuracy requirements. Total accumulated in a small amount compared with the overall sample survey should stand volume. If the cumulative deviation of ± 1 times $\sim \pm 2$ times because of the error check and correct cause, until the accuracy requirements meet. So far, where the cumulative deviation$> \pm 2$ times the allowable error, you should re-investigate.

5.2.2.3 Professional survey

In a professional survey, the investigation of the individual or several contents of soil, site conditions, growth, pests and diseases, consumption, wildlife and other forest resources projects are undertaken. Second investigation, select the project should be based on professional survey needs and technical conditions.

A region-wide survey with two types of professional survey, more independent enterprises, institutions and administrative units divided into units.

1) Common professional investigators

(1) Growth survey. Various forest growth, especially when an important basis for Volume Growth forest management decisions. The population dynamics of control over resources and reasonable arrangements for harvesting have a very important role. Growth investigations were dominant tree species, age class (group).

(2) Consumption survey. The survey mainly includes Felling of supplementary harvest, fuel wood removals and other kinds of production, the amount of wood consumed during life disaster.

This survey mainly independent of enterprises and institutions units.

(3) Forest protection survey.

(i) Investigation in this regard include the case of diseases, pests, and fire area.

Contents of fire investigation are kind of fire, time of occurrence, the number of continuation time, the reasons for fires, firefighting methods, and facilities.

(ii) Pest survey mainly refers to the situation of forest resources generate significant harm pests. The main contents of the survey include the type, quantity, degree of harm pests, because development occurs, and other losses, prevention methods, and measures.

(4) Site conditions survey. Growth forest habitats is a collection of environmental factors, good site conditions directly with the poor relation to various aspects of forest management, such as productivity, economic efficiency, harvest harvesting, silviculture direction, and speed.

Site condition factors including climate, soil, vegetation, topography and terrain and other factors. These factors play a role in a certain region of space, while the growth of trees produces more fixed corresponding impact, which is the impact of site conditions and reflect the production of land situation. Showing the merits of the site conditions and status indicators index level position in opposition to factor conditions such investigation has been completed, the status should be determined according to the relationship between class and status index trees age of trees, and tree age and tree height growth. It should also draw maps and forest soil type map site conditions for forest management operations purposes.

Site conditions survey management survey are not always carried out in the absence of major natural, climate, changes in crustal movement, the site conditions in a long time basically stable.

(5) The amount of material investigation. The main is to investigate a variety of species at a certain age, diameter, tree height, the number and proportion of forest reserves to produce commercial timber, which is an important work in timber operations.

(6) Nursery survey. The plant nursery is directly related to afforestation, artificial regeneration and artificial promotion of natural success.

Content nurserysurvey includes nursery area, species, breeding species, seedling growth status, annual seedlings, costs, benefits, equipment, and management system.

Nursery investigation is to permanently fix nursery-based, taking into account also the foreign seedlings and nursery stock output, supply and demand.

(7) Thinning investigation. Thinning forest management is an essential part, its main purpose is to better retain the tree growth, enabling the highest average annual productivity per unit area of woodland.

The content of thinning survey is the type of each species stands, thinning conditions, methods, strength, intervals, process, the amount of material and similar things.

2) Professional investigative manner

In the professional survey project, most projects using standard survey, woodplots, and standard parse, some projects survey need to be used in combination in several ways.

5.2.2.4 Multi-resources inventories

Forest resources in addition to forest area should also include space animal resources, vegetation resources, land resources, water resources, climate resources, recreation resources. In China, more resources survey means includes survey of wildlife, recreation, water resources, grazing and underground resources carried out. Forest resources are an organic in nature rather complex structure and function of many ecosystems. Forest resources and other environmental resources mutually influence each other.

More sustainable use of forest resources survey is sustainable timber from the forest to the multiple benefits sustainable transition gradually developed during the forest inventory project. States to investigate the type of multi-resource attribution is not the completely consistent world. In China's relevant regulations the investigation is still two tone multi-resource professionals in the scope of the investigation due to a wide range of resources and more involved in the investigation, more content, so in this book are listed separately.

1) Generation of multi-resource survey

Forests, human survival development and demand closely related to the sustainable use of forest multiple benefits in a very long time i.e. the use of wood and wood products. Therefore, the main objective of the survey conducted by forest, is the status of forest resources and how to harvest trees rational use of forest ecological and social benefits in the second low. Demand for forest functions and the number of people living standards and forests is closely related to quality.

As the human standard of living improve demand for forest functions is growing, especially in the global area, a significant reduction in the forest, their effectiveness has diminished significantly affect people's lives even more so.

The 1970s, the US Forest Service in the national forest resources survey launched multi-resource survey system. In the multi-resource survey in South Carolina, the main contents of the survey are ecological structure survey of forest vegetation, biomass survey, wildlife habitat survey, recreation investigation, soil and hydrographic survey, rangeland survey. US Forest Service in 1975 will be divided into six series of forest resources:

(1) Outdoor recreation and nature reserves resources.

(2) Wildlife and fish habitat resources.

(3) Pastures resources.

(4) Forest resources.

(5) Land and water resources.

(6) Human resources and social development.

After the 1980s, in many countries around the world have gradually launched an investigation and more resources. Investigation of forest resources in the past 20 years has been considerable development.

2) Multi-resource survey content and methods

Many different types of forest resources within each Ⅱ will investigate all the resources to investigate the reality is neither cut nor necessary. More resources survey project should be investigated in accordance with the necessary requirements for forest resources management. Now more resources on major research projects described below.

(1) Landscape resources survey. As people's living standards improve, urbanization, travel to natural areas, especially in the tourist resort of forest growing trend and forest recreation has become an important field of forest management.

In addition to forest resources include forest landscape vegetation. The landscape should also include a variety of natural landscape and cultural landscape of forest area, forest areas such as geographical landscape, the ancient landscape architecture, and folk customs.

(2) Economic plant resources investigation. Economic plants resource investigation is the investigation of forest wild plant with high economic value; or that in addition to providing the major forest products, but also provide other plant resources with a high economic value attached to the product.

(3) Water survey. Forest and water resources are closely related to each other. Many forests are located in the upper reaches of major rivers. The source, the total amount of water flow and quality of water are essential for stability. Water is not only important for the forest, it increases the attractiveness of the forest landscape and recreational areas. Also for the development of fisheries, farming and aquaculture play an important role in rural and urban domestic and industrial water important source. Our existing cities more than 600, of which 50% of the water shortage in cities, met a year of drought, water scarcity city 60%, thus maintaining water forests play an increasingly prominent role.

(4) Grazing resources survey. In many forest areas have grazing resources, as well as herbs. In addition to some shrub leaves, twigs and fruit, including some fruit trees. Herbs can be used for grazing resources distribution in the forest with the grass, low humidity, the shrub planting young plantations, rivers and some types of woodland.

Content of grazing resources survey are:

(i) Category name and the growing season.

(ii) Geographical distribution and counting.

(iii) Appliespastoralists.

(iv) Carryingcapacity of grazing resources.

(v) Wildlife resources survey.

Wildlife resources survey was carried out in the forest wildlife survey. The main contents are wild animal species, quantity, geographical distribution, habits, habitats, population dynamics of variation, and animal sex, age and the like.

5.2.3 Second class survey results

Main achievements:
1) Table book materials
(1) Stands description form. Forest inventory is based on the forest as a unit for forest resource information records and summary of the table, referred to as the survey book. The survey book consists of three parts:

(i) Cover. Summary of ' various types of land area' , stock volume and general situation of forest classes.

(ii) Closure. To record the status of the investigation factors of each sub-class, Linfen views of growth and management measures, the form of a small sub-class investigation card or computer code records.

(iii) Back cover. The forest classes within the small class changes in the operating records.

(2) Statistics of forest resources.

(i) Various land area statistics.

(ii) Various forest and forest area accumulation statistics.

(iii) Forest species statistics.

(iv) Trees forest area accumulation by age group statistics.

(v) Ecological public forest to statistics.

(vi) Red woods resources statistics.

(vii) Timber area accumulation by age level statistics.

(viii) Out material grade statistics.

(ix) The number of timber trees and the volume of each tree species in thenear mature forest.

(x) Timber forest and general public forest in the area of different age forest accumulation according to the ratio of large diameter wood table.

(xi) Forest commission meter.

(xii) Bamboo table.

(xiii) Shrub tables.

2) Figure materials

Major surface materials having basic maps, stock map, forest maps, the matic maps four categories, the above class diagram elements in the following Table 5-5:

(1) Base map. The base map is the basic figure is a woodland area, the compilation of stock map and a graph of surface materials for professional use. The main contents of basic diagram are administrative boundaries, roads, settlements, rivers, mountains, forests, the forest industry, forest notes, class notes, etc.

The basic map is based on the aerial plan or topographic map for the base map drawn, mainly in forest area as a unit. The scale is mostly 1 : 5000 or 1 : 25000.

(2) Stock map. Stock mapis based on surveys in Subcompartments, common charts for a draw in the forest.

Table 5-5 The main surface materials

Map name	Management and business units	Scale	Spatial scale/ha	Managing scale
Basic graph	Forest farm	1 : 5000 1 : 25000	10000 ~ 20000 in the North South < 10000 compartment General 3~20 North for 20	Sub-compartments
Stock map	Forest farm	1 : 10000 1 : 25000		Sub-compartments
Thematic maps	Forest farm/Forestry bureau	1 : 10000 1 : 25000		Sub-compartments
Forest distribution map	Forestry bureau	1 : 50000 1 : 100000	10. 15 million, 5. 1 million of boreal forest in the South of the North-South 100~200 <50	Forest tracts

Main contents are the small species, dominant species, age, size and other factors. The formal stock map is in color and representatives of various colors to distinguish the different types of classes and dominant tree species in sub-compartments.

The scale of forest phase map is 1 : 1000 and 1 : 25000 in China. The small class notes in the forest map generally use the fractional structure. The numerator is the small class number and the age class, the denominator is the status level (or status index, forest type, site type) and canopy density (or density).

(3) Forest map. Map is to reflect the Forest Bureau (county) the distribution of forest surface material, forest (township) units can also be drawn. Forest distribution map is based on the forest map is based on the reduction of the drawing; the figure reflects only Linfen and the above units of the resource situation.

(4) Thematic maps. Thematic maps are drawn on the basis of professional survey for professional use. The main types of thematic maps: soil map, the site map, the distribution of plant diseases and insect pests, wildlife distribution map, resource distribution etc. The matic maps for the forest (village, town) for the drawing.

3) Character material

Main including:

(1) Forest resource survey report.

(2) Professional investigation report.

(3) Quality inspection report.

5. 2. 4 The technologies for the second class forest resources management survey

5. 2. 4. 1 Space remote sensing technology

1) The development of Space remote sensing technology is widely used because of its advantages like big survey area, short survey period, and regular continuous monitor

With the development of space remote sensing, Quick Bird satellite image of Digital Globe in the USA, which resolution has reached 0. 61m on October 18, 2001, it met the need of decimeter accuracy.

On September 9, 2008, GeoEye satellite was launched and the resolution of GeoEye satellite image of has reached 0.41 m (panchromatic image) and 1.65 m (band 4 of the multi-spectral image). The WorldView-3 satellite which launched by Digital Globe company in August 2014, the resolution has reached 0.31 m (panchromatic image) and 1.8 m (band 8 of the multi-spectral image). China has launcher the ZY-3 satellite on January 9, 2014, its resolution of the image is 2.1 m (nadir view panchromatic image), 3.5 m (forward or backward view panchromatic image) and 5.8 m (4 bands of the multi-spectral image). This marks the China space remote sensing has entered the era of high resolution, acquisition of high-resolution remote sensing image has promoted the development of forest resources survey and monitoring.

2) Development of space remote sensing applies in forest survey

With the development of space remote sensing high resolution of big survey area in short survey, period has become the trend of image development. It takes technical support and update for forest resources survey and promoted real time dynamic survey and monitoring of forest resources.

At present space remote sensing technology mainly used in two aspects of the forest resources survey. One is the design of the project, including forest vegetation andland type classification and sampling with space remote sensing image, etc. For example, mechanical sampling in KM grid with space remote sensing image in National Forest Inventory; division sub-lot with remote sensing image and large-scale topographic map. In some province, high-resolution image and topographic map were used in systematic sampling and mechanical sampling in order to achieve the accuracy of sub-lot volume control and formed province plots, bureau plots, etc. It can enhance the accuracy of forest resources management survey.

Another aspect is an investigation of forest factors. The forest resources survey can be divided into two level: individual tree level and stand level, according to different objects. Investigation factors of the individual tree including stand tree height, diameter at breast height (DBH), the diameter of any transect, stem volume, crown width, biomass. Investigation factors of stand including stand average height, stand average diameter, stand volume, canopy density, stand biomass, etc. For space remote sensing image because of its resolution can't satisfy the requirement of individual tree investigation, so it's hard to use space remote sensing image to carry on the investigation of individual tree level. But according to the forestry professionals, scholars have long-term research found that the spectral information of images associated with stand volume, biomass, and other factors. Using the remote sensing image to retrieve these parameters, it has played a positive role in the rapid and accurate investigation.

From the 1970s, using the remote sensing image to retrieval stand volume and other forest investigation factors has become a hot spot. Parametric regression and nonparametric regression, neural network, image segmentation technology and so on are using in forest resources survey. The role of remote sensing technology in forest resources survey has been more and more recognized and used by people. Li Conggui (2006) using remote sensing and GIS information ground survey sample for the inversion volume estimation model. It is concluded that the canopy density played a very important role in volume inversion. Tong Huijie (2007) uses the band information of the remote sensing image and its derived data, terrain data and climate data, to establish the remote sensing inversion model of forest biomass. Ma Ruilan (2011) using SP3, SP (1-2)/(1+2), SP (2-3)/(2+3) three band or bands combination of SPOT5 image combined with spectral factor and slope gradient, slope, elevation established Forest stock volume estimation model of Huoditang , and their results are the relation between the number of 0. 782. Zheng G (2009) using the KNN method to retrieval forest volume based on sample sites level and stand level. He et. al use SPOTS remote sensing image, 1: 10000 terrain figure and ground survey data of image classification and evaluation, using BP neural network model built volume prediction model, and estimated the total volume of the vegetation of the Sun Yat-sen Mausoleum. Xu *et al.* using Quick Bird image to extract spectral factor, using DEM to extract terrain factors combined with ground survey factor, using BP artificial neural network method and the least square method. Fuzhou Minhou Baisha state-owned forest volume estimation model, accuracy could reach 88. 5% (BP neural network) and 85. 9% (partial least squares).

Estimation of biomass is another function of space remote sensing image. The traditional biomass measurement needs to establish the sample plots, sampling, weighing (including wet weight and dry weight), it takes a long time and high cost. With the development of RS technology, it provides a new method for biomass retrieval by means of remote sensing, especially for the research of biomass in a large area. The estimation of forest biomass using space remote sensing image and LiDAR is very precise and efficient. Hame *et al.* using Landsat-TM spectral information and ground biomass data to establish biomass estimation model. Dubayah *et al.* using LiDAR remote sensing technology for the estimation of biomass achieved good results. Since 21st Century, based on the mathematical methods, forestry workers have been using aerial remote sensing images and ground survey data to estimate and map the distribution of biomass.

In China, Zhanget. al proposed a model to estimate biomass by using the theory of statistical vegetation infrared radiation characteristics and the combined sensible heat exchange. Guo Z H (1997) through the forest in the west of Guangdong Province of coniferous and broad-leaved forest field measured volume, with seven bands of TM image data and the NDVI to analyze the correlation, the optimal spectral model is established using stepwise regression method, and estimate the forest volume. Li Jian (2005) according to the measured data of biomass combined with TM image vegetation data to establish the Poyang Lake biomass model, and in order to estimate the biomass. Tu et. al based on a former study of inversion factors, increased vegetation index factor, spectral information factors and meteorological factors such as to estimate yield model, for agro meteorological yield forecast, play a significant role. Huang G S (2005) based on MODIS image data to estimate the biomass of the three northeastern provinces. Li X analyzes the radar data of the Pearl River Mangrove Forest and combining the time series and the spatial variation characteristics of the estimated biomass. Gao Z H (2006) using TM image data to analyze the soil background spectrum based on the combination of the four vegetation index, the estimation of the biomass of low vegetation coverage was studied. Liu Z Y (2006) in Xilinguole Inner Mongolia natural grassland combined with multiple regression methods, the biomass model was estimated by hyperspectral remote sensing. Li et. al using TM images respectively, using the principal component linear regression, multiple linear regression methods and BP neural network method respectively in Beijing of coniferous forest and broadleaf forest to establish regression model to estimate the biomass of the entire city of Beijing.

In summary, aerial remote sensing technology is used for the inversion of volume and biomass with a large area, wide range, non-contact, and can be real-time dynamic monitoring. Reduce the workload and cost of the ground investigation in some degree. But the ground investigation is not completely abandoned at the present stage. So, how to use multi-source data more accurate quickly determined within a certain range of volume and biomass so that it can take the place of the ground survey and promotion in forestry field surveys, an aerial remote sensing technique in the survey of forest resources development direction.

3) The application of the traditional space remote sensing technology in forest resources management survey

Space Remote Sensing in the forest resources investigation and monitoring, mainly in the application of space remote sensing image, mainly reflected in:

Field survey sampling design. In the forest resources, survey sample has an important status, common kinds of the sample are fixed sample, random sample and so on. Sample selection is the priority for the sample survey, using the principle of stratified sampling, systematic sampling according to the principle of random sampling, combined with the topographic map based on space remote sensing image, for sample field survey prepared work.

Forestry division. Using remote sensing image, combined with the large-scale topographic map can divide forest region, compartment, and sub-lot. The smallest division of sub-lot can be of $2hm^2$.

Visual interpretation and area measurement. With the remote sensing image, land use types and forest types can be classified. It can also carry out area measurement, the relative accuracy of 96.8%.

Determination of some stands factors. Because of the limitation of space remote sensing, it can't estimate for all survey factors, but it has the advantage for the investigation and monitoring of some forest stand factors. Such as using the abundant spectral information for forest volume and biomass inversion, and then with ground survey test shows the effect is good, total volume sampling precision have reached the design requirements (85%).

5.2.4.2 Aerial photogrammetry technology

Aerial photography originated in 1849, Aime Laussedat a flight plan brought the emergence of this technology for all walks of life. The introduction of aerial photography to the forestry survey was introduced into the forestry survey in 1921, by Turski from the Soviet Union. Shilling low utilized by using aerial photographs of land when the military organization of the Commonwealth of forest resources survey, in accordance with a clear division of vegetation types, draw the boundaries of class to class diagrams and identifies the land type of use. With the development of computer technology, aerial photography technology developed steadily. In the seventies of last century, foresters began to draw on a variety of computer technology a study forestry thematic map. Kourt began the first step to establishing a complete rendering of the forest map. He began to study the use of aerial photographs in Canada to establish a three-dimensional model to identify species, stand tall, and canopy density and other factors and draw boundaries and compartments into a thematic map.

In the 21st century, with the rapid development of computers, the development of aerial photogrammetry technology, image technology, geographic information systems (GIS) and a variety of professional mapping software, the use of aerial photogrammetry techniques to obtain images of a higher spatial resolution can be analyzed based on the aerial images. Zheng et al studied individual tree crown structure and spatial structure using a color aerial images using the two-dimensional digital form on the spatial relationships. Gougeon F A proposed the use of a valley-following algorithm to identify the individual tree from aerial images with high spatial resolution, having a high accuracy. Erikson M studied on the canopy of trees segmentation method to extract information based on aerial images, presented a detection function simultaneously in the space domain and color domain. Separating the irregular shapes of the crown canopy to extract information, and applied this shape into species classification, proven an accuracy of more than 93%.

The development of the application of aerial photogrammetry technology in China is relatively late in the development of forestry. In 1953, China began to use aerial photogrammetry technology to investigate the forest resources. 1963 is the first time using aerial photography technology completed the national forest resource inventory and the mapping of the China's major forest distribution map and forest type map. 1973—1977 continued to use this technique, doing the second national forest inventory, inventorying forest stand volume of national, provincial, municipal, state and county forest area as well as drawing forest type map. 1977—1981 continued the third national forest resource inventories. The inventory work took the country as a unit and established a fixed sample inventory system. After the three National Forest Resources Inventory works, photogrammetry techniques are becoming increasingly sophisticated, have the advantages of updated timely, low cost, high accuracy, and others, overcomes requiring a lot of manpower, material and other defects.

The research technology becomes main technical means of monitoring the presentsituation of forest resources and their dynamic changes.

The resolution of the remote sensing image includes four aspects: spatial resolution, temporal resolution, spectral resolution and radiometric resolution. But in general narrow sense, it has an only spatial resolution. With the development of aerial image measuring technology of and the rapid development of digital image processing technology, the spatial resolution of aerial images is getting higher and higher. Forestry workers tried to use aerial images of forest areas for a variety of visual interpretation and visual interpretation. According to color, shape, and so most directly reflect the image of aerial images to measure factors out of the forest, forest and forest management based on the specification be divided into classes and draw the outlines. Traditional factors determining tree measurement method is the use of a magnifying glass, mirrors and other optical stereoscopic imaging instrument to expand the species, canopy, canopy density and the estimated accumulation amount and biomass, etc., or by means of three-dimensional measuring instrument, which is in a simulated three-dimensional environment based on tree height to measure tree height, average stand height, crown width, average crown width parameters.

For an aerial image, the most direct reflection of the forest is the information of canopy. The tree crown is the most important part of the tree and is the main place to obtain energy through photosynthesis. Crown size, shape, and area also affect trees growing and the biomass can be estimated through the canopy information. Therefore, the study of tree canopy is very necessary.

In recent years with the use of aerial photogrammetry technology to obtain the spatial resolution of the image data is constantly improved, the description of the features is more and more detailed.

But with improved spatial resolution, similar feature shown more complex texture features. Texture refers to the recurring images of surface features and characteristics of performance out of the rules, describing the spatial variation of the pixel brightness characteristics. With the development of computer technology and image recognition technology, the traditional texture feature extraction methods like GLCM, spatial autocorrelation wavelet multi-channel and similar are very common. Scholars at homeand abroad study a lot of information extraction canopy for the use of high spatial resolution aerial imagery. Cai W Fand Li F R established stereo pair's use of the 1: 10,000 aerial images through the digital photogrammetry software VirtuoZo. The use of IGS in the three-dimensional model was measured individual tree crown diameter things and North-South crown diameter. Verified crown diameter and crown diameter measured field correlation coefficient obtained by the measurement of digital photography can reach more than 85%. Wang Erli et al. (2011) use aerial images with a spatial resolution of 0.2m from Liangshui National Nature Reserve, which was extracted for regional pine, larch and spruce, adopting classification, visual interpretation method in conjunction with a circular model its crown shape. After the measured field data were analyzed, he used the method described above extracted crown diameter and crown diameter measured to establish a regression model, accuracies of up to 83.50%, 84.35%, and 82.26%. Wei et. al used light small aerial image data from Shangcheng of Henan and image parameters such as spectrum, texture on the surface of objects classification to multi-scale image segmentation, extracting individual tree crown diameter information, comparison and field-measured data. The extraction accuracy is up to 72%. Stand average diameter up to 87%. Wang Jia et al. aimed at the current situation of the main access to forest parameters characterized by a heavy workload measurement outside the industry caused by the low efficiency, took light and small airborne remote sensing system as a tool to get the LIDAR point cloud data. Preprocessed and extracted individual tree crown had been done combining with height and diameter from field investigation, taking Chinese pine plantations in the study. De-noising, classification, extraction process had also been done to obtain single tree height data. Aerial imagery data obtained pretreatment, matching stitching, segmentation, and extraction to obtain a single tree crown data, along with field sampling of individual tree height, diameter at breast height building regression models, and verify the accuracy of the model. The results showed: tree height extracted by LIDAR point cloud data has highly significant correlation with measured tree height. The model prediction accuracy established up to 97.5%. Diameter crown by measured and image extraction also has a very significant correlation, with a prediction accuracy of 91.6%, basically meet the requirements of forestry production.

Using aerial photogrammetry technology acquisition parameters compared with trees close-range photogrammetry, with the expansion of the area of photography, not only can get the entire plant standing forest within the tree measurement information but also increases the amount of information available. You can also obtain information stand as well as to estimate the entire forest accumulation, biomass, and other quantitative parameters, with broad application prospects in forestry.

With higher spatial resolution aerial imagery, many scholars did a lot of exploration and research on the use of canopy extraction from high spatial resolution image information, especially for single-crown-scale wood extraction and estimation and measurement. The basic obtained information to establish the crown diameter-crown model, combined with the establishment of a high tree stand volume models, measured to obtain tree factor rapidly and accurately, offering a new vision for the application of aerial remote sensing in forestry.

Application of aerial remote sensing technology in investigation of forest resources monitoring are two main areas:

Air Remote Sensing Technology in Forest Resources Monitoringthere are two major aspects:

(1) Sub-compartment divisions.

(2) Determination of partial stand factors.

Given the limitations of technology resolution aerial photography and other technologies, in areas with low canopy density, tree height, crown diameter, canopy density, stand density and other parameters can be directly measured, followed by modeling combined with the field survey data, used to calculate other related stand factors. In the high canopy density of stand canopy density of stand, estimation effect is good, the accuracy of the other factor is limited (Table 5-6).

Table 5-6 The forest survey factors air remote sensing technology can achieve

Parameter	Avg. relative accuracy/%	Technology and platform	literature
Tree height	95.95	Unmanned remote sensing aircraft	Li Yuhao, 2008
Volume	96.41	Unmanned remote sensing aircraft	Li Yuhao, 2008
Stand average height	83.00	Light small airborne LiDAR	Wei Xuehua, 2013
Individualtree height	88.00	Light small airborne LiDAR	Wei Xuehua, 2013
Number of trees per hectare	84.00	Light small airborne LiDAR	Wei Xuehua, 2013
Individual tree crown	72.00	Light small aerial remote sensing images	Wei Xuehua, 2013
Stand 87.00 average crown	87.00	Light small aerial remote sensing images	Wei Xuehua, 2013

5.2.4.3 Close-range photogrammetric technology

As a branch of photographic measurement technology, close range photogrammetry technology generally within one hundred meters of objects within the scope of photography to take pictures. Then collecting the image processing operation by changing the computer image processing technology combined with spatial mathematical model, which is analyzed the object taken shape, size and trajectory, ultimately, to obtain the required information. Photogrammetry has experienced nearly a century of development, from analog photography to measuring the development of analytical photogrammetry. Now, digital photogrammetry has also undergoing a change, but also from camera film camera development of today's digital cameras.

Currently, the camera for close-range photogrammetry measurement type is mainly divided into non-metric cameras and camera while both. Measurement type in the frame of the camera is provided with a specific standard frame, wherein interior orientation parameters are known or can be recorded directly, the focal length is divided into fixed focal length or adjustable focal length, lens distortion is small. Rather than measuring camera is not designed for close-range photogrammetry specially designed formulation, there is no specific measurement function of the camera, generally refers to an ordinary camera, interior orientation, focal length and other parameters are unknown, relative to the objective lens distortion larger. But with the deepening of close range photogrammetry technology development the interior orientation, focal length and other parameters are available through the camera calibration techniques to solve it.

Development of close-range photogrammetry mapping technology is used in the building of integrity ground photogrammetry photography rendezvous marked the beginning of the 18th century French proofreading A. Laussedat. 19th century, Alberhct Meydenbuaert first proposed "photogrammetry" term and became the beginning of photogrammetry technology development. In 1960, Japan adopted the method of measuring the photographic scene investigation to solve the problem of traffic accidents, this method not only by measuring various parameters of the photograph, but also as data was archived. In the 1990s, the development of three-dimensional reconstruction technology has become the focus of close-range photogrammetry, making the leap in pavilions, grottoes carved renovation and other construction and archaeological work has been qualitative digital photogrammetric system for landslide detection, the results obtained by the test method is lower than the electronic theodolite virtual photo law about 5 mm. Moreover, subsequent measurement and the processing of the measured shape of the housing industry car, check the processing quality and large mechanical parts and small parts assembly quality has been the development of bio-medicine dental, orthopedic correction technology also has been mature development.

Domestic close range photogrammetry technology development started late, from before the 1970s really began to study photogrammetry techniques. Wang Z Z for Academician in his monograph "photogrammetry" a book on the relative orientation of foreign experts formula is not ideal for mountain place a higher precision formula. The book represents the level of Photogrammetry theoretical work, that has a great influence on both inside and outside. Wang G H *et al.* proposed a no fixed station's close-range photogrammetry. The method that is not necessary to determine a fixed station, nor need to set as the control point but need to set a few basic surface measurement scale, then on the subject matter and arbitrary scale and photographed. Then the pixel coordinates of the image side by modified P-H algorithm into an object space coordinates, to finally determine the spatial coordinates of the object being photographed.

With "precision forestry" and proposed to promote and speed up the trees to obtain accurate information about the development of technology, and therefore requires a new, fast, accurate measurement of forest trees factor approach. A digital camera as a measurement tool into a forestry worker's horizons, which is a non-contact measurement, not only can it based on the theory of photogrammetry and image recognition technology to extract the standing image information and solver stumpage various parameters but can also shape the image study tRUNK research. The image data archiving reserved for future dynamic monitoring data as the most solid foundation.

Close range photogrammetry technology in forestry in 1952, Marsh first proposed the use of terrestrial photogrammetry tree trunk diameter measurement method, although there are some preliminary results, but the results were not satisfactory, the error was larger. In 1963, Grosenbaugh used remote computer technology the dashboard and the standing timber volume meter readings and calculated binding volume. The results showed that within ± 4% volume error obtained. Ecrielle Tbaof, rB. S using non-metric camera for loblolly pine stereo photogrammetry to study the three-dimensional crown variation, the results show that the measurement accuracy is better. In 1998, Juujarvi proposed ordinary digital camera to measure the diameter of the plant standing method, and by measurement of Scots pine, the results show that the method is effective and feasible. Wang X M (2001) developed a set of advanced photo processing forest survey system. Notebook systems in the field does not need to be required to make any measurements of forest surveys, just ingested a certain number of photos can be obtained tree canopy diameter and other information, after testing the accuracy of the comparison result was better. Wu X L (2008) proposed the use of close-range photogrammetry techniques can be non-destructive monitoring of standing tree, and the study of stem form provides a graphical basis, and laid a solid foundation for the dynamic monitoring of standing trees survey information lossless files. Clark use of close-range photogrammetry technology to develop automatic measuring trees System (TMS) model tool, which is a tool to quickly capture dimensional standing information. We can fully and efficiently collect tree measurement factor information for Image Processing in the "precision forestry" application to lay a solid foundation. Yang Huachao (2005) proposed a method suitable for measuring the plant standing on 2D imaging equation method and this method can be felled without precise calculation of plant standing wood volume. Cao M L has proposed the use of 3D camera measurement information obtaining individual standing, standing for plant and tree DBH high average accuracy can reach 97. 88% and 93.09%, this method of close-range photogrammetry in forestry development step on a new level.

5.2.4.4 Traditional ground surveys

Traditional Forest Resource Inventory, after the sub-compartment divisions need to field surveys, its methods are:

1) Plot survey

Within the geographical scope of the investigation, the use of sampling to sampling, and then sub-compartment. Among them, there are random sampling, mechanical sampling, stratified sampling and other samplings. Likewise shape of round, rectangular, square, etc. Found in the plot of the Stand survey factor, according to mathematical and statistical methods used to estimate the overall value of the measured sample corresponding values calculated after sampling error, accuracy, and other factors. Sampling may choose reliability, observed changes in the value of the coefficient, a measure of error factor requirements.

2) The standard method of investigation

Within the geographical scope of the investigation, the results of the standard method of investigation estimated the overall value. Standard ground generally rectangular shape, ribbon and so on. Prior to the establishment of standards, and then measured, with the results estimated accumulation amount and other units within the sub-compartment size and the entire range of sub-compartment factor.

The key to reduce the error using a standard survey method is selecting criteria as possible " standard" that is the standard way to have the status of the overall survey average. In addition, the area of standards must be guaranteed a certain amount, if the area is too small it is difficult to ensure that the selected samples representative. Then it will have a large deviation when estimating overall. However, the standard area of increased workload over the General Assembly to increase the investigation time andoperating costs, also the standard layout should try to avoid systematic errors.

3) Visual investigation

In the state of forests relatively simple case, the choice of investigations by investigators with this side assessment.

4) Angle gauge investigation

Use basal area per hectare by factors such investigations angle gauge with high work efficiency, sub-compartment perspective in good condition, investigators in the case of relevant experience, can use this method. Angle gauge constant should be selected depending on the size of the trees. Angle gauge layout should follow the principle of random points, avoid systematic errors and errors forest edge.

Sub-compartment in the field survey, the problems of these methods:

Measurement visual method randomness, and with the investigators and different experience, required more people, so in the field survey, the visual method should be used with caution.

Great angle gauge operations and the observer's experience, habits, visual relationships, more particularly common errors caused by the use of angle gauge.

5.2.5 Second class survey precision

Class II quality measured by the accuracy of the outcome of the investigation and the precision error values can be used to evaluate. The error is smaller, the higher the precision. Accuracy based on forestry production requires the establishment of a standard precision. How to compute, measure, to improve the applicability of forest inventory information is a very key issue?

5.2.5.1 Area precision

1) Accuracy of the sub-compartment area

Sub-compartment survey is an important content of forest resource investigation, statistical analysis of forest resources is based on sub-compartment area accuracy. So, how to assess the area of sub-compartment to meet the requirements?

When measuring sub-compartment area determination accuracy, usually measured sub-compartment area as the standard. Only the measured compartment when an error is smaller, in order to obtain reliable results and measured in sub-compartment area of the main sources of error include errors, such as quadrature error. Practice shows that the compass wire measured small border area of the calculated error is relatively small, 100-90 are in sub-compartments area error value is less than 1%, can be used as assessment criteria.

Verification found that relative error of sub-compartment area with sub-compartment size the size of the change, the larger relative error smaller. Use zoom in map calls directly into maps: a compartment in the 30 hm^2 above, the relative error is within ± 10%; compartment 15~30 hm^2, relative error is within ± 15%; compartment below the 5 hm^2, relative error exceeds ± 20%; when you use the aerial tune into maps, precision can be increased a three.

2) Forest area of the main class accuracy

Issued by the Ministry of Forestry of the technical regulations for forest resources investigation of the 18th article, based on the farm unit, predominantly in the area of forestry land are three levels of accuracy: class a is 95%, b is 90%, c is 85%. The above requirements are suitable for several sampling methods to determine the area of land throughout the whole class. When doing the sampling method within a population all over the area, first determine the number of sample units. To ensure the accuracy of such estimates are required, with a minimum overall percentage to determine the number of sample units of the class second class area around the estimated. Sample around the class unit weight and the ratio of the total number of sample units is calculated around the area of class; the third is to determine estimates of error, errors in the application of the sampling error of the formula of land area.

Major forest forestry land in the area should meet the three requirements for accuracy. Percentage of sampled population is not a forest but when within the forest is divided into more than one general, respectively, in the forest area of a major class of accuracy.

5.2.5.2 Stock accuracy

Forest stock volume measurement and visual survey methods, measured and divided into standard and sample survey. Factors for the determination of diameter at breast height, tree height, check volume table to get individual standing tree volume. Tree accumulation per plant per unit of area and volume per unit area can be obtained, multiplied by the small area out of the total volume in sub-compartments. Therefore, the determination of standing tree volume error factor tree diameter at breast height, tree height measurement and volume table errors and the first two elements by the operator effort. As stipulated in the technical measure excluding gross and systematic errors, errors still exist but the volume table volume is a volume table of error per unit area, is the second accuracy per unit area. Relative accuracy, the relative error is more than 1/100 per unit area, less impact on the volume is negligible.

1) Standard to test the accuracy of volume

Those commonly used standard method of test volume are set in a forested class measurement with a standard volume of the three-wide and small volume. Major errors of this method are standard size error, sub-compartment area error. Based on error propagation theory, measurement error in the function calculation formula is:

$$m_v = \pm \sqrt{m_x^2 + m_y^2 + m_p^2} \qquad (5\text{-}12)$$

Error of $m_x = \pm 3\%$, standard $m_y =$ area error $\pm 1\%$, sub-compartment area error $m_p = \pm 10\%$, into the formula to be $m_v = \pm 10.5\%$. That is stand volume accuracy of 89.5%.

2) Sampling method for determination of stand volume accuracy

After 70, class II investigation to forest volume are determined by the sampling method, precision implementation units as fields or forest. On the existing technical regulations are replaced with the sampling accuracy of stand volume, but can instead of the actual error of the volume still needs discussion. In order to further explore the sample volume and the difference between truth and error should be taken into account in calculating the volume accuracy volume table, area error or sampling error. Error is:

$$m_v = \pm \sqrt{m_x^2 + m_p^2 + m_s^2} \qquad (5\text{-}13)$$

Error $m_x = \pm 3\%$ volume table, area error $m_p = \pm 10\%$ $m_s = \pm 10\%$ volume sampling errors, bring into the equation $m_v =$ available 14.4%. That is stand volume accuracy of 85.5%.

If forest units should also calculate the precision of forest tree farm for 5 overall, each overall error of $\pm 14.4\%$, calculated using the following formula:

$$m_v = \sqrt{\frac{m_l^2}{n-1}} \qquad (5\text{-}14)$$

Into the available $m_v = \pm 16.2\%$.

3) Volume accuracy evaluation

From the above calculation shows that accumulation of errors per unit area cannot be considered as the accumulation of errors in the overall area, you must include the size difference error volume. Stand volume error, including tables, size error, a difference of about ± 5%. In order to ensure that stock volume is accurate to 85% ~ 90%, rises the area and sampling accuracy ± 5%, in order to achieve the desired effect.

5.3　Forest manager's three kinds of investigation

5.3.1　Forest manager's three kinds of investigation overview

Three types of the survey are also called job design survey. It is the basic unit of forestry to meet the needs of cutting area design, cradle cut design of the investigation. The volume of trees and the quantity of material is planted to make accurate measurement and calculation. In the process of investigation wood cutting need to be registered, according to the size of the survey area and the homogeneity of forest stand can be measured using the whole forest or standard survey method.

Forest design survey research is used to determine field operations and to obtain the theoretical methods of forest resource information. In fact, it is to work for forest resources in the field of quantity and quality evaluation. How to obtain reliable information on forest resources is very important, because the forestry production must be carried out according to the substantial law and to achieve the purpose of reasonable development, management, and sustainable utilization. Forest management is continued during doing job design investigation. There are three kinds of the survey in forest resource survey, job design and investigation. It is on the forest resources investigation (i. e., the forest manager survey), on the basis of it is the basic unit of forestry to meet cutting area design, cradle cut design, and other services. Forest type of job design investigation is final felling, tending felling, forestation, and promote regeneration and low yield transformation such as job design survey. The aim of forest operation investigation is to use the lowest cost, for obtaining the fast reliable and sufficient information, in order to meet the needs of the job design or production decisions. Forest operation design research task is to find out, the forest on the precision work site quality and quantity of change characteristics and its change rule; investigation result can objectively reflect the real situation on the ground, at the work site of forest resources to carry on the comprehensive evaluation, put forward the comprehensive accurate forest resource survey data, design data and drawings. Survey materials including working in the cutting area and updating sub-compartment resources current situation, the condition of plantation etc, according to the material in the field technology design of cutting area.

Design research field work is mainly to obtain the working in the cutting area and the condition of sub-compartment resources, the condition of the plantation and update the information to meet the requirements of job design. Job task is to draw the design all kinds of cutting area in the industry of floor plan, statistics, and calculation of resources.

The characteristics of forest operation design investigation require high precision, cutting area accuracy requirement for 1/200, a sub-compartment is 1/100. It is necessary to carry out the survey in sub-compartment with the 95% reliability of volume estimation in sub-compartment ($t = 1.96$), 90% accuracy requirements in order to meet the requirements of forestry and timber production. The forest design survey has an important role in the rapid and accurate access to forest resources information, forestry production decision-making is the basis for the correct decision-making of forestry production, rational forest management, development, utilization and to achieve sustainable development.

The forest operation design investigation is the first step in forestry production site construction, are the bases of reasonably organized production, it is the important mean of checking the management effect and forestry management. Through an understanding of the characteristics and laws of forest resources, the potential productivity of un-used forest production grasps the change and its law of forest quality.

5.3.2 Significance and characteristics of three types of investigation

The purpose of the three types of survey design is to investigate the forest resources in a cutting area or area having some tending operations to investigate the amount of timber, growth conditions, forest structure and laws. Three types of investigation are the grass-roots investigation in the forestry sector to complete the task of cutting work, development of forestry material, logging targets. The design of the three types of the survey is based on the forest cutting quota, timber production, tending, low conversion, plantation, cutting and afforestation tasks approved by the competent forestry authorities at the national and higher levels. Based on the recent nationally approved forest management program and the 'Forest Law' and other laws, regulations and operating procedures, the qualified professional forest investigation team conduct an investigation. The three types of surveys are designed to serve timber production and silviculture in the grass-roots forestry bureaus and have the legal effect given by the state.

5.3.3 Methods of forest job design investigation

The forest operation survey design is a pre-job survey of forest production, which aims to obtain pre-operational data for rational design and construction. At present design survey of most, the forest areas in China are mainly is the whole forest measurement method or 10% of the strip plots survey method. The former is mainly used for the final felling job design, which is mainly used for raising logging. This method is of high volume, cost. In order to improve the work efficiency, job design survey can be used in the aerial photograph and ground combined sampling survey method.

5.3.4 Kinds and content of forest operation design survey

Operations of survey design survey can be divided into two categories, namely the final felling operation design and job management design. See below specific classification (Figure 5-2).

Forest job design on the basis of the theoretical investigation on deforestation and update, follow on the basis of forestry sustainable utilization policy, combining with the current rules, regulations and other relevant provisions of design. The design method of general adopts the method of aerial photograph combined with the ground.

An aerial photograph is mainly used for cutting area planning, design, and measurement. They are an important part of the conduct cutting operations in design survey. Before field in indoor according to the principle of regionalization and the necessary technical data, in the stereo images of aerial photograph export cutting area, area, sub-compartment boundary line and branch line, skid trail lines, and determine the position of the loading field.

The interior districts after division design field. It is an area indoor design results, application of aerial photograph accurately calibrated to the ground. In situ division, main content is made the cut lines and buried, and so on. Simultaneously in situ survey and mapping are used to determine

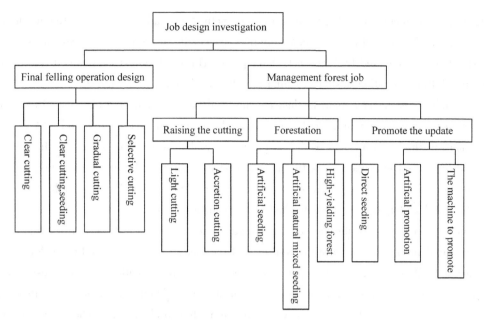

Figure 5-2　Job design investigation content

the sectional area.

In situ districts after job design investigation. Survey method mainly adopts the whole forest every wood survey method, the ribbon sample plot method. For small plantation age, average height, diameter, crown density and update, gradient, slope direction, slope position stand factor must be a detailed investigation, so that the design in the industry and provide a scientific basis for a production decision.

5.3.5　The development of forest operation designs survey technology

Job design survey technology development is with the development of forestry production and other subjects and developed. Job design from countries around the world, forest survey technology development process can be divided into three stages:

1) Visual inspection stage

Before the 18th century, the development of capitalism in the early sales of mountain forest, this period of forest investigation did not form a complete system, only visual investigation, and give priority to in order to estimate forest volume. By the 18th century, visual inspection of the whole forest volume method is to survey area is divided into several partitions, with a visual method to estimate volume per unit area, and the sample wood cut down the measured the volume of the earth to correct visual survey results. The advantage of this method is rapid and low cost.

2) The measured phase stage

In the early 19th century, the tree measurement technology is developing rapidly, the forestry workers began using tree height, DBH, shape and volume of the relationship between tree species stocking table is established. As the demand for forest products from fuel wood to timber, visual stand calculates the whole forest volume small area already cannot satisfy the production practice, the visual measurement method by the measured method instead. At present, the whole forest measurement method is still used in some countries, especially to the precious forest of job design. The method is characterized by high cost, slow speed. The whole forest every wood questionnaire survey final felling operation design in China is still used in most of the forest.

The whole forest measurement method can be used for the small area of forest. But for large areas of forest volume the cost is high, more than the typical standard earth and ribbon sample measurement method. To investigate the proportion of real measure, the greater the higher precision is considered. The proportion of the measured size at the time by experience judgment, such as Europe and the United States has banned measured by 5% ~ 10%, China had adopted overripe 5% of the standard for the measured method belongs to this period.

From the visual stage to the measured phase, accuracy has improved a lot. But subjective decision measured proportion increase the workload of the measured, usually need to explorea perfect survey method. With the rapid development of the theory of mathematical statistics, especially its application in forestry investigation makes this contradiction. Mathematical statistics theory provides the optimal design of forest survey, with minimum effort to obtain the highest accuracy, make the minimum workload, the highest precision, it marks the forest survey has entered the phase of sampling survey.

3) The sampling stage

Sampling survey of forest resources began in the 1920s, the first application in Norway, Sweden, and the United States, 1930 – 1950, gradually spread to other countries. In 1970s sampling method was applied to the job design investigation. Mainly used to simulate the system sampling survey method, in the cutting area of the main and tributary according to certain distance setting sample plot (or analog) used to estimate the small volume method.

Sampling survey method compared with the standard ground survey, avoid the head deviation and sampling scheme is simple, with the continuous development of sampling theory and method and improvement of increasing forest survey precision, more diverse research methods.

The end of the 19th century, with the development of the forest aerial photography, a German forester's bolt put the balloon to gain aerial photograph. In the 1920s, aerial photography is applied to the forest survey; promote the development of forest survey technology. After gradually spread to the United States, Africa, Canada, Australia and New Zealand and other countries and it was improved. Investigate the application of aerial photograph can be obtained by various categories of land area. Were recorded in a considerable aerial photograph provides a feature image feature in photography moment live, reflects the seen and features of live. From photographs can interpretation on all kinds of forest survey factors, such as level of trees, forest age, position, crown density and other factors, and the use of an aerial photograph of the figure, all kinds of surface materials.

After the 1960s, due to the development of the computer, aviation remote sensing and statistical science, forest survey technology development to a new level, formed in forest survey, computers, and the combination of remote sensing investigation method, the formation of the forest survey automation system.

In the early 70's the forest survey process automation development direction and forest survey automation system gradually formed. In the early 80's the forest operation design of survey data processing in the industry development in the direction of the automation was occurred. In the late 90's forest job design automation system in the field was introduced. Data processing in the industry from manual calculation to the computer automatic processing, greatly improve the quality of the forest operation design investigation and speed. This method greatly reduces the cost, reduce the labor, improve the quality of the survey, make the job design investigation has reached the standardization and normalization in the industry, promote the development of the forestry production.

Are the theoretical basis of forest operation design survey of mathematical statistics, it is based on probability theory to design work investigating the best solution, with the lowest cost to achieve the highest accuracy, or accuracy of the established can lower the cost. In accordance with requirements of the production and research, based on the principle of design and guiding ideology the development of forest operation design sampling design process is as follows: the forest every wood gauging to fixed area of the sample area, point to 3P sampling, sampling simulation sample.

5.3.6 Sub-compartment cutting material design

The traditional method needs to be considered in determining the material rate of different tree species the planted out material rate. Material rate refers to the quantity of material is planted the material of the ratio of the corresponding total volume. With the skin material rate refers to the quantity of material is planted the material of the ratio of the corresponding total volume with the skin. In the forest, survey works it usually needs to take economic material yield grade into consideration. The material level according to the material volume often determines the percentage of stand volume level. But in order to simplify the determination of work, can also according to the material inside the forest tree number and forest determine the percentage of the total number of grades. Studies have planted out material rate can be determined planted out material structure regularity, right quantity and quality of comprehensive evaluation of forest resources, reasonable forest management to provide an important basis for accurately.

But traditional indeed set material ratio method to determine material, in the actual operation will produce a lot of inconveniences, therefore put forward a kind of stand out without considering material can be obtained rate method is necessary. Put forward a kind of to set the angle theta control sampling observation about 10 trees infer that stand out the material rate method, do not need to consider the different tree species, as only a single material yield of wood can calculate the stand.

The value of wood is that it can be used to directly use the wild plant resources value, it can be directly exchanged as a commodity in the market of the product value of resources, therefore, to study how the wood value maximization has particularly important practical significance. Certain value of wood is the main standard of surveying, according to the standard results of the survey, the diameter order statistics each tree species (different tree species with different diameter order level the market price) in the forest community types of strains, and use the local volume table Chad a each tree diameter order and calculate the volume, according to the formula to calculate value.

This method by super station instrument for the tool, select about 10 in the cutting area of trees, large, medium and determined by measuring the path of material on the number of material, build the price of the corresponding matrix, realize the maximum value of wood cutting.

5.3.6.1 Stand out material ratio calculation principle

In mini superstation instrument, choose a vantage point in the cutting area sub-compartment, select 10 or so tree as counting wood of sampling observation. The count number, diameter at breast height (DBH), tree height, and diameter at any point, we can also calculate the whole stand out rate of material and waste rate (Figure 5-3).

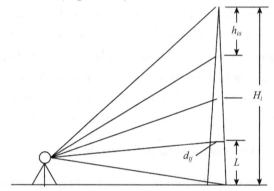

Figure 5-3 Measurement principle diagram

In the cutting area of the forest far away from the forest edge that can represent the average state point as the observation point. To set the angle theta control selected 10 or so tree as counting wood of sampling observation, as $I(I=1, 2, \cdots, r;$ general r 10 or less) and to measure the count of wood DBH $d_i(d_i > 5$ cm, otherwise can't counting as wood); For $k = (50 \text{ theta}) 2$, the density of wood for the selected count:

$$n_i = \frac{4k \cdot 10^4}{\pi d_i^2} \tag{5-15}$$

Thus calculate the stand density (unit: plants/ha) as follows:

$$N = \sum_{i=1}^{r} \tag{5-16}$$

Put the MINI superstation instrument in observation point, first measure the height of each count wood tree H_i again from the base of standing tree and up according to the material length L integer times $(L, 2L, 3L, \cdots, sL)$ in turn up small head diameter measuring each log section d_{ij} ($j = 1, 2, \cdots, s$); according to different small head diameter d_{ij}, wood section can be divided into large size order material (d_{ij} 26 cm or higher) and pitch diameter order material (20 d_{ij} < 26 cm or less), trail order (6 d_{ij} <or less diameter 20 cm), and statistics the order material number; When d_{ij} 5 cm or less (dis) for small diameter at this time, regardless of the material the wood and tree trunks above according to the waste treatment, to calculate the wood pitch the height of the tree.

According to the measured data, we get the quantity of material stand always V the general calculation formula is:

$$V_{total} = \sum Nad_i^b \tag{5-17}$$

Type of b and a for volume equation coefficient; will waste some approximate cone, forest waste volume V waste calculation formula is:

$$V_{waste} = \frac{\pi}{12} \sum Nd_{is}^2 h_{is} \tag{5-18}$$

According to the quantity of material and waste quantity, get the calculation formula for material rate Q stands for:

$$Q = \frac{V_{total} - V_{waste}}{V_{total}} \cdot 100\% \tag{5-19}$$

W, forest waste rate calculation formula is:

$$W = 1 - Q \tag{5-20}$$

5.3.6.2 Timber value calculation principle

The income of different tree species according to the diameter size is divided into large diameter d_1, large diameter d_2, d_3, d_4 trail order in diameter, build price matrix:

$$\begin{array}{c} \begin{array}{cccc} d_{j=1} & d_{j=2} & d_{j=3} & d_{j=4} \end{array} \\ \begin{array}{c} i=1 \\ i=2 \\ \vdots \\ i=m \end{array} \begin{pmatrix} C_{11} & C_{12} & C_{13} & C_{14} \\ C_{21} & C_{22} & C_{23} & C_{24} \\ \vdots & \vdots & \vdots & \vdots \\ C_{m1} & C_{m2} & C_{m3} & C_{m4} \end{pmatrix} \end{array}$$

Type C said the price, I said trees species (1 I m or less) or less, j said categories (1 j 4 or less) or less;

Through observation records N_{ij} density of different tree species and DBH d_{ij}, build order matrix density-tree-diameter:

		d_1	d_2	d_3	d_4	Σ
$i=1$	N_{11}^0	N_{12}^0	N_{13}^0	N_{14}^0	N_1^0	
$i=2$	N_{21}^0	N_{22}^0	N_{23}^0	N_{24}^0	N_2^0	
...	...	...	...	...	...	
$i=m$	N_{m1}^0	N_{m2}^0	N_{m3}^0	N_{m4}^0	N_m^0	
Σ	N_1	N_2	N_3	N_4	N	

Through the model $V_t = \sum_{i=1}^{m} \sum_{j=1}^{4} C_{ij} N_{ij}^0$, computing stand maximum value of the timber and according to the observation of DBH d_{ij}, get DBH growth Δd_{ij}, build price matrix-tree-diameter and density matrix, maximum value calculation next year's wood V_t+1, at this time, $\frac{(V_{t+1} + V_t)}{V_t} \leq l$, 1 said the annual benchmark interest rates.

5.4 Forest parameters inversion through remote sensing image pair

Limited by vision, angle, shade, non-metric Camera can only be used in a small piece of forest based on terrestrial photogrammetry technology. It is difficult to meet the rapid for investigation and management of the large area. With the development of aerial photogrammetry and thecompletion of relative hardware and software products, aerial photogrammetry technology is widely used in forestry. It has a lot of advantages such as quick mapping technology, high precision, high spatial resolution and time resolution, not limited by climate and season etc. At present, many scholars have researched the application of aerial photography technology in forestry and resulted in many modes and methods for forest resource investigation. Nowadays, the spatial resolution of aerial remote sensing image can reach 0.16 meters, it is relatively accurate to obtain the crown, tree height, stand density, and other factors in relative high-precision level. This technology could replace ground compartment division and forest investigation, and can greatly improve the efficiency of the forestry survey.

Zhou S F by used high-resolution aerial images with the method of multi-scale segmentation, according to different sexual set different segmentation method and scale information to extract the crown. Xu Li Gong using cold water nature reserve of the aviation image of the 193 aerial triangulation and carries on the orientation is corrected for the 1 meter of the DOM. From the forest through the interpretation to the class division in images and use the orthogonal projection extraction tree as advanced research. Aerial photograph of the three-dimensional model is utilized to extract high tree, comparing with the ground experimental data and the extraction accuracy can reach 93.96%. Wu Aibin area, the peak bears a natural park of 106 small to test and use of RC plane carrying a digital camera to obtain high spatial resolution image to extract the corresponding angle of small-scale, mixed degree index, etc. Crown competition for the extraction of stand spatial structure of high resolution remote sensing image to provide a reliable basis. Zhao Jian used the cool water of aerial photograph of canopy and tree height were the extraction, extraction accuracy can reach above 70%. Air has a very high spatial resolution remote sensing image, image features and to class is clearly visible, can according to image interpretation of color, shape, texture, and the forest quarter divisions and Sub-compartment divisions. According to the spectrum and texture of images to extract the crown of a tree, the tree Height, tree parameters and stand density, average crown breadth of parameters, such as, whether can replace the forest resource survey requirements. This chapter will be taken by aerial photography system to obtain high spatial resolution aerial images, studies its application in forestry in automatic extraction method of forest parameters.

5.4.1 Extraction of Individual Tree Height Using Aerial Image

Tree height is an important indicator of forestry investigation management that reflects the individual trees and looks like the whole forest health. Fast accurate measurement of height is for the entire forest resources investigation has important practical significance and application value. In field investigation to determine the tree height, often can meet no precision measuring equipment, have experience of forestry workers are usually use visual method or by using a stick and trees to form a similar triangle method under test to measure tree height, or using the shadow or the length of the shadow of similar ratios calculated tree height. But the above methods require workers experience, it is a convenient and quick method, but the error is bigger also. To accurately determine the forest tree height, in addition to the traditional compass, Feng Z K(2007) put forward an instrument by using total station through traditional trigonometric elevation, tree height measuring principle to measure tree height and electronic measuring tree gun to measure tree height. Methods like electronic theodolite, total station and electronic angle gauge, and other precision instruments is introduced in the investigation of forest resources, get accurate height. With the development of remote sensing technology, has been widely used in forestry, for independent tree height was measured using LIDAR data, and using single camera based on angle measurement, displacement measurement to calculate the tree height, and so on a number of methods.

In field on-the-spot investigation work, however, is time-consuming, laborious, especially in some more complex terrain conditions areas cannot suitable measurement instruments and equipment. Along with the development of the aerial photography technology, a large area of forest resources investigation begins to be three-dimensional model is established by means of aerial photographic technologies to investigate, for our country's forest resources investigation provides a new investigation method.

Dimension and relative to the aerial images are used to establish a three-dimensional model, in the area of crown density is relatively low, selection can be more clearly observed on an aerial photo of the outline of buck's 50 strains sample wood. Respectively using digital photogrammetry and the measured height measuring tree gun, calculate the difference.

5.4.2 Forest stand mean height Estimation

Forest stand mean height is main factor investigation in forest resource survey, it is the sign of the trees stand height. Commonly used in forestry average diameter of trees to calculate the average stand height. However, due to the height and diameter at breast height of trees, there is a close relationship curve, so the traditional computing by average tree height is by calculation arithmetic average height and diameter at breast height of sectional area total product divided by the sum of DBH area. The above methods are needed in the field every wood gauging, tree height and diameter at breast height measured to calculate the average tree height of the sample area.

The method of using aerial photography is difficult to get directly to the diameter at breast height, only can be observed through three-dimensional relative tree height. So with the measured angle gauge point the coordinates of the point as the center of each sample to do 20 m buffer. Under the environment of the three-dimensional model, the three-dimensional cutting plant buck crown and the root part of the ground, the difference is the tree height. Within each round sample extraction 5-10 plant sample wood, its calculation arithmetic mean as the sample of the stand high on average.

5.4.3 Extraction of canopy

The canopy of trees for photosynthesis is the most important photosynthetic area and volume reflects the growth of trees. High(tree) affect the growth of trees in the vertical distribution and horizontal distribution growth(crown). In forest group, upper canopy also directly affect the growth of the lower trees and entire ecosystems. The canopy is also an important predictor of tree growth. Thus determine the size of the canopy of the forest growth and change is of great significance to the dynamic trend. This year in forest resource survey determine the size of the canopy is also an important factor. But how to quickly extract canopy information is the focus of the scholar's study. With high spatial resolution, remote sensing satellite images and aerial images quickly and accurately extract the canopy information is more and more attention from scholars. The most intuitive reflect the forest information in the remote sensing image is the canopy information, the remote sensing image on the canopy size, shape, texture and spectral information and other factors is the important basis of extraction of canopy information. The high spatial resolution of satellite remote sensing image to achieve decimeter level, in 2001 theQuickBird satellite image spatial resolution of the panchromatic band is 0.61 meters, 2008 launch of GeoEye satellite image spatial resolution of the panchromatic band can reach 0.41 meters. Aviation remote sensing image spatial resolution has can reach 0.1 meters. In such a high resolution remote sensing image forest canopy. If you want to extract the image of the canopy, by far the most commonly used method is visual discrimination. According to the reading of Ariel photograph interpretation generally divided into direct and indirect interpretation. Direct interpretation is according to the actual situation of ground objects, according to the features inherent in the image features, such as images, such as shape, size, color and shadow. On forestry in order to obtain more comprehensive information, detailed close to direct interpretation is not enough, still need to combine relationship between features, such as terrain, forest, tree species distribution and the change rule of auxiliary factors such as social status to interpretation. Due to the ground and air measurement observation direction is inconsistent, tend to cause some error. In ground measurements are often the most peripheral of dusty dry were measured in, but in the aerial photograph measurement is based on edge detection auto-

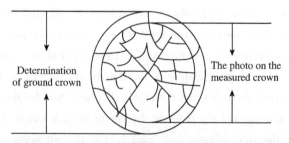

Figure 5-4 Crown measurement schematic drawing

matic tend to ignore some small branches. So often cause Ariel photograph measurement value is less than the ground measurement values(Figure 5-4).

Despite the current commonly used visual interpretation canopy information extraction method can achieve a certain effect, but just rely on this method to extract the canopy on the high-resolution remote sensing image information is not enough, both are time-consuming, inefficient, and the precision of the extracted and operators a direct link between the operation of the technology. How to pick up fast and accurate tree crown is still a problem. With the development of earth observation technology and aerial photography, technology is widely used in forestry and mature application of computer image recognition technology. Aviation remote sensing image with high spatial resolution canopy information extraction method proposed the new challenge. Use computer recognition technology of automatic or semi-automatic extraction tree crown, is expected to be an efficient method to replace the manual interpretation, and the extraction accuracy will be close to or even more than artificial interpretation.

Tree crown automatic extraction research at home and abroad in recent years have been done, the use of methods mainly include local maximum method, the bottom tracking method, contour line scanning method, template matching method, threshold mask method, seed region growing method. The local maximum method is to the point by canopy spectral reflection as a theoretical basis for seeking the biggest canopy spectral reflection value. The climax of the canopy, crown center and then further tofind the edge of depicting the border and then extract the crown area. Valley tracking method is based on spectral emission strongest area crown center theory, and some other outside edge will be dark, then find the spectra of the minimum as the boundary of the canopy and then extracted.

Based on in a variety of methods, forestry work the authors to start to do a lot of trial and research, expect by computer image recognition technology and various extraction model to rapid, efficient and accurate information to extract the canopy. Seewu (2010) the vegetation in QuickBird image area the spectrum threshold value method is used for primary segmentation, and then to the secondary with the improved algorithm based on edge segmentation, finally get to extract the overall accuracy of 84.67%. Cui S W respectively using QinckBird remote sensing image using seed region growing method for GuLiMu and crown density is more dense forest areas to extract canopy, GuLiMu extraction accuracy can reach 93.56%, but crown density between 0.5 to 0.8 at the time of the extraction accuracy can reach 89.74%. Fong W using aviation remote sensing image using object-oriented multi-scale segmentation method which can realize to plant trees or open canopy extraction accuracy is higher, but for the thick forest stand extraction accuracy is relatively low. Wang Erli (2011) the childlike yuan classification method, the national nature reserve of liangshan aviation remote sensing image of Korean pine, larch dragon spruce and fir canopy information extraction, respectively by the comparison with experimental data and the canopy extraction accuracy reached 83.50%, 84.35%, and 83.50% respectively. General tree species in the forest canopy shapes vary, but most are high around the low form in the middle. The broad-leaved forest canopy is generally high middle low around the circular arch form, the coniferous forest is high low points form around in the middle. Both broad-leaved forest and coniferous forest canopy generally have the highest point, canopy area is to point to the center, to edge bounded close to a plane of the circle.

In forest resource survey, the crown is also a very important factor investigation, generally in the measurement of east-west and north-south direction two direction measurements. But in the field in the forest canopy complexity also tend to have a variety of forms; canopy extraction cannot only use a kind of method applies to thewhole forest, only fit with the same characteristics of forest areas. Can generally in the area of low canopy density or canopy can clearly distinguish the area of a piece of extracting effect is better. But in the canopy structure is relatively complex, or have obscured under the condition of extraction accuracy is poorer. Sometimes need human intervention, for visual interpretation. This is because the self-imaging features of the remote sensing image. Therefore, this study for the canopy extraction preferable area as the research on visibility Guizhou area. Aviation remote sensing image is utilized to extract and then get the crown canopy contour information, can directly reflect the size of the crown and shape, save the ground survey, time-consuming. The extraction of canopy of research after years of development and with the efforts of several scholars although there are some achievements, for some particular stand living forest category with the extraction algorithm and the model of a mature, but for complex forest form and lack of many fundamental research and mainly displays in the low degree of automation.

The texture feature and spectral characteristics of forest canopy can be extracted as follows:

(1) First, make the orthogonal projection with projection information by pre-processing the aerial image, and then analyze the land types of aerial images. In this survey, land types of the study area cover main forestland, grassland and bare land, buildings and other places (Figure5-5), the author set up 987 sampling points in 5 different land types as a tree, shadow, bare land, grassland, building. Then of the sample function of Spatial Analyst Tools in ArcGIS to respectively extract the spectra value and texture information of various land types sample(Table 5-7).

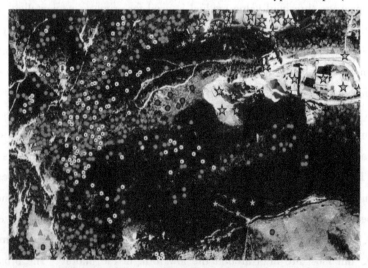

Figure 5-5　Point sampling of the different land type

Table 5-7 Spectral and texture information of different land type Statistics

Type	Tree	Shadow	Bare land	Grassland	Building
r	68.64	94.70	117.35	47.69	39.24
G	94.28	109.89	116.26	56.00	56.10
B	60.96	83.13	97.09	58.31	46.19
R_var	22.99	20.89	14.13	8.31	39.24
R_sm	97.90	88.80	97.17	133.54	93.90
R_mean	75.53	85.01	102.13	61.77	157.21
R_hom	178.32	181.39	187.13	212.08	159.95
R_entro	147.34	156.06	147.48	116.77	156.31
R_diss	58.80	56.89	50.70	34.92	81.19
R_corr	249.43	249.01	253.87	254.92	240.60
R_con	25.42	23.67	17.57	12.23	50.14
G_var	24.17	21.83	15.17	7.92	39.10
G_sm	97.90	88.80	97.17	133.54	93.90
G_mean	79.28	84.67	103.22	64.00	155.43
G_hom	178.45	181.06	184.78	217.38	164.05
G_entro	145.12	150.06	143.65	95.44	147.69
G_diss	60.50	59.07	53.30	32.00	80.76
G_corr	247.32	244.81	255.00	254.85	239.00
G_con	26.45	24.81	17.78	11.62	50.95
B_var	24.21	20.47	14.00	10.15	39.21
B_sm	99.24	101.09	106.87	129.62	94.88
B_mean	58.25	64.10	79.09	52.00	138.50
B_hom	180,36	184.06	196.43	201.77	158.86
B_entro	146.63	144.30	139.87	120.08	153.90
B_diss	59.62	55.50	46.65	43.23	83.17
B_corr	245.47	246.29	251.61	253.62	232.86
B_con	26.61	22,41	16.52	50.79	26.61
			14.38		

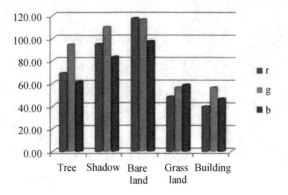

Figure 5-6 Red, green and blue spectral reflection characteristics of different land category

Eight texture factor including Mean, Variance, Entropy, Second Moment, Contrast, Homogeneity, Correlation, and Dissimilarity. Entropy reflects the richness of image texture, more entropy means more rich texture. The Second Moment reflects the degree of image texture thickness. These texture factors can be analyzed and quantitatively express. Calculating the 8 texture factors form7 * 7 window of R, G, B three band in the aerial remote sensing image, can get 24 (3×8) gray texture feature layer.

(2) Image segmentation. Image segmentation is the process of dividing an image into a number of sub-regions according to the pixels with the same attributes. In general, there are two ways of image segmentation: one is based on the boundary of the segmentation; the other way is based on region segmentation. In recent years high spatial resolution image is constantly emerging, the quality of high spatial resolution image segmentation will affect the precision of subsequent processing. However, the method faces high spatial resolution is less. Domestic and foreign scholars have made much researches: Liu Y X (2004) determine the basic unit of standard farmland extraction of remote sensing information using edge extraction and edge growth segmentation method. Zhang X Y (2007) using multi-scale segmentation method to extract urban green space information. Chen Z (2006), improved the classification accuracy of high-resolution remote sensing images by fitting multi-scale segmentation and multi-classifier. Liu S (2012), improve the research of SVAM and proposed SVAM + PCNN integration-segmentation model of automatic semantic image l, the result is superior to the traditional SVAM model. Feature Extraction tool of ENVI 5.0 was used in canopy extraction, the tool based on edge segmentation calculation methods, it mainly based on crown's texture features and spectral reflection, brightness, color, shape in the image pixels to extract. Segmentation effect is different according to different segmentation scale. Smaller segmentation scale, more detailed segmentation result for each image, there is an optimal segmentation scale which is not suitable for all classes. In order to achieve high accuracy of information extraction best segmentation scale is necessary which can be obtained by trial and error according to the characteristics of the canopy. For high crown density region, segmentation scale commonly set a little higher, otherwise, there will be many spare figure spots(Figure 5-7).

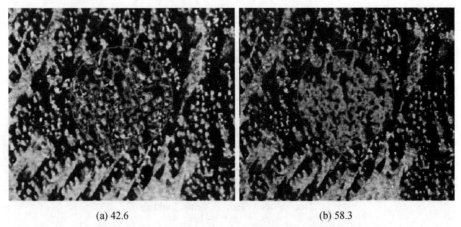

(a) 42.6   (b) 58.3

Figure 5-7 Preview effect of two kinds of segmentation scale

(3) Featureextraction. Feature extraction is achieved by supervised classification, rule classification, and direct vector output. Rules of classification set numbers of rules according to the classification of the various spectrums and texture feature, each rule can be described by several attribute expression. This research extracts single tree crown information, setting rules according to a crown factor of spectrum and texture feature. From visual effect, canopy extraction effect is good, but for medium and high canopy density forest, the signal crown is difficult to separate from the group.

5.4.4 Estimate the average crown

With high spatial resolution aerial images popularized in forestry, the use of the image can be relatively easy to extract powder raw wood or crown density lower canopy of opening information. But meet high canopy density forest canopy structure is complex or simple only using high spatial resolution image is hard to calculate, you need to combine with the spatial statistics analysis method to extract by the average crown. But in actual production, tend to choose themost simple method, direct use of forest land area and visual interpretation of the number of trees ratios to estimate the average crown. This method can directly affect the use of high spatial resolution image to obtain forest coverage area and the number of trees, simple and easy to operate, reduces the spatial statistical analysis and its various calculations.

The specific operation is as follows: continue to choose to calculate forest crown density ofPuShan forest region of 40 pieces of sample plots and jiufeng also 30 pieces of the sample, respectively. The circular sample plots in the crown area, visual interpretation in the sample area of the number of trees, such as canopy structure more complex can invoke the three-dimensional model, under the environment of three-dimensional interpretation and finally based on the ratio of the two for the block sample average of the crown.

5.4.5 Forest crown density estimates

Forest canopy density and survey of forest resources in a very important factor is also one of the most basic parameters on behalf of the film characteristics of forest which reflects the crown cover ground, namely, in terms of the forest to forest tending selective and forest management are extremely important indicators. According to the different canopy density generally fall into the forest land (>0.7), moderate self-cultivating forest (0.2—0.69) and opening (0.1—0.2). There are many ways to surveyabout crown density and application at present most is the survey method and the use of an aerial photo of the canopy area is calculated, combined generally divided into visual, outlets, crown density canopy projection diaphragm comparison method and measurement method. Or a visual method of field investigation, but the traditional method are generally not random sampling, has a small range, have no common representative defects. At present, the high spatial resolution aerial images can be easy to get, according to the image to get the distribution of forest and the spatial pattern of the area, then extracted according to the image of tree canopy area is relatively easy to extract factor regional large-scale crown density estimate can be realized. Morales (2008) using high spatial resolution IKONOS images of tropical drought in the north of Hawaii to evaluate forest canopy density, finally according to the image extraction of crown density and field measured the crown density of the correlation coefficient is 0.86. Wan H M(2011), using QuickBird image on the lower reaches of Tarim River, popular and Blame willow extracted canopy information, using the average crown canopy density and number of trees, and the relationship between the sample size to estimate crown density, results of forest canopy density estimation accuracy up to 69.45%.

Forest canopy density calculation method is a classic formula calculation, use of canopy projection area and the area ratios to estimate. Concrete implementations have become: first of all in ArcGIS, selected as the center of the circle the center of the sample. Angle gauge to do 20 meters outward buffer, to get round samples and then according to the circular sample plots in the crown area and the area of the whole circular sample area ratios as sample crown density of this block.

5.4.6 Forest volume estimate

As spatial resolution aerial images are more and higher, many scholars with high spatial resolution image to extract the canopy information made a lot of exploration and research especially for the extraction of the crown. Individual tree and based on the measured model a crown diameter at breast height, crown information recombination establish tree volume model, rapid and accurate for measuring tree factors further, forthe application of remote sensing in forestry provides a new perspective.

For stand volume and another measurable factor, there is a certain relationship between that can quickly give average height, average crown diameter and crown density or number to get to the volume. Can make use of aerial photography technology rapid access to the sample of the stand average height, average crown diameter and crown density or number, volume and their relationship model is set up, to estimate the stand volume quickly.

Stand volume as the dependent variable and then according to the above method by average height of stand, crown, crown density (or number) for the parameters, such as independent variables, set up the stand volume of multivariate linear:

$$M = a \cdot \overline{C^b} \cdot \overline{H^c} \cdot N \tag{5-21}$$

$$M = a \cdot \overline{C^b} \cdot \overline{H^c} \cdot P \tag{5-22}$$

$$M = (a \cdot \overline{CH} \cdot b\overline{C} + c) \cdot N \tag{5-23}$$

Type, a, b, c as the regression coefficient, stand high on average, for the canopy interpretation average diameter, N for interpretation strains, P for crown density.

For different stand conditions, select the different formulas to estimate forest volume. Better interpretation on the low canopy density, the number of regions (5-21) and (5-22) is adopted to establish the number, average crown breadth, the average tree height and accumulation model; Such as crown density is bigger, can't read a good number canopy density, average crown breadth, high average and accumulation model, such as type (5-23).

5.5 Forest parameter inversion by UAV image pair

Forest resources dedication in each period of human development has played an irreplaceable role. For forest resources quantity, quality, distribution and dynamic change of investigation and monitoring also come with the full stage of human development and utilization of forest. Nowadays, remote sensing technology application in forest resource survey and monitoring has very wide. In aerial remote sensing, 1954, the original forest aerial survey of the Ministry of Forestry investigation brigade in northeast China was carried out by a large number of forest aerial survey research. The development of the airborne remote sensing has laid a solid foundation. Unmanned aerial vehicle (UAV) for its characteristics such as low cost, easy operation, short cycle, is bound to in the forest resources investigation and dynamic monitoring plays a great role.

5.5.1 UAV photogrammetry principle and method

In aviation photogrammetry technology camera platform on a plane is used for photogrammetry. Flight direction and range of the ground are important factors as well as set up the stand volume and pictures extraction factor, the model of the relationship between selection factors including average height, crown density, and crown diameter, gradient, slope direction are also important.

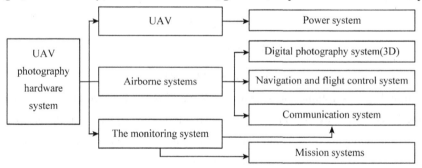

Figure 5-8　UAV aerial system integration

UAV aerial system integration (Figure 5-8) is as follows: UAV aerial photography hardware system includes UAV, airborne systems by aviation platforms, airborne digital camera, the control system composed of three parts. Control system including airborne flight control and the ground sensors.

5.5.2 Classification and characteristics of unmanned aerial vehicle

According to the different classification standards, UVA can be divided into different types. UAV system according to the power supply system can be divided into two major types i.e. fuel engine system and battery power. The price of fuel engine power system is high, heavy load and large size, the main advantage of fuel engine UAV is that it can fly to far distances but also have serious consequences of the accident. In battery power UAV system price is low, the load is light, small volume, cruise, generally, it will not be too painful to crash.

According to the characteristics of the wing, UAV can be roughly divided into fixed wing UAVs and rotor unmanned aerial vehicle (UAV). Fixed-wing unmanned aerial vehicle (UAV) has two types: one is the runway for take-off and landing; another is by the ejection device to take off and then landed by parachute. Fixed-wing UAV'S take-off require relatively open area or field is suitable land. Rotor UAV can be divided into two kinds: common rotor unmanned helicopter and unmanned aerial vehicle (UAV) the type of unmanned aerial vehicle (UAV) easy to carry and flight stability, suitable for small scale low aerial mission.

5.5.3 UAV forest parameters inversion

Forest tree height and the crown have close relation with the amplitude factor and volume. The traditional binary air volume model tends to be on the photo interpretation of tree height, crown width and ground of the measured volume to establish regression model, which are established for the single standing tree, but often needs in actual production the estimated volume of large area, each wood gauging is unable to do it. By aerial photograph interpretation average height, average crown diameter and crown density (visibility good area available on the number of per unit area) and other factors to establish the relation between the measured volume of ground, the commonly used equations as follows:

$$V = a \cdot K^{\bar{b}} \cdot H^{\bar{c}} \cdot N$$
$$V = a \cdot K^{\bar{b}} \cdot H^{\bar{c}} \cdot P \qquad (5\text{-}24)$$
$$V = (\overline{KH} \cdot b\overline{K} + c)N$$

In the formula, V is volume; K is crown density; H is the average high; N is stand density.

Individual tree height extraction can be used the following equation, using the pictures of the exterior orientation elements to calculate baseline weight B_x, B_y photography and B_z:

$$\begin{cases} B_x = X_{s2} - X_{s1} \\ B_y = Y_{s2} - Y_{s1} \\ B_z = Z_{s2} - Z_{s1} \end{cases} \qquad (5\text{-}25)$$

Determine each image point as auxiliary space coordinates:

$$\begin{bmatrix} X_1 \\ Y_1 \\ Z_1 \end{bmatrix} = R_1 \begin{bmatrix} x_1 \\ y_1 \\ -f \end{bmatrix} \quad \begin{bmatrix} X_2 \\ Y_2 \\ Z_2 \end{bmatrix} = R_2 \begin{bmatrix} x_2 \\ y_2 \\ -f \end{bmatrix} \qquad (5\text{-}26)$$

Then can be calculated by the projection coefficients on the ground of the unknown target feature point photogrammetry coordinates:

$$\begin{cases} N_1 = \dfrac{B_x Z_2 - B_z X_2}{X_1 Z_2 - X_2 Z_1} \\ N_2 = \dfrac{B_x Z_1 - B_z X_1}{X_1 Z_2 - X_2 Z_1} \end{cases} \quad \begin{cases} X = X_{s1} + N_1 X_1 = X_{s2} + N_2 X_2 \\ Y = Y_{s1} + N_1 Y_1 = Y_{s2} + N_2 X_2 \\ Z = Z_{s1} + N_1 Z_1 = Z_{s2} + N_2 X_2 \end{cases} \qquad (5\text{-}27)$$

Forest volume estimate can be combined with remote sensing inversion theory, with angle gauge sample measurement volume as the dependent variable, the UAV aerial obtain sample average crown, crown density, forest plant information as independent variables. Through the SPSS statistical software for curve fitting establish Jiufeng region sample volume estimate regression model, as follows:

$$V = -276.54 + 14.73 \cdot K + 163.48 \cdot \rho + 0.076 \cdot N + 21.33 \cdot \left(\sum_{i=1}^{n} \frac{H_i}{n} \right) \quad (5\text{-}28)$$

In the formula, V is sample volume; K is the sample average crown; P is sample crown density; N is the number of trees per hectare; H_i and n, respectively sample individual sample tree height and a number of trees. Equation fitting regression coefficient was $0.001 < 0.05$, meet the requirements.

5.6 Ground photography tree measurement technology

5.6.1 Binocular camera vertical baseline photography tree measurement technology

Binocular camera vertical baseline photogrammetric system, by using the fixed long baseline fixed two identical cameras simplifying or leave out the relative. Absolute orientation process will coordinate calculation in the space as the auxiliary coordinate system, shooting mode selection or integrity to photography, on the premise of its vertical fixed on a tripod, integrity photography with around the rotation of the baseline by the rotation of the space as the auxiliary coordinate system to complete the sample investigation.

Two cameras photography center of the specific steps the attachment is the baseline direction as the X-axis direction, with a camera D center of photography as the origin pass through the origin D and perpendicular to the plane of the baseline for YOZ plane. With camera D photography di-

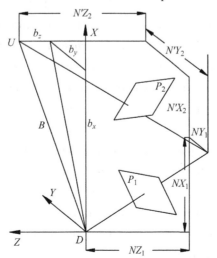

Figure 5-9 Binocular camera vertical baseline photography sketch map

rection in *YOZ* plane for the Z axis (photography direction is negative direction) coordinate system is established (Figure 5-9).

In auxiliary stereo pairs at the completion of binocular camera coordinate system after the orientation, the first thing we need to be the same point in stereo pairs of relative coordinates calculation. Namely stereo pair space forward intersection, the specific method is based on forward intersection point projection coefficient and the use of the front of the collinear equation intersection method.

$$N = \frac{b_x Z_2 - b_z X_2}{X_1 Z_2 - Z_1 X_2} \quad N = \frac{b_x Z_1 - b_z X_1}{X_1 Z_2 - Z_1 X_2}$$

$$\begin{bmatrix} X_1 \\ Y_1 \\ Z_1 \end{bmatrix} = \boldsymbol{R}_1 \begin{bmatrix} x_1 \\ y_1 \\ -f_1 \end{bmatrix} \quad \begin{bmatrix} X_1 \\ Y_1 \\ Z_2 \end{bmatrix} = \boldsymbol{R}_2 \begin{bmatrix} x_2 \\ y_2 \\ -f_2 \end{bmatrix}$$

$$X = X_D + NX_1 = X_D + b_x + N'X_2$$

$$Y = Y_D + NY_{YD} + b_y + N'Y_2 = Y_D + \frac{1}{2}(NY_1 + b_y + N'1Y_2)$$

$$Z = Z_D + NZ_1 = Z_D + b_z + N'Z_2 \tag{5-29}$$

The experimental results show that the use of space forward intersection calculation based on the projection coefficients on the round diameter at breast height and the actual measurement of the average absolute error is 0.5 cm, with an average diameter at breast height measurement of absolute error is 1.3 cm, the average relative error is 1.8%. USE space forward intersection calculation based on the collinearity equation of diameter at breast height and wheel measurement of average absolute error is 0.4 cm, with a diameter at breast height measurement of the average absolute error is 1.2 cm, the average relative error was 1.8%.

5.6.2 CCD arbitrarily tree measurement technology

Arbitrarily photogrammetric system is mainly composed of CCD camera, CCD lens, tablets, photogrammetry software. This system has the function of individual tree height, average height, DBH, volume, stand density, sample volume and 3D modeling and other functions.

Specific stepsare near the area under test to select two points, with a fixed focus camera to measure the target for nearly horizontal integrity photography, obtaining a stereo pair, reuse of collinear equation mathematical model and calculated six or more like a point with the same camera relative exterior orientation elements. Near the object under test upright, two fixed-length flowers stem respectively. After the two spots to photography calculated by fixed length rod endpoint object space coordinates, using the mathematical model and point coordinates, work out the relative elements of the exterior orientation of the camera. Near the object under test upright a fixed-length flower stem, and then in the photograph known more than three known points, using fixed-length flower stem length and three known points, calculate the relative exterior orientation elements of camera photography (Figure 5-10).

Figure 5-10 Schematic diagram of CCD trees measurement

5.6.3 Monolithic photography tree measurement technology

Monolithic photography tree measurement technology is a traditional tree measurement technology. Monolithic tree photography test technology in forestry survey also plays a bigger role. The system consists of a CCD camera/smartphone/tablet and photogrammetry software component. According to the mathematical model of the breakthrough image decoding classical theory, implement on the ground of uniform slope, that breast height line and has a fixed length rod or other parts of the image geometry information known three-dimensional calculating single photo (Table 5-8).

Table 5-8 Three kinds of photogrammetry coordinate, height, distance calculation method

	Binocu larvertical line	Arbitrary image pair	Single chip photography
X	$X = \dfrac{L}{q} u$	$X = (N_1 u + N_2 u'' + \cdot x)/2 + X_{S_1}$	$X = \dfrac{Z_0}{U_0} u$
Y	$Y = \dfrac{L}{q} f$	$Y = -N_1 + Z_{S_1}$	$Y = \dfrac{Z_0}{U_0} f$
Z	$Z = \dfrac{L}{q} v$	$Z = N_1 v + Y_{S_1}$	$Z = \dfrac{U_1}{V_2 - V_1} h$
H	$H = \dfrac{L}{q} \cdot V$	$H = \sqrt{(x_2 - x_1)^{\cdot 2} + (Y_2 - Y_1)^{\cdot 2} + (Z_2 - Z_1)^{\cdot 2}}$	$H = \dfrac{Y}{f} \cdot V$
$d_{1.3}$	$d_{1.3} = \dfrac{L}{\cdot q} \cdot u$	$d_{1.3} = \sqrt{(x_4 - x_3)^{\cdot 2} + (Y_4 - Y_3)^{\cdot 2} + (Z_4 - Z_3)^{\cdot 2}}$	$d_{1.3} = \dfrac{Z_{1.3}}{V_{1.3}} \cdot u$

5.7 Modernization of ground forest observation

5.7.1 Modern forest intensive method

Observations h_i, d_i (4 or more) and H plots fine measuring, meet the requirements for a class survey, the steps are as follows using MINI superstation followed by measuring up from the 1.3 m woods h_i anywhere, d_i and tree height H, standing timber volume can be obtained as follows:

$$V = \frac{10^{-4}}{4} \pi [1.3 d_{1.3}^{\ 2} + \frac{1}{3} \sum_{i=1}^{3} (d_i^{\ 2} + d_{i+1}^{\ 2} + d_i d_{i+1})(h_{i+1} - h_i) + \frac{1}{3} d_4^{\ 2}(H - h_4)]$$

(5-30)

In the formula, V for Standing timber volume (m^3); d_i for anywhere in diameter at breast height (cm); H for tree height (m); h_i for anywhere in the tree height.

3D angle rules plots fine measuring, meet II class survey requirements, specific steps following, in sample plot to selected sampling points. This sample points for angle rules points to MINI superstation instrument tool for around measuring. According to angle rules count principles for count measuring and records the distance from angle rules points to count tree L_i, count wood azimuth angle T_{0i}, count tree high H_i, and DBH d_i, crown and species. Electronic angle rules function technology says that when the viewpoint and the formation of the baseline on the screen angle diameter it is cut with the adjacent tree around, it will counts as 1, but while count is 0. Electronic angle gauges feature specific steps: press the ⬜ key until |I·I| display in the upper left corner of the display; press the 🔍 key, switch angle coefficient; press the 🔢 key, counting an increase of 1; press the [CLEAR OFF] key, you are prompted whether to end the current measurement, end the count data to zero(Table 5-9).

Table 5-9 3D angle measurement solution like the goblin

Modern forest intensive method	Stand average height	Stand average diameter at breast height	Stand density	Survey ingstands volume
3D angle gauge plots the precision measurement	$\overline{H} = \dfrac{\sum_{i=1}^{k} H_i g_i}{\sum_{i=1}^{k} g_i}$	$\overline{d} = \sqrt{d_i^2}$	$N = F_g \sum_{i=1}^{n} \dfrac{1}{g_i} n_i$	$G = F_g Z$ $M = G(\overline{H} + 3)f_{1.3}$

In the formula, d_i is tree diameter at breast height; H_i is tree height; g_i is tree basal area; k is the wood counted number; F_g is the angle coefficient; n is the wood counted number; G is the number hectares of basal area; Z is the angle count of wood; $\overline{H}$ is the weighted average tree height; $f_{1.3}$ is the number of tree at breast height.

Sentinel square method to measure, meet II class survey requirements, the steps are as follows, using MINI superstation in point P in order to determine recent n trees of diameter d_i ($i = 1, 2, \cdots, n$), H and the surrounding trees to the horizontal distance of the point P D_i;

$$N_i = \frac{2 \times (i - 0.5)}{(d_n + 2D_n)^2} \times 10^4 \tag{5-31}$$

$$g_n = \frac{\pi \times (0.5 d_n^2 + \sum_{i=1}^{n-1} d_i^2)}{16(d_n + 2D_n)^2} \tag{5-32}$$

If $\dfrac{(g_n - g_{n-1})}{g_{n-1}} \leq 5\%$

$$M = g_n f_{1.3}(\overline{H} + 3) \tag{5-33}$$

$$\overline{d} = \sqrt{d_i^2} \tag{5-34}$$

In the formula, g_n is total stand basal area of trees; $\overline{H}$ is the weighted average tree height; $\overline{d}_i$ is the average diameter at breast height; $f_{1.3}$ is the tree at breast height form factor.

Growth volume sample measuring meet forest business survey requirements specific steps following. When consider a forest points of site conditions, forest according to the different sites divided into 5~7 classes. According to different species determine count sampling wood, that this forest point has $i = 1, 2, \ldots, m$ species, each species has $j = 1, 2, \ldots, n$ tree counts sampling wood; using MINI superstation instrument collect more than 40 observation points data, and on count sampling tree for more than consecutive two years of observation. Getting sampling tree diameter d_{ij}, tree height for H_{ij}, density for N_{ij}; known this stand the mixing ratios of K, the average height is H, average diameter D, total density N, mathematical model of diameter at breast height and tree height growth of established:

$$\cdot d_{ij}(\cdot H_{ij}) = aN^b H^{-c} D^{-d} K^e N_{ij}^f H_{ij}^g d_{ij}^h \qquad (5\text{-}35)$$

In the formula, $\cdot d_{ij}$ is annual growth for diameter at breast height; $\cdot H_{ij}$ is annual growth of trees; observations into mathematical models, solution 8 coefficients a, b, c, d, e, f, g, h;

In the cutting area fine measuring, meet three class survey requirements, specific steps following, in the cutting area away from the edge of selected representative of the forest- average status point to vantage point, with fixed angle θ control 10 or so tree is selected as sampling observation of counted wood, remember for i ($i = 1, 2, \ldots, r$; general $r \leqslant 10$), and measurement count wood of DBH d_i(requirements $d_i > 5$ cm, or cannot as count wood); makes $k = (50\,\theta)^2$, is by selected count wood of density $n_i = \dfrac{4k \cdot 10^4}{\pi d_i^2}$, which calculation out forest points of density (units: strains/hectares) for: $N = \displaystyle\sum_{i=1}^{r} n_i$; in observation points placed MINI superstation instrument, first the measurement of each tree count wood of height H_i, from the state tree roots as L followed by measuring each piece of wood up small head diameter $d_{ij}(j = 1, 2, \cdots, m)$. According to head diameter d_{ij} of different, tree section is divided into big diameter order material ($d_{ij} \geqslant 26$ cm), and in the diameter order material ($20 \leqslant d_{ij} < 26$ cm) trail order material ($6 \leqslant d_{ij} < 20$ cm) statistics the diameter order material number. When $d_{ij} \leqslant 5$ cm, wood section regardless of material, this wood and the above trunk are by waste material processing, calculation get this wood section from trees of height h_{is}; according to above data, calculation forest points of total out material volume:

$$V_{\text{total}} = \sum Nad_i^b \qquad (5\text{-}36)$$

Waste material volume:

$$V_{\text{waste}} = \frac{1}{12}\pi \sum Nd_{is}^2 h_{is} \qquad (5\text{-}37)$$

Forest points of out material rate for:

$$Q = \frac{V_{\text{total}} - V_{\text{waste}}}{V_{\text{tptal}}} \times 100\% \qquad (5\text{-}38)$$

Scrap rate is:

$$W = 1 - Q \qquad (5\text{-}39)$$

5.7.2 Modern forest measurement method in detail (two or three)

Block conditions observation sample for details measuring, meet II class survey requirements, specific steps following. By using MINI full station instrument measurement sample to within trees tree high H, and trunk two at diameter d_1 and d_2 and correspond to ground of distance h_1 and h_2, then with into to mathematics model $d_i = a(H - h_i)^b$, which, H for trees of height, d_i for trees a height at of diameter, h_i for the height, a and b for constants coefficient. Solution get constant coefficient a and b, to get anywhere in the trunk of the tree diameter and height of the convergence model and characteristic parameters of model a and b for the trunk. For much of the region the species tree for model fitting can be obtained anywhere in the region the trunk of the tree diameter and height of the convergence model. So as to achieve the level of smart-phone chip straight photography to extract characteristic parameters of a region of a tree trunk, and in the case of trees blocking, to estimate the tree height.

Binary model of detailed investigation meets II class survey requirements, primarily using MINI superstation measured tree diameter at breast height of d and tree height h, and volume model of binary volume. Binary volume table based on tree diameter at breast height and tree height two factors prepared and disclosed volume table of standing timber volume. China national dual volume model in the form of $V = aD^b H^c$. There are 180 species of 197 000 sample trees throughout the country were used 56 binary volume table information compiled, 35 of coniferous trees, broad-leaved 21, series general system error at plus or minus 1%, few in between positive and negative 1% and 2% (Table 5-10).

Table 5-10 Dual model for volume

$V = a_0 + a_1 D + a_2 D^2 + a_3 DH + a_4 D^2 H$	Mai ye
$V = a_0 + a_1 D^2 + a_2 D^2 H + a_3 H + a_4 DH^2$	Meng Xianyu
$V = a_0 + a_1 D^2 + a_2 D^2 H + a_3 H^2 + a_4 DH^2$	Nasi lunde
$V = a_0 D^{a_1} e^{a_2 H - \frac{a_3}{H}}$	SiQidu
$V = a_0 D^{a_1} H^{a_2}$	Yamamoto and the Tibetan
$V = a_0 (D + 1)^{a_2} H^{a_3}$	Ka Song
$V = a_0 + a_1 D^2 H$	Siboer
$V = D^2 (a_0 + a_1 H)$	Aogaiya
$V = \dfrac{D^2 H}{a_0 + a_1 D}$	Kazuhiko Takada
$V = a_0 D^{a_1} H^{3-a_2}$	De Witt
$V = a_0 (DH)^{a_1}$	Siboer
$V = a_0 + a_1 D^2 + a_2 D^2 H + a_3 H$	Stott
$V = a_0 D^2 H$	Siboer
$V = a_0 D^2 e^{a_1 - \frac{a_2}{H}}$	SiQidu
$\lg V = a_0 + a_1 \lg D + a_2 (\lg D)^2 + a_3 \lg H + a_4 (\lg H)^2$	German linke
$V = a_0 D^2 H + a_1 D^3 H + a_2 D^2 \lg D$	Zhao Kesheng

5.7.3 Modern forest survey act (two or three)

Using a general surveyof model-like, meet II class survey requirements, primarily using MINI superstation measured tree diameter at breast height d, then a volume model for volume calculation. Single entry volume table in the application area can give good results. Because of the single entry volume table only considers the relationship between volume and diameter at breast height. Even in the same area, diameter at breast height of the same tree, the tree height and volume may vary greatly. So local unitary volume table, its use should not be too large. If you do not know the suitability of single entry volume table in the region, should first be randomly a number of sample trees, the applicability of the measured material volume tables on a test if the error exceeds 5%, be renumbered table(Table 5-11).

Table 5-11 Meta-model for volume for example

$V = a_0 + a_1 D^2$	Parking his-Richard Engelhardt
$V = a_0 D + a_1 D^2$	Disaisi ku-Mai ye
$V = a_0 + a_1 D + a_2 D^2$	Covered Nader
$V = a_0 D^{a_1}$	Bokehuote
$\lg V = a_0 + a_1 \lg D + a_2 \dfrac{1}{D}$	Bulinake
$V = a_0 \dfrac{D^3}{D+1}$	Lu Ze
$V = a_0 \dfrac{D^3}{D+1}$	Zhongdaoguangji

Using common angle gauge plots a general survey, meet II class survey requirements, the steps are as follows, choosing: Site far away from the forest edge (50 m) within the forest pick one point, to this point as the center of rotation, around the measuring and counting. Around the counting measuring method: Trees that are cut with the angle of sight of count 1, apart tree count as 0. Stand basal area per hectare: $G = F_g \times Z$, F_g angle gauge sectional area coefficient; Z is around the measured total count. Around measuring: angle gauges by strain measurement and count the number of trees. Critical tree: tangent to the angle of sight of the trees. Stand weight: $M = G(\overline{H} + 3)f_{1.3}$. Stand density: $N = F_g \sum_{i=1}^{n} \dfrac{1}{g_i} n_i$.

Point to a general survey sample square method meets II class survey requirements, the steps are as follows, using MINI superstation in point P in order to determine recent n trees of diameter d_i ($i = 1, 2, \ldots, n$) and the surrounding trees to the horizontal distance of the point P D_i.

$$N_i = \frac{2(i - 0.5)}{(d_n + 2D_n)^2} \cdot 10^4 \tag{5-40}$$

$$g_n = \frac{\pi \cdot (0.5 d_n^2 + \sum_{i=1}^{n-1} d_i^2)}{16(d_n + 2D_n)^2} \tag{5-41}$$

If $\dfrac{(g_n - g_{n-1})}{g_{n-1}} \leqslant 5\%$,

$$\bar{d} = \sqrt{\overline{d_i^2}} \qquad (5\text{-}42)$$

In the formula, g_n is n total stand basal area of trees; $\bar{d}$ stand average diameter at breast height; $f_{1.3}$ is tree at breast height form factor.

Circular plot a general survey meets II class survey requirements, the steps are as follows, using MINI superstation, after the calibration of circular plots in the forest, by measuring the diameter of each tree, tree height, standing tree volume(5-40). Circular sample plot volume formula (5-41), diameter ratio calculation formulas (5-42), stand average height stand inside a circular plot of arithmetic average formula, see (5-43).

$$V_i = g_{1.3}(h_i + 3)f_\varepsilon \qquad (5\text{-}43)$$

In the formula, $g_{1.3}$ for each tree diameter at breast height of basal area; h_i for the height of each tree; f_ε for the experimental form factor, experimental form factor selected $f_\varepsilon = 0.415$.

$$M = \frac{\sum_{i=1}^{N} V_i}{S} \ (\text{m}^3/\text{m}^2) \qquad (5\text{-}44)$$

In the formula, N is the trees inside a circular plot number; S is the circular plot area (m^2).

$$n_i = \frac{N_i}{N} \ (\%) \qquad (5\text{-}45)$$

In the corresponding N_i diameter i stand number; N is the trees inside a circular plot number.

$$\bar{H} = \frac{\sum_{i=1}^{N} H_i}{N} \qquad (5\text{-}46)$$

In the formula, H_i is the height of each tree; N is the effective standing inside a circular plot line.

A general survey of the polygonplots meets II class survey requirements. Specific steps following, using MINI superstation instrument, take an as common point for measuring points, in measuring points around select 1 tree away from measuring points as center tree (figure in the tree 1); the center tree for coordinates original points, in East, West, South, and North four direction formed four quadrant. Each quadrant selected away from a center tree recently of two trees for observation tree. If a recent tree happens to be on the axis, then the tree is counted as two adjacent the quadrant is a quadrant with a smaller counter-clockwise angle so that there are nine central and observational trees. The calculation model of polygon sample measurement parameters is as fol-

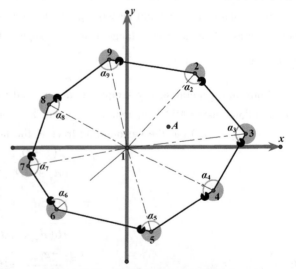

Figure 5-11　Polygon sample plot survey

lows(Figure 5-11):

Polygon sample to measure parameter calculation model is as follows:

Volume M (unit: m^3/hm^2) (5-47), which, f_θ for the experimental form factor, f_θ look-up table of main tree species are available, but taking into account the operation of the instrument, take the average form factor $f_\theta = 0.472$; $P_i = \dfrac{\alpha_i}{2\pi}$, α_i $i-1$ number of $i+1$ strains and strains posed by angle (measured in radians), (5-48), in which when you $i=2$, $i-1=9$, when $i=9$, and $i+1=2$. S is the polygon area (m^2), (5-49), in which when you $i=2$, $i-1=9$, when $i=9$, and $i+1=2$. H_i for the sample to be measured tree height (m).

$$M = \frac{1}{4}\pi f_\theta \sum_{i=1}^{n} P_i D_i^2 (H_i + 3)/S \tag{5-47}$$

$$\begin{cases} S_{i-1,i} = \sqrt{(x_i - x_{i-1})^2 + (y_i - y_{i-1})^2} \\ \alpha_i = \arccos\left(\dfrac{S_{i-1,i}^2 + S_{i,i+1}^2 - S_{i-1,i+1}^2}{2S_{i-1,i}S_{i,i+1}}\right) \quad (i = 3, 4, 5, \cdots, 7, 8) \end{cases} \tag{5-48}$$

$$S = \frac{1}{2}\sum_{i=1}^{n} x_i(y_{i+1} - y_{i-1}) = \frac{1}{2}\sum_{i=1}^{n} y_i(x_{i+1} - x_{i-1}) \tag{5-49}$$

Stand average height $\overline{H}$ (m) (5-50) P_i calculation method with volume calculation formulas.

$$\overline{H} = \frac{\sum_{i=1}^{n} P_i H_i}{\sum_{i=1}^{n} P_i} \tag{5-50}$$

Stand density N (strain/hm^2) (5-51) is obtained, P_i and S calculated as above:

$$N = \left(\sum_{i=1}^{n} P_i/S\right) \times 10^4 \tag{5-51}$$

Diameter ratio N_j (%) j the proportion of stem diameter, (5-52), and met $\sum_{j=1}^{n} N_j = NP_i$ calculation method with (5-52).

$$N_j = \frac{P_j}{\sum_{i=1}^{n} P_i} N \tag{5-52}$$

5.8 Mobile/tablet/GNSS forest survey counted measuring system

Nowadays smartphone/tablet refers to as personal computers, having an independent operating system, independent space can be used by users to install software, games, navigation and another third-party service. Providers to provide programs and can be operated via mobile communications network wireless network access type in general.

Use of smartphone/tablet has been reached all over the world but not everyone knows the use of smartphones with the excellent operating system. Many features in smartphone like free installation of various software (Android only), full-screen, touch-screen characteristics make this technology superior and the end of full keyboard cell phones a few years ago. Google (Google), Apple, Samsung, Nokia and HTC (HTC electric) are five big brand in world most widely known while millet (Mi), Huawei (HUAWEI), MEIZU (MEIZU), Lenovo (Lenovo), ZTE (ZTE), Cool Sent (Coolpad), One Plus Phone (Oneplus), GIONEE (GIONEE) and Tianyu (Tianyu, K-Touch) brand in China are also famous.

Forest survey measuring system built on phone/tablet is convenient, light weight, good network characteristics included in the Android/IOS system, forest records table construction procedure, fast standing timber volume query evaluation procedures further provide a more intelligent platform for forestry information collection technology.

5.9 Forest map drawing and GIS spatial analysis

The pictures for forest management investigation mainly include basic maps, forest maps, and thematic maps. These are most widely used and basic pictures materials for forest management. The traditional methods of drawing are complex and not easy to carry. Secondly, these were difficult to update. Now with the involvement of the GIS, these problems have been resolved and forestry data can be easily analyzed statistically.

1) Basic map

The basic map is mostly used to calculate forest area, basic drawing data compilation i.e. forest boundaries administrative divisions, roads, settlements, rivers, mountains, bound compartments, forestry sector, compartments annotation, annotation and other small groups. The basic map is a plan view of aerial photos or topographic mapping, mostly drawn by using a scale of 1 : 5000 and 1 : 25000 (Figure 5-12).

2) Forest map

Forest map is based on sub-compartment investigative material of the forest. The

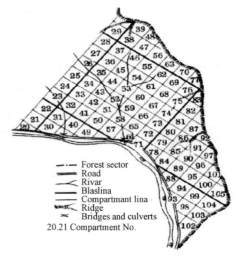

Figure 5-12 Basic chart

main contents of forest map are a division of each sub-compartment to class, dominant tree species, age group, area and other factors. Official forest maps are colored; In sub-compartment different classes and dominant tree species are used with a variety of different colors. The depth of colors is used to represent different stand age groups, i.e young forest shallow color, slightly dark color for the middle-aged forest, deep color for the mature forest.

Forest map scale mostly used in our country is 1 : 10000 and 1 : 25000. Score structure is used to a small note of stock map, the molecular level for sub-compartment size and age, the denominator to position (or status index, forest type, site type) and crown density (or density) (Figure 5-13).

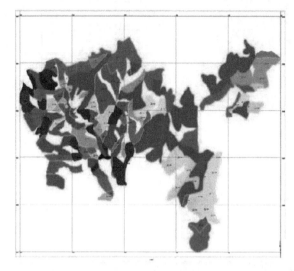

Forest facies map of Guanghua Temple forest area

Figure 5-13 Forest map

3) Forest distribution map

Distribution map is a picture material which reflects the forest distribution in Forest Bureau (county). In forest distribution map the mostly used scale is 1 ∶ 50000 and 1 ∶ 100000.

4) Thematic map

The matic map is a professional map based on professional survey. Thematic map is of different major types depending on its purpose like the soil profile, site type, distribution, elimination of pests, distribution of wildlife, by-product resource distribution, etc. Thematic maps are drawn in forest farms (township and town) as the unit. In practice, a map which project entirely by forest management needs(Figure 5-14).

5) GIS technology to realize forestry information intellectualization

In recent years, GIS technology is increasingly applied to the mapping of forestry and forestry system thus the rapid realization of information inquiry, statistics and analysis(Figure 5-15).

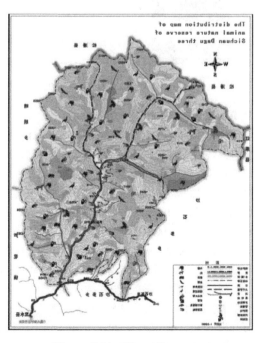

Figure 5-14 Thematic map

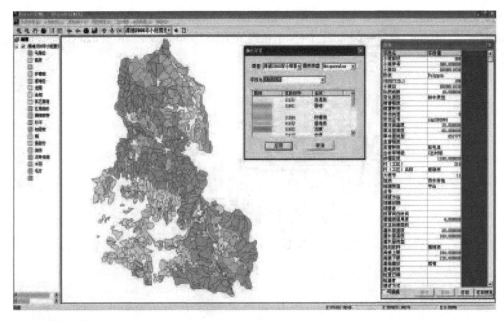
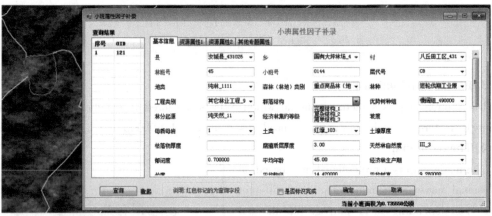

Figure 5-14   GIS technology to realize forestry information intellectualization

References

蔡文峰,李凤日,2010.基于 VirtuoZo 系统对林木冠幅信息的提取[J].森林工程,26(02):4-7.
陈忠,赵忠明,宫鹏,2006.一种快速高分辨率遥感影像分割算法[J].计算机应用研究(10):154-155,185.
冯仲科,隋宏大,邓向瑞,等,2007.三角高程法树高测量与精度分析[J].北京林业大学学报 (S2):31-35.
高志海,李增元,魏怀东,等,2006.干旱地区植被指数(VI)的适宜性研究[J].中国沙漠(02):243-248.
郭志华,李琼婵,1997.广东省植被潜在生产力的估算及其分布[J].热带亚热带植物学报(01):45-52.
黄国胜,夏朝宗,2005.基于 MODIS 的东北地区森林生物量研究[J].林业资源管理(04):40-44.
李崇贵,赵宪文,2006.聚类分析在监测区域样地分类中的应用研究[J].遥感学报(02):256-262.
李健,舒晓波,陈水森,2005.基于 Landsat-TM 数据鄱阳湖湿地植被生物量遥感监测模型的建立[J].广州大学学报(自然科学版)(06):494-498.
仝慧杰,冯仲科,罗旭,等,2007.森林生物量与遥感信息的相关性[J].北京林业大学学报(S2):156-159.
万红梅,李霞,董道瑞,等,2011.塔里木河下游林地树冠 QuickBird 影像信息提取与分析[J].西北植物学报,31(09):1878-1885.

王国辉,马莉,杨腾峰,等,2005. 手持普通相机监测隧道洞室位移的研究与应用[J]. 岩石力学与工程学报(S2): 5885-5889.

王秀美,曾卓乔,2001. 数字摄影测量技术在森林调查中的应用研究[J]. 林业资源管理(01): 31-35.

杨化超,邓喀中,2006. 利用2维DLT和共线方程分解相机外方位元素[J]. 测绘科学技术学报(03): 232-234.

张秀英,冯学智,刘伟,2007. 基于多分类器结合的IKONOS影像城市植被类型识别[J]. 东南大学学报(自然科学版)(03): 399-403.

郑刚,彭世揆,戎慧,等,2010. 基于KNN方法的森林蓄积量遥感估计和反演概述[J]. 遥感技术与应用,25(03): 430-437.

Akay A E, Oguz H, Karas I R, et al., 2009. Using LiDAR technology in forestry activities [J]. Environmental Monitoring and Assessment, 151 (1-4): 117-125.

Alvarez E, Duque A, Saldarriaga J, et al., 2012. Tree above-ground biomass allometries for carbon stocks estimation in the natural forests of Colombia [J]. Forest Ecology and Management(267): 297-308.

Armstrong A, Fischer R, Huth A, et al., 2018. Simulating forest dynamics of lowland rainforests in eastern Madagascar [J]. Forest, 9(4): 214.

Bakula M, Przestrzelski P, Kazmierczak R, 2015. Reliable technology of centimeter GPS / GLONASS surveying in forest environments[J]. IEEE Transactions on Geoscience and Remote Sensing, 53(2): 1029-1038.

Bartelink H H, 1996. Allometric relationships on biomass and needle area of Douglas-fir[J]. Forest Ecology and Management, 86 (1-3): 193-203.

Bartelink H H, Mohren G M J, 2004. Modelling at the interface between scientific knowledge and management issues. Towards the sustainable use of Europe's forests [J]. Forest Ecosystem and Landscape Research Scientific Challenges and Opportunities, 49: 21-30.

Cao L, Coops N C, Innes J L, et al., 2016. Estimation of forest biomass dynamics in subtropical forests using multi-temporal airborne LiDAR data [J]. Remote Sensing of Environment, 178: 158-171.

Carrer M, Castagneri D, Popa I, et al., 2018. Tree spatial patterns and stand attributes in temperate forests: The importance of plot size sampling design and null model [J]. Forest Ecology and Management, 407: 125-134.

Charaey N D, Babst F, Poulter B, et al., 2016. Observed forest sensitivity to climate implies large changes in 21st centur north American forest growth [J]. Ecology Letters, 19 (9): 1119-1128.

Clawges R, Vierling L, Calhoon M, et al., 2007. Use of a ground - based scanning lidar for estimation of biophysical properties of western larch (*Larix occidentalis*) [J]. International Journal of Remote Sensing, 28 (19): 4331-4344.

Liu S, He D, Liang X, 2012. An improved hybrid model for automatic salient region detection [J]. IEEE Signal Processing Letters, 2012, 19(4): 207-210.

Liu Y X, Liu L, Li X, et al., 2008 High quality prime farmland extraction pattern based on object-oriented image analysis[C]. 16th International Conference on GeoInformatics and the Joint Conference. International Society for Optics and Photonics, 2008.

Liu Z Y, Huang J F, Xiang R, 2008. Characterizing and estimating fungal disease severity of rice brown spot with hyperspectral reflectance data[J]. Rice Science, 15(3): 232-242.

Ma R, Li W. Selection of factors for estimating forest volume based on SPOT5 remote sensing images: A case study of Huoditang Forest Station[J]. Journal of Northeast Forestry University, 2011.

Wang J, Yang H, Feng Z, et al., 2013. Model of characteristic parameter for forest plantation with data obtained by light small aerial remote sensing system[J]. Transactions of the Chinese Society of Agricultural Engineering, 29(8): 164-170.

Wang Z, 1996. On the renaming of the discipline 'photogrammetry' [J]. Isprs Journal of Photogrammetry & Remote Sensing, 51(1): 1-4.

Wu J, Bauer M E, 2012. Estimating net primary production of turfgrass in an urban-suburban landscape with QuickBird imagery[J]. Remote Sensing, 4(4): 849-866.

Xiaolan W U, Gao L, Kan J, 2009. Extraction of information of standing tree images based on region growth and edge detection[J]. Science & Technology Review, 27(22): 44-47.

CHAPTER 6 System Design of Forest Precision Manangement and Manangement Platform

6.1 Forest precise management system design

Forests are an important part of the earth ecosystem and the foundation of human society that sustained and healthy development. China has built a fairly complete set of National Forest Inventory (NFI) Survey System technology with efforts of 60 years in order to grasp the present situation and dynamic of forest resources. Eventually we made forest resource inventory that results in a five-year survey period. At the same time in our country there is a second category of 10-year survey forming a unique and comprehensive survey of forest resource monitoring and management systems at the local level and also at national level.

Forest precise management system design includes forest planning, forest observation, and forest management. The main contents are shown below (Figure 6-1).

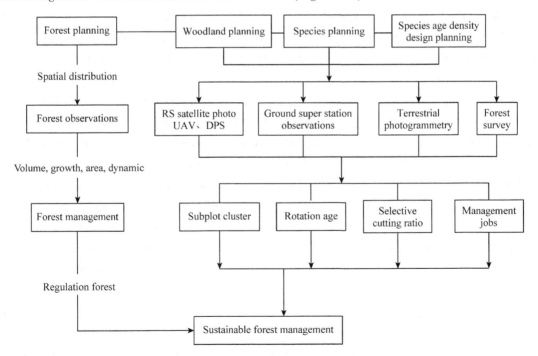

Figure 6-1 Forest precise management system design

The key issue:

(1) What is the growth level at different sub-plot in the forest? (forest for young, middle, nearly mature, mature or over-mature).

(2) How to run various types of woodland? (forest for young, middle, nearly mature, mature or over-mature).

(3) What is the best rotation age?

(4) Whether the business model tending selective cutting the mature orover-mature forest (especially natural forests)?

(5) What are the best-mixed forest, age differential (DBH ratio) and the density of the sub-plot? Maximum growth? The maximum volume?

(6) What is tending selective cutting ratio? Which order correspond the selective cutting? What position of selective cutting gives the best forests output?

Forest planning and design for land use/land cover change (LUCC) is a forest spatial planning that is done by mean of 3S technology. The target of ecological environment and energy savings is to balance and coordinate the regional economic and social development. On the basis of the optimization design, forest management workers to assume the task of designing forest species, forest types to design coordinated with the regional environmental, economic, social, fully embodied the human spirit.

Forest sub-plot is the most basic business unit in the inventory of forest resource management and basic statistics. Forest management must first solve the precision about the growth of sub-plot i. e. what is the level of growth through the investigation and observation, such as 5 trees method and 3D angle gauge observation. These methods are used to determine whether the forest is young forest, middle-aged forest, near mature forest, mature forest, over-mature forest category. According to the internal characteristics, business purposes and adopted operating system technology we have divided forest into the different management plan. Specific forest management is based on forest tenure, management objectives, species, site quality and stand origin, etc. Depending on forest's amount, quality, the forest coverage rate, growth rate and structure of forest tree species to determine forest resources development goals. Plantation is classified separately according to commercial forest or ecological forest. The artificial commercial forest is managed through observation and investigation of species, habitats, and growing conditions stand density, management techniques, etc. Based on the average growth or accumulation of volume or stand of trees has reached the maximum, to determine whether the stand or the number of mature trees. According to market demand and stand structure parameters for different wood species, to determine whether mature technology.

Although China has implemented a ban on logging policies for natural forest. As a large country, there is wood demand and consumption; we still need to study the utilization of harvesting and processing of timber, such as plantations harvest quota, mature forest (especially natural forests). How tending type selective cutting operation would done? We need to seriously study the survey design and the timber harvesting rate statistics and related technologies, methodology establishment, such as plantations harvest quota, mature forest (especially natural forests) tend type selective cutting operation. Detailed design of the management plan, primarily regulating mixed in sublot, age distribution, diameter distribution, especially in the design of optimum density. How to design density to achieve the most selective cutting of timber in sub-lot for a period of time with the growth of the largest, best environmental effects? To tend selective cutting, through research and to determine the proportion tending harvesting more selective cutting which a path rank? Which selective cutting position to make the best forest output?

6.2 Evaluation of forest

6.2.1 Value of woodland

Detailed and careful study assessment is used to determine the object's meaning value or condition. The evaluation process is a complex process includes evaluation and determination of object through comprehensive analysis, calculation, observation and counseling methods.

Forest resources not only include forest land and wildlife on it but also the survival of other plants, microbial resources, as well as forest landscape resources. Sustainable development of forest resources in natural forest is essential and for an effective economic evaluation of forest resources, the protection of forest resources and environment groups is required. Evaluation system for the classification of forest resources is based on different elements. In practice, the evaluation method of forest resources for different purposes like sale, exchange, division, addition to the merger, loan etc., forest pests and diseases suffered losses from natural disasters is also evaluated.

Determining compensation for expropriation of woodland and forest land use right of the lift in the case of having a mortgage assessed value of the security; when the implementation of the approved amount of forest loss and the insured amount, forest taxation standards to determine that. In addition it can also be applied to evaluate the assets and management of forestry enterprises calculate price last almost no contact algorithms etc., especially in recent years with special use forests, such as various protective forest, natural parks, reserved forests and other landscapes a comparison between the horizontal and vertical elements of the economic effects and ecological effects of different regions and different species are brought to be involved. There are some restrictions in special forest evaluation of rootstock. Forest resource value evaluation of wildlife and microbial of different areas both require extensive knowledge of the forest resource assessment.

Forest resources is for material connotation of assets, it is only with part of the economic resources of the nature of the assets of forest resources; it is forestry enterprise survival and development of material basis. The means of forest production and the main source of forest resource assets are the main natural resource assets in the part, is a kind of regenerative capacity of natural resources assets. Compared with other resource, assets assessment of forest resources is very complex. Forest resources with the general characteristics of profitability, liquidity, and possession, has the following 5 characteristics.

The first one is the sustainability of operations. Forest assets belonging to renewable resource assets, forest resource assets consumption can be through the reasonable operation, according to the characteristic rules and regeneration of forest growth the scientific management and utilization of forest measures be compensated. Forest resource assets are affected by natural disasters and man-made destruction. The scientific and reasonable operation is not depreciation problem. Every year if we sell some assets of forest products the total amount of forest resources assets remained unchanged or increased slightly.

The second one is the long-term regeneration of forest resources. According to the forest growth, the products needs many years for its growth and after maybe 10 to 15 years or more than that we will able to sell this product and get revenue.

The third one is the vast distribution. The forest is widely distributed; usually, the forest resources assets management is widely distributed to tens of thousands of hectares.

The forth one is the diversity of function. The complex structure of the forest resources assets, shapes, have their own characteristics, some components of the forest resources assets in addition to the commodity attribute of value can be exchanged. Also has some value that is difficult to measure like ecological public welfare effect, esthetic value, land degradation, etc.

The fifth one is harder to manage. Forest resources assets are distributed all along different terrains, aspects, wild vast woodlands, no go areas the security guard of these assets is very difficult, like fire, insect, illegal logging and other man-made or natural disasters are difficult to control, so we can say that forest resources have possibility increased the risk of loss. The forest resources assets management must introduce the risk mechanism, in order to make the development of the socialist market economy. Forest asset evaluation is different from the general procedure, more complex and difficult. In addition considering the value of the forest also consider the status and role of the forest.

6.2.2 Forest resources evaluation

Forest land resources are used in the regeneration of forest resources, forest regeneration is the most basic means of production. It is an important part of forest resources and an indispensable factor of production and non-renewable.

1) Present cost method

Woodland resource evaluation system including location, type, slope, soil erosion, pH value, organic matter, buried underground water depth, salinity atmospheric environment, and other elements. These elements are most costly and they cannot be reliably measured. In the case of forest land transactions or events, we can evaluate with the current cost of forest land of forest resources as the basis.

Current woodland cost refers to the current market conditions for forest assets; the current price is required to pay cash or its equivalent.

You can use the current cost method to the trading behavior of forest land resource evaluation mainly include usufruct of forest land in the market and the relationship between the change of business ownership after payment of cash and reliable measurement of equivalents. The lessee and the rent, mortgage and mortgage party woodland rent or mortgage relations after expenses.

2) Present income value method

Forest income approach is the assessment of future value forest as well as the distant future earnings, all discounted at present value. This is calculated essentially expected land price calculation method. As a result of selective cutting in the uneven-aged forest, the profitability of forest trees is continuous, cannot be broken down. We can calculate the expected price formula and the result is a composite price of forest trees of the forest floor. To determine the value of forest it must separate the value of the land and the value of its trees. Usually divided manner proportional coefficient method is used the formula is as follows:

$$B_u = A_u / (1+p) n (n^{th} \text{ power}) - V/P, \quad (6-1)$$
$$B = B_u \times K \quad (6-2)$$
$$K = k_1 \times k_2 \times k_3 \quad (6-3)$$

Where: B_u is present income value; A_u is selective cutting income; V is the management and protection expenditure; p is the interest rate; n is selective cutting cycle; K is the adjustment factor (where k_1 is the site quality factor, k_2 for the location level adjustment coefficient, k_3 for the price index); B for the land price.

k_1: Is based on site quality score table and can be drawn from the ratio of forest woodland and other site type class IV, in order to adjust the other site types of forest land.

k_2: Standing tree price at the time of the actual stand cutting / reference to standing price at the time of main cutting.

k_3: Due to the principle of the timeliness of cash, in order to ensure the value of the assessment can be effective in a long period of time so that the price of forest and land prices and the same period between the price of material products remain stable to ensure the basic forest production and expanded reproduction process, as well as in various forestry economic activities in an independent position and play a regulatory role. Therefore, the price of forest land and government rent must be changed dynamically. The value of which is based on official data published by the National Bureau of Statistics.

Current woodland cost method and income present value method is a kind of macro forest land value evaluation method. It can be used for general analysis of the policy. Unable to make reasonable analysis, each element of forest land carry on comparative analysis to the impact of various factors.

3) The monitoring method of spatial data

The monitoring method of spatial data is pointed by using 3S (GPS, RS, GIS) technology by monitoring record of tree species in different forest growth situation and the mathematical model to adopt integration technology setting up a corresponding model to be analyzed. It contains main forest land suitability evaluation, tree forest productivity evaluation, woodland forest site type and the microscopic evaluation between various factors.

Forest productivity includes natural productivity and economic productivity. Natural productivity refers to the absence of any human material input. The forest that relies on natural resources may be developed by organic matter, biotic factors naturally. Economic productivity refers to the interference of human input, which must be the sum of the number of labor after the formation of the organic matter. VNPP (vegetation net primary productivity) estimation model mainly includes climate-related model, light energy utilization model and process model.

Woodland forest site type classification evaluation refers to different site conditions which are responsible for the growth of the trees. Different methods, technical measures are used; each type of site quality is suitable for one specific type of tree species. Now we are using 3S and RS technology to obtain variety of aerial images and information pluralism in relief maps, soil maps, geological maps, etc. Thematic maps and other site factors rasterized on the basis of the analysis carried out stacked on GIS platform by means of mathematical techniques related to the formation of visual expression, forest resources evaluation system easy to use. This is also the assessment of forest resources and cutting-edge trends.

6.2.3 Forest ecological social benefit evaluation

1) Forest resources evaluation

Forest resource assets evaluation is the most important content. Most of the forest resource assetswere trading forest resources assets. Use of assets can be divided according to forest resources i.e. timber, forest with ecological forests. Forest resources assessment is a long process of accumulation which is affected by various factors which can be assessed by special risk assessment. These are the followings risks like disaster risk, interest rate risk, management risk, policy risk, price risk, risk and alternative species risk. Therefore as we discussed that the valuation theory forest resources assets appraisal has a very important practical significance. Assets evaluation for different forest types is different; we select the appropriate assessment methods and forest quality adjustment factor assessment to estimate evaluation methods which are the following:

(1) Market act.

(i) Price-back: Market price pour algorithm is made after timber being evaluated by timber harvesting market total sales revenue and deducting the cost of the timber business consumption (including related taxes) and deserved profits remaining forest assets as part of the appraised value. The formula is:

$$E_n = W - C - F \qquad (6-4)$$

In the above formula: E_n is forest assets evaluation value; W is total sales revenue; C is timber operating costs (including the cost of harvesting, sales expenses, administrative expenses, financial expenses and related taxes); F is timber operator's reasonable profit.

(ii) The prevailing market price method: The prevailing market price method is based on the same or similar current market price of forest assets as a basis for comparison. The estimation method is evaluated to assess the value of forest assets. The formula is:

$$E_n = K \cdot K_b \cdot G \cdot M \qquad (6-5)$$

In this formula: E_n is forest assets evaluation value; K is stand quality adjustment factor; K_b is price index adjustment coefficient; G is reference accumulation unit trading price (RMB/cubic meter); M is being evaluated forest stock volume of assets.

(2) Income approach.

(i) Income net present value method: Net present value method is used to assess the forest assets net income for each year of the next operating period at a certain discount rate to present value and the income to the cumulative sum method to assess the value of forest assets. The formula is:

$$E_n = \sum_{i=n}^{u} \frac{(A_i - C_i)}{(1+p)^{i-n+1}} \tag{6-6}$$

Where in: E_n is the forest assets evaluation value; A_i is i annual income; C_i is i annual costs; u is operating period; p is discount rate (determined in accordance with specific local forest average investment earnings); n is stand age.

(ii) Harvesting present value method: Harvesting present value method is the use of the discounted value of the harvest table to predict forest assets at felling net income. The difference between the discounted value of forest production cost after deducting the expenses during the assessment to the felling of trees as a way to assess the value of assets. The formula is:

$$E_n = K \frac{A_u + D_a(1+p)^{u-a} + D_b(1+p)^{u-b} + \cdots}{(1+p)^{u-n}} - \sum_{i=n}^{u} \frac{C_i}{(1+p)^{i-n+1}} \tag{6-7}$$

Whereas in formula: E_n is forest assets evaluation value; K is stand quality adjustment factor; Stand in net income A_u is standard felling time (refer to timber sales revenue after deducting harvesting costs, selling expenses, management fees, finance charges, taxes and fees, some reasonable profit after the timber business); D_a, D_b are standard forest -a, b of the thinning net income; C_i is i year silvicultural production costs; u is operating period; n is stand age; p is rates.

(iii) Annuity capitalization approach: Pension benefit capitalization method is stable income will be evaluated annually forest assets as capital investment, according to an appropriate investment yield estimate the value of forest assets evaluation methods. The formula is:

$$E_n = \frac{A}{P} \tag{6-8}$$

Where in the assessed value: E_n is forest assets; A is average annual net income (net of rent).

(3) Cost method.

(i) The number of workers required sequence method: The number of workers required in sequence method is based on the existing man-days and the cost of production forest asset management method of the average number of workers in each forest workers estimated replacement value of assets. The formula is:

$$E_n = K \cdot \sum_{i=1}^{n} N_i \cdot B \cdot (1+P)^{n-i+1} + \frac{[R(1+P)^n - 1]}{P} \tag{6-9}$$

Where in: E_n is forest assets evaluation value; K is stand quality adjustment factor; N_i is the required number of workers in year i; B is the production cost in terms of working days at the time of assessment; P is interest rates; R is rent; n is stand age.

(ii) Replacement cost method: Replacement cost method is based on the existing level of production and wages to re-create a similar assessment with the cost of stand desired phase of forest assets. It is evaluated as a method to assess the value of forest assets. The formula is:

$$E_n = K \cdot \sum_{i=1}^{n} C_i(1 + P)^{n-i+1} \tag{6-10}$$

Where in formula: E_n is forest assets evaluation value; K is stand quality adjustment factor; C_i is i year at current production levels and wages for the standard calculation of production costs, including wages for each year of investment, consumption of materials, rent, etc; n is stand age; P is rates.

(iii) The historical cost method of adjustment: Historical cost adjustment method is based on the cost of inputs when the basis of the method is evaluated to determine the value of forest assets evaluation and assessment of price index changes according to the time investment. The formula is:

$$E_n = K \cdot \sum_{i=1}^{n} C_i \frac{B}{B_i}(1 + P)^{n-i+1} \tag{6-11}$$

Where in: E_n is the assessed value of forest assets; K is stand quality adjustment factor;

The actual cost of the i-th input; B is assessment of the price index; B_i is input price index when; P is interest rates; n is stand age.

(iv) Different types of the forest resources valuation methods

Timber (including firewood) forest resources valuation methods: Selected timber forest assets evaluation generally based on the type of forest managementage groups. The young forest is generally used in the current market price method, replacement cost method and a serial number of workers required method. The middle-aged forest is generally used in the current market price method the present value of the harvest method. When using the present value method must be harvested energy.

Harvest reflecting the growth table or tables of the local production process. In the absence of these data, tables can also be used in the investigation of local materials, fitting the local forest average growth process, in order to obtain the predicted value. In near mature forest, the main choice of the current market price method in the market price pour algorithm. Timber forest assets evaluation should give full attention to convergence between the various age groups of the assessed value.

Forest resources valuation methods: Forest assets are generally used to assess the prevailing market price method, the income approach, and replacement cost method. Considering the economic life of the forest business income approach should be chosen forest residual differences between growth stages of forest product yields and cost the end of economic life.

Shelterbelt forest resources assets evaluation methods: The main purpose of the shelterbelt forest is to act as wind and sand breaker, homeland security, improve agricultural production conditions and other protective functions. Shelterbelt asset valuation including the value and benefits of forest ecological protection of assessment estimated market value which is generally used in forest valuation income approach and replacement cost method.

2) The forest landscape resources evaluation

Evaluation of forest landscape is an important part of the forest resources management system.

With the development of forest tourism, landscape evaluation has become the main content of the forest landscape evaluation. Forest landscape biological resources are mainly divided into landscape resources, water resources, cultural landscape, sky landscape resources and its formation factors. Forest resources assessment through landscape resources evaluation method depends on size, structure, composition, function of landscape resource. The economic development potential of the forest landscape resources depends largely on the quality of the landscape and its development and utilization conditions.

The forest landscape resources evaluation mainly includes experts, psychophysical school of thought and cognitive school of thought. Psychophysical faction is mainly based on the introduction of psychophysics signal detection method to evaluate the quality of the landscape and scenery aesthetic landscape it as a stimulus-response relationship. This specific approach is based on public attitudes to the aesthetic landscape scenery obtain a reaction mass scale and then establish a direct mathematical relationship to determine the scale and composition of a landscape. Among them, the method proposed by Daniel and Boster and the comparative analysis proposed by Buhyoff *et al.* are the most effective methods for the application of forest landscape assessment. The beauty degree judgment method (SBE) judges the ideal representative value of beauty degree by analyzing the value of SBE as the judgment result, which is the result of the combination of perception and judging standard of the beauty of the observer. Comparative discrimination (LCJ) is the first to judge the landscape and then comparative analysis.

The cognitive school and the psychophysical school also judge the overall quality of the scenery by measuring the composition of the landscape, but the human school of thought is more focused on the aesthetic process of the landscape as a living space.

The phenomen ological school is the person in the aesthetic judgment of the competent role of the absolute height. It regards what distinguishes the landscape as the expression of human personality and its cultural and historical background, ambition and taste. Based on the method of expert school, this portion of the book introduces the establishment of forest landscape evaluation system from the perspective of aesthetics. Evaluation index is divided into novelty, diversification, naturalness, mystery, scientific value, historical value, harmony and coordination to establish a systematic evaluation system.

(1) 'Forest landscape rrsources evaluation system for aesthetic value' 'Central south forest inventory and planning'. In the Table 6-1 A-1, A-3 applies to a single forest landscape evaluation while A-2 is suitable for evaluation of a scenic or scenic.

Forest landscape novelty composite score can be calculated as:

Table 6-1 Forest landscape aesthetics novelty factor classification table

Programme	Distance	Scale	New species	Score range
No	A-1	A-2	A-3	
I	Closer	Very small		10~20
II	Close	Small	Little	20~40

III	Far	Big	Many		40~60
IV	Farther	Bigger	More		60~80
V	Far away	Biggest	Most		80~100

$$A = \sum_{i=-1}^{3} a_i \eta_i \qquad (6\text{-}12)$$

Where in formula: A is novelty score; a_i is a weighting factor score; a factor i is a weighting factor.

(2) The degree of diversification. These five factors can be graded and score as previously described, in the same manner, the extent of forest landscape diversity score, the formula is as follows (Table 6-2):

Table 6-2 Forest landscape aesthetics of diversification factor classification table

Grade	Number of species	Seasonal variation	Individual variation	Age changes	Structural complexity	Score range
	B-1	B-2	B-3	B-4	B-5	
I	Smaller				Very simple	0~20
II	Small	Little	Little	Little	Simple	20~40
III	Big	Many	Many	Many	Complex	40~60
IV	Bigger	More	More	More	Very complex	60~80
V	Biggest	Most	Most	Most	More complex	80~100

$$B = \sum_{i=-1}^{6} b_i \eta_i \qquad (6\text{-}13)$$

The natural forest landscape of mystery and aesthetic evaluation calculated using the following formula (Table 6-3):

Table 6-3 Type natural forest landscape and aesthetic mystique factor classification table

Grade	Human disturbance degree	Origin	Deep depth	Opening	Score range
	C-1	C-2	C-3	C-4	
I		Virgin forest		Very wider	80~100
II	Little	Secondary forest	Little	More wider	60~80
III	Many	Natural regeneration	Many	Wider	40~60
IV	More	Artificial promoting Regeneration	More	Occlusion	20~40
V	Most	Artificial forest	Most	Very occlusion	0~20

$$C = \sum_{i=-1}^{4} C_i \eta_i \qquad (6\text{-}14)$$

Where: C is a factor score; $d = a_3 N^{b_3}$ is weighted.

(3) Scientific and historical value. Besides recreation, learning is a great motivation in travel. Thus forest landscape has scientific value also affect the aesthetic quality of landscape, scientific value refers to the representation of vegetation, and whether there is scientific research, teaching

and so on. The historical value of forest landscape in the process of its formation is accompanied by historical events, and historical legends (Table 6-4), etc.

Table 6-4 Forest landscape aesthetic value of historical sciences factor classification table

Grade	Typical representative	Value of scientific research	Historical background	Score range
	D-1	D-2	D-3	
I	Rare	Higher value	More	80~100
II	Seldom	More value	Most	60~80
III	Un-common	Moderatevalue	Common	40~60
IV	Very un-common	General value	Seldom	20~40
V	Common	Hardly useful	Little	0~20

The scientific historical value score is calculated as follows:

$$D = \sum_{i=-1} d_i \eta_i \qquad (6-15)$$

Where: d_i is a weighting factor; η_i is a factor score.

(4) Harmonious coordination. Harmony forest landscape mainly refers to the internal and external harmonious coordination, identification of this factor mainly from the color line in combination with each other as well as the structure and aesthetics of the aforementioned factors. There is no single factor can represent that such a stand have harmonious coordination. Generally, may be carried out according to the following classification criteria.

Level I: Very harmonious and coordinated. The landscape with the surrounding environment and internal structure of lines has a natural transition, good color combination, and reasonable structure.

Level II: Coordinated landscape with the surrounding environment is in the form of similar or compatible, no obvious color contrast changes.

Level III: More coordinated landscape in the form of structure and the surrounding environment there is a clear transition, little color change.

Level IV: Uncoordinated. Landscape and surroundings transition is not natural, there are tough undermine the overall lines and color contrast style inconsistent.

Level-V: Landscape surrounding undermines the overall effect of the presence of other environments. Dazzling color contrast did not match the style change, seriously affecting the viewer's psychological balance, low aesthetic value.

Respectively, A, B, C, D, E weight.

Score in 0~20 is rated as general landscape area;

Scores in 20~40 is rated as more beautiful landscape area;

Scores in 40~60 is rated as the United States landscape area;

Scores in 60~80 is rated as the beautiful landscape area;

Scores in 80~100 is rated as a wonderful landscape area.

The evaluation system of each factor criterion more intuitive and easy to master, although it is dependent on the expert scoring method, but in addition to quiet depth, openness and coordination is difficult to master, even if other factors is the landscape evaluation workers is also common to master and use, and not because the evaluation personnel are not the same and have great difference(Figure 6-2).

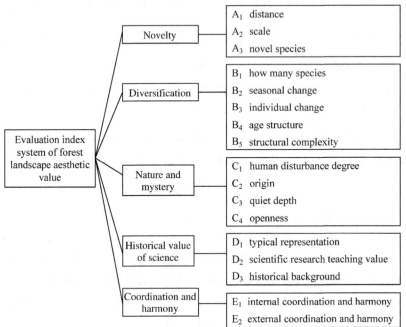

Figure 6-2 Standard evaluation system

3) Forest resources assessment other flora and fauna

Wildlife resources and plants in the forest are the main part of the evaluation and are the sum of wildlife resources including both flora and fauna. Wildlife resources belong to the natural biological resources, which refers to free survival of animals in the wild and habitats in natural vegetation. China is the among world's wildlife-rich countries, China's use of wildlife is more reflected in the drugs, traditional handicrafts, food, leather products, etc. In addition aesthetic wildlife ornamental value, the value of bio-genetic, biological research value, is not negligible.

Wildlife as an important economic, ecological resource also plays an indispensable role in promoting economic development and maintaining the ecological balance. It will use according to the following wildlife resources are a monetary expression that is a measure of the price proposed specific assessment methods.

(1) Wildlife valuation concept. Wildlife resource value is the wildlife resources to meet people's property and capacity needs as production and living materials, or namental objects, objects of scientific research the value of wildlife. Valuation wildlife resources assessment and refers to estimates of the needs of people and wildlife resources related to the value carried out.

(2) Characteristics value of wildlife resources. The preciousness of wildlife resources can be largely determined by the scarcity, utility, energy and demand of wildlife resources. First of all the

wild animal and plant resources because of human want only killing and cutting, its use have been far more than the wildlife resources regeneration. The contradiction between economic development and the protection of wild animals and plants has become increasingly acute. In many areas despite the sustainable and long-term development of the economy, only the immediate and temporary interests have been focused and due to this, the natural resources are destroyed. According to incomplete statistics in the world, every day 100 species wild animals and plants come near to extinction.

Wildlife resources play an important role in economic development and ecological balance, and wild animals and plants attract people who are busy in the social environment all day long to watch. As wild animals and plants of important tourism resources, wild animals and plants ornamental value cannot be ignored. In addition, the wildlife scientific research value is more and more people have been recognized.

(3) The method of valuation of wildlife resources. Use according to wildlife resources, the wildlife resource value assessment methods are divided into three major methods. Evaluation Products Act respectively, as well as corporate law for non-productive sectors.

(i) Products Act: First when the wildlife resources as a direct product, the product yield and multiplying the unit price, then the value of the sum of all kinds of products, a total value of wildlife products. This method is suitable when wildlife resources as a direct product exchange, the calculation formula is as follows:

$$S = \sum_i^n P_i O_i \qquad (6\text{-}16)$$

In the formula: S is total price of wildlife; P_i is prices of i-th product; Q_i is number of i-th product; n is product category number.

(ii) Corporate law: This valuation method following a wild animal and plant resources used in processing and manufacturing enterprises. When evaluated on wildlife resources for this purpose, usually manufacturing companies is seen as a whole, the specific approach is to assess production of various products in each category multiplied by the unit price of the product and then summed. The formula is:

$$S = \sum_i^n P_i O_i \qquad (6\text{-}17)$$

In the formula: S is overall value of the enterprise; P_i is i-th product prices; Q_i is number of i-th product; n is production products.

(iii) Wildlife resources for the evaluation method of non-productive sectors: When wildlife resources for the leisure industry, especially tourism, at this time belong for-profit, non-material production sectors. Since the profit-making activity occurred when required expenses are derived from its income, therefore wildlife scenery, tourism value equal to all those engaged in unit revenue industries is combined.

In addition, the wildlife resources will be used for scientific research. When wildlife for scientific research, wildlife resources at this time will not be used for profit, they are usually used for maintenance of the public environment, education, culture, health, social welfare and scientific research, these industries usually not only not profitable, the state will spend a lot of burdens. Thus the value of wildlife resources at this time to be its research department, management department should pay for the costs associated. When wildlife resources for this type of non-profit, non-material production sectors valuation formula is: total value wildlife resources = daily expenses of all wildlife research, administration expenses + fixed assets + fixed virtual asset depreciation expense.

(iv) The value of forest resources and value of ecosystem services: According to the 'forest ecosystem services assessment specification' can be calculated the following ecosystem services value.

Vale of water conservation: According to the method to measure the value of forest as water conservation, first find the total amount of water conservation forest, the total forest impoundment of water i. e. is the amount of precipitation and evaporation forest areas and other poor consumption, namely:

$$x_3 = 0 \qquad (6-18)$$

Where: Y is amount of water intercept by forest; A is rain forest impoundment area (hm^2); P is precipitation (mm/a); E is evapo-transpiration (mm/a); C is surface runoff (mm/a), because of the forest surface runoff few negligible.

In the 2011 data, for example, Beijing 2011 annual precipitation is 721.1mm, ET is about 60% (in 1994, mainly Mentougou data)

Regulation of water value:

$$U_{tune} = 10CA(P - E - C) \qquad (6-19)$$

Where: U_{tune} is stand on regulation of water value (RMB ¥/a); C-library is reservoir construction unit capacity investment (re-location compensation, cost and maintenance costs, etc.), (RMB ¥/m^3); A is eleven stand area (hm^2); P is one precipitation (mm/a); E is stand evapo-transpiration (mm/a); C is runoff (mm/a) because of the forest runoff few negligible.

The value of forest storage and precipitation is equivalent to the value of equal-volume reservoir, accounting for the construction cost of the reservoir with 1 m^3 of water. According to the survey 1 m^3 of flood storage reservoirs, the construction costs is about 0.5 RMB ¥/m^3

Water purification value: Due to the existence of forest vegetation the streams and rivers have the dissolved oxygen (DO), which have a great impact on pathogens and chemical properties of water. In the course of water flow through the forest i. e. the canopy layer, the litter layer under the canopy and soil layer all these filters the water from various pollutants, absorption and purification, thus purifying the water quality. State Planning Commission and Ministry of Health jointly issued standards for drinking water standard, for Industrial Enterprises and Standards for Farmland Irrigation i. e the water pH, volatile phenols (DO), arsenic, mercury, cadmium, chromium, lead and other 14 indicators in full compliance with drinking water standards. The water flowing out of the forest basin is cool and rich in dissolved oxygen (DO). Because the main heat source of the forest stream is direct solar energy on the surface of the stream the sunshine of the canopy surface is less, the water temperature is lower and the dissolved oxygen (DO) content is higher because of the shading effect of the forest canopy. Forest soil has a good aggregate structure, suitable for microbial growth in the temperature and humidity conditions, the entire ground layer making the forest than the open forest has a stronger purification. In short, the living matter and litter layer retention and filtration, microbial decomposition of compounds ground objects on the ion uptake the physical adsorption of soil particles, soil chemical adsorption of metal elements such as precipitation and so on through the water quality of forest ecosystem is obviously improved.

$$U_{water} = 10KA(P-E-C) \tag{6-20}$$

In the formula: U is the value of regulated water; C is library—reservoir construction unit capacity investment (land demolition compensation, project cost, maintenance costs, etc.), (RMB ¥/m^3); A is 1 stand area (hm^2); P is precipitation (mm/a); E is stand evapo-transpiration (mm/a); C is surface runoff (mm/a); U is water quality – decontamination value of water purification (RMB ¥/a); KA is water purification costs (RMB ¥/t)

Soil conservation valuation:

$$G_{earth\ soil} = A(X_2 - X_1) \tag{6-21}$$

In the formula: $G_{earth\ soil}$ is stand soil volume (t/a); A is stand area (hm^2); X_2 is the soil erosion modulus [t/(hm^2·a)]; X_1 is woodland [t/(hm^2·a)];

$$G_N = AN(X_2 - X_1) \tag{6-22}$$

Type: G_N is reduced the loss of N (t/a); A is stand area (hm^2); N is soil nitrogen content (%); X_2 is Soil erosion modulus [t/(hm^2·a)]; X_1 is soil erosion modulus [t/(hm^2·a)].

$$G_P = A_P(X_2 - X_1) \tag{6-23}$$

Type: G_N is reduced the loss of P (t/a); A is stand area (hm^2); N is P content in soil (%); X_2 is soil erosion modulus [t/(hm^2·a)]; X_1 is the soil erosion modulus [t/(hm^2·a)]–woodland.

$$G_K = A_K(X_2 - X_1) \tag{6-24}$$

Type: G_N is reduced the loss of K (t/a); A is stand area (hm^2); N is K content in soil (%); X_2 is soil erosion modulus [t/(hm^2·a)]; X_1 is soil erosion modulus [t/(hm^2·a)].

Because there is no relevant experimental data, soil data in the survey of forest resources in Shandong Province in 2008 as a reference and required follow-up of more detailed investigation and sampling) forest soil, N was 0.370%, P was 0.108%, K 2.239%. The annual soil erosion modulus was 36.85 t/hm^2

The value of solid earth:

$$U = AC_{solid\ earth\ soil}(X_2 \cdot X_1)/P \tag{6-25}$$

Where: $U_{Solid\ soil}$ is solid stand on the amount of soil value (RMB ¥/a); A is stand area (hm^2); C_{soil} is digging earthworks and transport unit volume required fee (RMB ¥/m^3); X_2 is no soil erosion modulus [t/(hm^2·a)]; X_1 is soil erosion modulus [t/(hm^2·a)]; P is soil bulk density (t/m^3).

Fertilizer value:

$$U_{fertilizer} = G_N C_N + G_P C_P + G_K C_K$$

Where: $U_{fertilizer}$ is stand in fertilizer value (RMB ¥/a); G_N is reduced N Loss(t/a); G_P is P loss reduced(t/a); G_K is reduced k Loss(t/a); C_N is ammonium bicarbonate price(RMB ¥/t); C_P is superphosphate price(RMB ¥/t); C_K is price potassium(RMB ¥/t).

Adjust the temperature, clean up the environment:

The value of forest absorption of SO_2: Sulfur dioxide is the most harmful gas and most widely distributed. Trees have a certain sulfur dioxide absorption capacity. Cost alternative method is required to mitigate the harm of sulfur dioxide to use the value of forest resources to absorb sulfur dioxide. According to the Nanjing Institute of Environmental Science, State Environmental Protection Administration a report named 'China National Research Report on Biodiversity', evaluate the value of forest purification SO_2 in China by using the average management fee of SO_2. Its basis is: the absorption capacity of forest to SO_2:

Broadleaf forest: $\quad q_1 = 88.65 \text{ kg/(hm}^2 \cdot \text{a)}$

Coniferous forest: $\quad q_2 = 215.6 \text{ kg/(hm}^2 \cdot \text{a)}$ (6-26)

The value of forests to absorb fluoride: According to the Beijing Municipal Environmental Protection Science Research Institute for the determination of emissions of hydrogen fluoride in the vicinity of the enamel plant, white poplar, Canadian poplar, *Robinia pseudoacacia*, locust, ash and other hardwood the fluorine absorption capacity is the highest reaching up to 4.65 kg/hm². Chinese pine and in other evergreen the fluorine absorption capacity is 0.5 kg/hm². Forests absorb hydrogen fluoride using the average price of coal-fired furnace air pollutant emission fees about 0.16 RMB ¥/kg.

The value of forests to absorb nitrogen oxides: According to the study, the annual capacity of forests to absorb nitrogen oxides is 380 kg/hm². China's air pollution charges according to the average of standard pollutants are namely 1.34 RMB ¥/kg; calculate the value of the value of forests to absorb nitrogen oxides.

The value of forest carbon sequestration of oxygen: Vegetation carbon sequestration:

$$G_{\text{vegetation carbon sequestration}} = 1.63 R_{\text{carbon}} \quad (6\text{-}27)$$

Where: $G_{\text{vegetation carbon sequestration}}$ is carbon sequestration in vegetation (t/a); R_{carbon} is carbon dioxide in the carbon content of 27.27%; A is stand area (hm²); B is the stand net productivity (hm² · a/t).

The release of oxygen:

$$G_{\text{oxygen}} = 1.19 AB \text{ oxygen years} \quad (6\text{-}28)$$

Where: G_{oxygen} is stand release of oxygen (t/a); A is stand area (hm²); B is the stand net productivity [t/(hm² · a)].

The value of carbon sequestration:

$$U = B\ C\ A 1.63 R \text{ years} \quad (6\text{-}29)$$

Where: U is stand on carbon sequestration value (RMB ¥/a); R is carbon dioxide in the carbon content of 27.27%; C is solid carbon price (RMB ¥/t); A is stand area (hm²); B is the stand net productivity [t/(hm² · a)].

Oxygen value:

$$U = C_{\text{oxygen}}\ 1.19 AB \text{ years} \quad (6\text{-}30)$$

Where: U_{oxygen} is oxygen stand on solid value (RMB ¥/a); R_{oxygen} is carbon dioxide in the carbon content of 27.27%; C is solid carbon price (RMB ¥/t); A is stand area (hm^2); B is the stand net productivity [t/($hm^2 \cdot a$)].

According to the purchase price of oxygen is 2.4 RMB ¥/kg basis.

6.3 The forest management precision sub-compartment cluster analysis

6.3.1 Sub-compartment cluster analysis based on site condition

Sub-compartment cluster analysis is based on many research fields including data mining, statistics, equipment learning, and pattern recognition and so on. As a function of a data mining, cluster analysis can be used as a stand-alone tool to obtain the distribution of data and summarizes the characteristics of each cluster or focus on the further analysis of the specific clusters.

Data mining is a prominent feature of handling the huge, complex data sets, this clustering technology presents special challenges like that require scalable algorithms, processing capability of different types of property, clusters of arbitrary shape, high-dimensional data processing capability. According to various potential applications, data mining clustering analysis put forward different demands.

The input of the cluster analysis can be represented by a set of ordered pairs (X, s) or (X, d), X represents a set of samples while S and D respectively is a measure of similarity or dissimilarity between samples (distance) of the standard. The output of the cluster system is a partition, If $C = \{C_1, C_2, \cdots, C_k\}$, in which $C_i(i=1, 2, \cdots, K)$ is a subset of X, and is satisfied

$$\begin{cases} C_1 \cup C_2 \cup, \cdots, \cup \quad C_K = X \\ C_1 \cap C_2 = \varnothing, \quad i \neq j \end{cases} \quad (6\text{-}31)$$

C member in $C_1, C_2, \cdots, C_K$ is known as classes or clusters (cluster), every class or cluster is described by a number of features usually said through their centers or distant relations (boundary) represents the spatial point for a class, using clustering tree in the graph nodes represents a class.

The goal of cluster analysis of is the formation of clusters of data and to satisfy both of the following conditions, i.e. the data within a cluster of possible similarity (high intra-class similarity); try not to have different clusters of similar data (low inter-class similarity).

Cluster analysis is a method based on the characteristics of the thing itself, the purpose of the study is to classify the similar things. Its principle is that the individuals in the same category have great similarity and the individual in different categories have very large differences. This approach has three features firstly it is not suitable for classification without prior knowledge. If there is no prior experience of some international, domestic and industry standards the classification will appear to be arbitrary and subjective. At this time as long as the set of perfect classification variables, you can get a more scientific and rational classification by the method of cluster analysis. For example, to be based on consumer purchases size classification is relatively easy, but if during data mining, according to the requirements of various indicators of consumer purchases, household income, household spending and age to classify are often complicated and poly class analysis can solve these problems. Cluster analysis is an exploratory analysis, the ability to analyze the characteristics and the inherent law of things, and according to the principle of similarity grouping of things, is commonly used as a data mining technique.

6.3.2 Main principle and the calculation method of cluster analysis

According to the main idea of the clustering analysis algorithm, it can be summarized as follows.

1) Partition method

Classification of data based on certain standards. The data set is divided into K groups each group contains at least one data record and each data record belongs to a group each of which is composed of a class. Change the grouping by an iterative method, so that each time after the improvement of the grouping scheme is better than the previous one. The standard is the closer the record in the same group, the better, and the different groups in the record as far as possible. Use this basic idea of the algorithm: K-means、K-modes、K-prototypes、K-medoids、PAM、CLARA、CLARANS etc.

2) Hierarchy process

For a given set of data objects, hierarchical decomposition is generally divided into 'bottom-up' program and the 'top-down' program. 'Bottom-up' program: each data alone and as a group, through iterative method, those adjacent to each other and combined into a group, until all of a group of records or a certain condition is satisfied, on behalf of algorithms: BIRCH algorithm, CURE algorithm, CHAMELEON algorithm. 'Top-down' program: including the K-MEANS algorithm, K-MEDOIDS algorithm, Clara algorithm, Clarans algorithm.

(1) K-MEANS algorithm. choose k objects from n data objects as the initial cluster centers and for the rest of the other objects, according to their similarity (distance) of these cluster centers, assign them to their respective most similar (cluster centers represented) clustering; calculated for each cluster center obtained a new cluster (cluster mean that all objects). Repeats this process until the beginning of the standard measurement function converges. It is generally used as the standard deviation measurement function. K clustering has the following characteristics: each cluster itself as compact as possible, but as much as possible between each cluster separately.

(2) K-MEDOIDS algorithm. K-MEANS has the drawback that the size of the generated class is not very large, nor very sensitive to the data. K-MEDOIDS for its improved algorithm we select an object called mediod to replace the role of the center of the above such a medoid identifies the class.

(3) Clara algorithm. K-medoids algorithm is not suitable for a large amount of data calculation. The idea of Clara algorithm is to replace the whole data with the sampling of the actual data and then use the K-medoids algorithm to get the best medoids algorithm on these samples data. Clara algorithm from the actual data to extract more than one sample in each of the samples are u-

sing K-medoids algorithm to get the corresponding $(O_1, O_2, \cdots, O_i, \cdots, O_k)$, then select one of the smallest E as the final result.

(4) Clarans algorithm. The efficiency of the Clara algorithm depends on the size of the sample and is generally less likely to get the best results. On the basis of Clara algorithm, the algorithm of Clarans is proposed. Clara algorithm is different from Claran algorithms in certain way that in the Clara algorithm to find the best medoids in the process the sample remains the same but in Clarans algorithm during each cycle of sampling used is not the same. Different from the process of finding the best medoids in the last class, it is necessary to limit the number of cycles.

3) Other methods

In addition to the above methods, there are some other methods which include density-based methods. These methods are based on the data object connected density evaluation. In grid-based methods, the data space is divided into a finite element of the grid structure, based on the grid structure for clustering. In model-based methods, it is supposed to be a model for each cluster and then look for the data set that can be very good to meet the model.

4) Small class clustering data processing method

(1) Total standardization. Respectively the corresponding data of the cluster elements is the sum of the data the elements of the data divided by the sum of the elements of the data:

$$x'_{ij} = \frac{x_{ij}}{\sum_{i=1}^{m} x_{ij}} \quad (i = 1, 2, \cdots, m; j = 1, 2, \cdots, n) \tag{6-32}$$

The new data obtained by this method is satisfied:

$$\sum_{i=1}^{m} x'_{ij} = 1 \quad (j = 1, 2, \cdots, n) \tag{6-33}$$

(2) Standard deviation. The new data obtained by this method is the average value of each factor is 0, the standard deviation is 1, that is, there is:

$$x'_{ij} = \frac{x_{ij} - \bar{x}_j}{s_j} \quad (i = 1, 2, \cdots, m; j = 1, 2, \cdots, n) \tag{6-34}$$

$$\bar{x}_j = \frac{1}{m} \sum_{i=1}^{m} x'_{ij} = 0 \qquad s_j = \sqrt{\frac{1}{m} \sum_{i=1}^{m} (x'_{ij} - \bar{x}'_j)^2} = 1 \tag{6-35}$$

(3) Maximum standardization. After this new data, the maximum value of each factor is 1, and the rest is less than 1.

$$x'_{ij} = \frac{x_{ij}}{\max_i \{x_{ij}\}} \quad (i = 1, 2, \cdots, m; j = 1, 2, \cdots, n) \tag{6-36}$$

(4) The range of standardization. After this new data, the maximum value of each factor is 1, the minimum is 0 and the rest is between 0 and 1.

$$x_{ij} = \frac{x_{ij} - \min_i \{x_{ij}\}}{\max_i \{x_{ij}\} - \min_i \{x_{ij}\}} \quad (i = 1, 2, \cdots, m; j = 1, 2, \cdots, n) \tag{6-37}$$

Common distance calculation: According to the axiom of the distance the four conditions of self-similarity, the minimumity, the symmetry and the triangle inequality are required when the distance measure is defined. Commonly used distance functions are as follows:

Absolute value distance:

$$d_{ij} = \sum_{k=1}^{n} |x_{ik} - x_{jk}| \quad (i, j = 1, 2, \cdots, m) \tag{6-38}$$

Euclidean distance:

$$d_{ij} = \sqrt{\sum_{k=1}^{n} (x_{ik} - x_{jk})^2} \quad (i, j = 1, 2, \cdots, m) \tag{6-39}$$

Minkowski metrics:

$$d_{ij} = \left[\sum_{k=1}^{n} |x_{ik} - x_{jk}|^p\right]^{\frac{1}{p}} \quad (i, j = 1, 2, ..., m) \tag{6-40}$$

Chebyshev distance:

When the Minkowski distance $p \to \infty$, there is:

$$d_{ij} = \max_{k} |x_{ij} - x_{ij}| \quad (i, j = 1, 2, ..., m) \tag{6-41}$$

The small growth-class situation can directly determine the growth status and growth potential of forest and it has importance for its significant in forest diversity management. In order to determine the growth process of small class accurately and timely, the cluster analysis was introduced into the study of the dominant factors of small class growth. Based on the data of small-scale forest resources investigation, the dominant tree species with the same management type and site type were selected and cluster analysis was carried out on several sub-classes and growth status of the investigation and identification of forest sub-plot was determined to provide appropriate measures. Several forest sub-compartment leading factor of poly class analysis and automatic clustering of each cluster (per class) for survey and identification of growth to provide corresponding measures for forest sub-compartment and management.

Cluster analysis method is the correlation between the dominant factors of forest size and the clustering method system studied by K-means algorithm and so on. Classification of forest classes was obtained on the basis of the comparison and analysis of the variance or the difference between different attributes of the same dominant factor among different forest classes. The results are analyzed so that the distance between two data objects belonging to any different class is greater than the distance between two data objects in the same class. Which can be applied to determine the growth status of small groups of trees, to provide technical support for speculating forest sub-class management measures and to provide a theoretical basis for scientific and effective development of forest management program.

5) The establishment of clustering analysis algorithm

In cluster analysis objects in one cluster are 'similar' but not 'similar' to objects in other clusters. The 'quality' of clustering can be expressed by 'diameter', and the diameter is the maximum distance between two arbitrary objects in a cluster. The centroid distance is another measure of clustering quality, which is defined as the average distance between the cluster centers (mean 'object', or the average point in the cluster space) to each clustering object.

Cluster analysis is divided into four-step process, which is data preprocessing, feature selection & extraction, clustering algorithm design, and evaluation of clustering results. The main goal of the cluster analysis algorithm is to use the clustering algorithm to achieve the dominant factor of small forest growth.

The experimental area in Huangnihe Forestry Bureau, based on sub-compartment data is located in the northwest of Dunhua City, Yanbian and Jilin Province. The geographical coordinates of the experimental area are east longitude 127°40′~128°42′, north latitude 43°25′~44°06′. south and Southeast is bounded by Dunhua City Forestry Bureau; Its west and northwest part are connected with the territory of the Jiaohe Baishishan Forestry Bureau and Jilin province forestry bureau Jiaohe experimentation area management; The north is adjacent to the Heilongjiang Province Forestry Bureau. Forestry bureau is located in the territory of Huangnihe town of Dunhua City, 35 km from the Dunhua area, about 196596 hm^2 (hectares).

The quality of data preprocessing is directly related to the authenticity and accuracy of the data analysis in the later stage. The preprocessing of data includes four steps. i. e. data cleaning, data integration, data transformation and data reduction. Data cleaning is usually defined by filling the gaps, smoothing the noise data, identifying the outlier to 'clean' data. Data integration is to combine data from multiple data sources in data storage. Data transformation is to transform the data into a form suitable for analysis and processing. Data reduction is the original data values with interval or a high-level conceptual replacement. Only in this way can be in close proximity to maintain the integrity of the original data based on, simplify the operation of data and produce approximately the same results.

For the growth of forest sub-compartment data, 2005 Huangnihe Forestry Bureau contains survey data from Okawa, Weihu River, Weihu ridge, North Mission, seedling of Peking University, Northwest Branch, birch bark, plant launched, bald Dingzi, EMU, loose spirit, Jin Gou, Tuanshanzi, are Ling, white, green hook, the malugou, Tara station. A total of 18 forest farms, 17048 sub-compartment. After removal of the missing values of the indicators, select the type of business consistent, site conditions consistent with a total of 2521, the growth of its dominant factor analysis.

It should be emphasized here that the various data preprocessing methods mentioned above are not independent of each other but are interrelated.

6) Selection and design of clustering algorithm

In the selection of clustering algorithm, the type of data, capacity and purpose of clustering should be considered. In general, the clustering algorithm is based on the idea of the algorithm and can be divided into a variety of forest sub-compartment clustering, based on site conditions analysis, dominant small growth, and factor such as slope (level), age, canopy density, per hectare, the volume factor (HD^2) etc. Several forest sub-compartment leading arbitrary factor A or several combined as a feature to distance based clustering analysis.

Distance-based clustering is a clustering method based on the distance from the boundary of the sample to the center of the sample. For a given sample, first create an initial partition number k, k value is generally required for a given, and then use the iterative re-positioning technology, through the object in the class to improve the classification results. In order to optimize the classification results, the partition based clustering may be exhaustive of all the division. At present, most of the clustering applications use *K*-Means and *K*-Medoids which are divided into two types. That is to say after the standardization of data processing, the software automatically generates the initial aggregation center, after the *N* step of the iterative process of the cluster center to complete the convergence, success will be successfully divided into forest class *K*.

Clustering is to find the potential of the data in the group, therefore the evaluation of the results of the cluster, i. e. the discovery of the cluster is an important part of the cluster analysis. When evaluating the clustering results, the optimal number of clusters is an important aspect of evaluating the effectiveness of clustering. Especially for multi-dimensional feature vector model, we cannot directly see the clustering effect, need to use a variety of indicators to measure it.

Clustering has two basic principles, i. e. compactness and separation. From that point of view of the distance between the objects in the cluster required to be as close as possible and it requires that the cluster should be as far away as possible. On the basis of this principle according to the research field and the goal the evaluation index of the cluster effect can be selected. The following are some of the commonly used effectiveness evaluation indicators:

(1) Distance matrix of cluster center. Cluster center vectors are provided, $i = 1, 2, \cdots, k$, the matrix $Z(i,j) i, j = 1, 2, \cdots, k$, the distance matrix between cluster centers. The greater the distance between the cluster centers the higher will be the quality of clustering.

(2) The distance variance between sample and cluster centers in each cluster domain is that the smaller the variance the better the clustering effect. We can see that these two indicators the former emphasizes the degree of separation between clusters while the latter emphasizes the tightness of the cluster.

(3) Dunn index. This is a clustering validity index proposed by Dunn, which aims to find a close and separate clustering. Where nc is the specified number of clusters, $I = 1, 2, \cdots,$ nc is the representative point of each cluster domain. Obviously, the Dunn index value indicates that the data set contains compact density and good separation degree.

$$D_{nc} = \min_{i=1,\wedge nc} \left\{ \min_{j=1+1,\wedge nc} \left[\frac{d(c_i,c_j)}{\max\limits_{k=1,\cdots,nc} diam(ck)} \right] \right\} \qquad (6\text{-}42)$$

In addition to the above cluster validity index, the experts and scholars also study and define a number of other evaluation indicators. It can be seen that all kinds of evaluation methods have their advantages and disadvantages and scope of application. For the same clustering algorithm, different evaluation functions are likely to be inconsistent results. Moreover, the data types of data objects are mostly numeric classes, and the evaluation of the classification attribute data and other complex data types should be further studied.

In this paper we select data from 2521 sub-compartments of Huangnihe forest, five forest sub-compartment factor obtained after finishing, respectively: the slope (level), age, canopy density, per hectare, volume factor (HD^2). Specific data for each case mentioned in the Table 6-5.

CHAPTER 6 System Design of Forest Precision Manangement and Manangement Platform

Table 6-5 Basic information description

	Number	Range	Minimum value	Maximum value	Mean		Standard deviation	Variation number
					Statistical information	Standard error		
Grade	2521	3	1	4	2.3	0.014	0.723	0.522
Age class	2521	5	0	5	2.56	0.023	1.143	1.307
Canopy density	2521	1	0	1	0.627	0.0041	0.2083	0.043
Per hectare	2521	5000	0	5000	1026.35	16.591	833.017	693917.932
Volume factor(HD^2)	2521	1.3328	0	1.3328	0.213126	0.0042533	0.2135581	0.046
Effective number	2521							

Using SPSS 22.0 data standardization processing, the software automatically generatesthe initial cluster center, after 16 step of the iterative process of the cluster center complete convergence, success will 2521 classes are divided into five categories, five kinds of forest sub-compartment factor the clustering of significant analysis(Table 6-6).

From table 6-6 we can know that the differences between the five small forest types are very significant. In addition, the degree of importance of each variable on the clustering results as follows: HD^2 > Age class> Per hectare> Canopy density> Grade (grade).

Table 6-6 ANOVA table

	Gather		Error		F	Significant
	Mean square	df	Mean square	df		
Zscore (Grade)	182.634	4	0.711	2516	256.785	0
Zscore (Age class)	485.031	4	0.23	2516	2104.483	0
Zscore (Canopy density)	425.395	4	0.325	2516	1307.753	0
Zscore (Per hectare)	436.544	4	0.308	2516	1419.37	0
Zscore (HD^2)	500.087	4	0.207	2516	2421.278	0

For a more detailed analysis of the meaning of each cluster, the original data contained in each cluster is described in the original data, see Table 6-7.

From table 6-6, we can see that this experiment selected five forest small size factors, the difference is very significant in each cluster, so the five factors selected in this paper are the dominant factor of the forest. As table 6-7 shows, By this method, Huangnihe Forestry Bureau forest sub-compartment is divided into five categories, namely: young forest, middle age forest and near mature forest and mature forest, over-mature forest and thinning area.

Table 6-7　Raw data description of The clustering small class

		Number	Range	Min value	Max value	Mean Statistical information	Mean Standard error	Standard deviation	Variation number
Class 1	Grade level	843	3	1	4	2.78	0.018	0.531	0.282
	Age class	843	2	1	3	2.22	0.021	0.609	0.371
	Canopy density	843	0.6	0.4	1	0.717	0.003	0.0869	0.008
	Per hectare	843	2400	800	3200	1632.72	15.251	442.81	196080.303
	Volume factor(HD^2)	843	0.384	0	0.384	0.102066	0.0028926	0.0839852	0.007
	Effective number	843							
Class 2	Grade level	190	3	1	4	2.04	0.052	0.715	0.512
	Age class	190	1	0	1	0.15	0.026	0.355	0.126
	Canopy density	190	0.4	0	0.4	0.043	0.0077	0.1056	0.011
	Per hectare	190	5000	0	5000	2461.05	40.694	560.927	314639.098
	Volume factor(HD^2)	190	1	0	1	0	0.003	0.038	0.001
	Effective number	190							
Class 3	Grade level	364	3	1	4	1.91	0.037	0.71	0.504
	Age class	364	2	3	5	4.16	0.021	0.402	0.162
	Canopy density	364	0.6	0.3	0.9	0.635	0.0054	0.1029	0.011
	Per hectare	364	1600	0	1600	544.71	21.157	403.65	162933.429
	Volume factor(HD^2)	364	0.9488	0.384	1.3328	0.619885	0.0078117	0.1490378	0.022
	Effective number	364							
Class 4	Grade level	440	3	1	4	2.43	0.031	0.651	0.424
	Age class	440	3	0	3	1.98	0.031	0.645	0.417
	Canopy density	440	0.9	0.2	1	0.615	0.0093	0.1947	0.038
	Per hectare	440	1500	0	1500	119.03	14.673	307.789	94733.837
	Volume factor(HD^2)	440	0.29	0	0.29	0.0713	0.00331	0.06936	0.005
	Effective number	440							
Class 5	Grade level	684	3	1	4	1.9	0.022	0.582	0.339
	Age class	684	2	2	4	3.17	0.02	0.513	0.263
	Canopy density	684	0.8	0.2	1	0.682	0.0037	0.098	0.01
	Per hectare	684	2300	0	2300	720.47	21.286	556.7	309914.645
	Volume factor(HD^2)	684	0.5004	0.018	0.5184	0.283173	0.0039579	0.1035117	0.011
	Effective number	684							

6.4 The cluster analysis of forest in small class based on forest type

For young and middle-aged forest tending measures taken include tillage, weeding, intercropping, irrigation, fertilization, drainage, pruning, under story planting and other work. Furthermore, it should choose a relatively flat area planted, so that growth is better and faster; proper thinning or final felling updates for the nearly mature forest, mature forest, for the over-mature forest; seen from Table 3. The next major work of harvesting in Huangnihe Forestry Bureau sub-compartment with class 1 and class 5-based that is the majority of the forest in sub-compartment are middle age forest to mature forest stage,

Classification unit of forest community is named as forest type. It is in accordance with the internal characteristics of the community the external characteristics and the dynamic law of the division of the same forest lots. The purpose of classification of forest types is for forest surveying, afforestation, management and planning design provides a scientific basis and takes different silvicultural measures of different types. The classification principles, units and systems of forest communities and the disciplines of the growth, development and spatial distribution of the forest communities under certain geographical conditions are called the forest types.

In the ecological school of thought, the unity of the stand and the habitat is emphasized. Habitat changes much more slowly than the stand, the interaction between the two determines the composition, structure, productivity and stand characteristics stand, etc., the amount of forest habitat conditions caused by its composition and productivity of qualitative change. This school of thought is an indicator of site conditions that is the same site conditions the differences of the forest are mainly affected by the soil factors there may be several forest type. The classification system is the type of site condition (forest plant condition type), and the basic unit of classification is the general term of soil nutrient and water condition.

Forest-type is the type of site condition of lower level units under the same conditions of water and fertilizer in soil due to differences in climate formed different forest types. The similar climate of forest and non-forest area collectively known as a forest type, site conditions is the type of climate variations. Stand type is the lower unit level of forest type namely under the same soil water, nutrients, and climatic conditions, dominant tree species in the similar stand form a stand type. Stand type can be divided according to age, canopy density, and productivity. So the cluster analysis based on the site condition is necessary.

This paper mainly focused the lower class in a kindergarten of Miyun County on the coniferous forest, broad-leaved forest for the work of data mining. Considering the hybrid mixed forest tree species, site conditions are not unified and sparse forest sub-compartment study of little significance, we do not place these two classes for data mining work. Specific to the class distribution is shown below (Table 6-8).

Table 6-8 Miyun forest subcompartment class statistics

Land type	Number offorest	Percentage of total/%
Coniferous forest	2009	54.1
Broad-leaved forest	900	24.2
Mixed forest	756	20.3
Sparse forest land	51	1.4
Total	3716	100

In Coniferous forest data mining the use of K-means clustering algorithm in this paper set the number of clusters for the 5 categories. The average index of stand growth can be seen and the second kind is better than other kinds of growth. The features of the second type are as follows: mean elevation in 776.99 m; average slope at 17.67°; soil average thickness in 26 cm; Bardo thin pine (Ⅳ 2) the site types are; almost all the forest is water conservation forest; the whole area is sloppy; the north (shady) slope is in the largest proportion; landform belongs to Zhongshan, almost all the cinnamon soil, the soil parent material is shale type, soil texture is the majority of loam; the vast majority of dominant tree species were larch (40.62%), which was significantly different from other classes, the second is Chinese pine (37.5%), followed by *Platycladus orientalis* (21.88%) (Table 6-9 and Table 6-10).

Table 6-9 Average growth index of all kinds of forest statistics

Class code	Average height of stand/m	Average DBH/cm	Canopy density	Timber Volume/(m^3/hm^2)	Vegetation coverage/%
1	5.982095	12.53394	0.54183	423.7326	83.89665
2	10.71125	15.49063	0.709375	1204.794	90
3	6.552174	11.90725	0.571014	272.8123	79.49275
4	6.993487	12.31454	0.619436	432.7526	81.06083
5	5.798784	10.6666	0.565347	364.0213	78.99614

Table 6-10 The characteristics of the second class

Salient features	Number, mean, or attribute code	Salient features	Number, mean, or attribute code
The number offorests	32	slope position	* 7 (50%)
Altitude	776.989	slope direction	* 4 (28.12%)
Slope angle	17.667	soil texture	* 2 (46.88%)
Soil thickness	26	soil parent material	* 2 (78.12%)
Site type	* 42 (31.25%)	soil name	* 30 (68.75%)
Forest	* 111 (78.12%)	landforms	* 3 (100%)
Dominant tree species	* larch (40.62%) * pine 21.88% orientalis 37.5%		

6.4.1 Optimal rotation age in small clear felling

So far the best way to determine artificial timber rotation is mainly by the realoption method. Real option method is based on the real option theory, i.e the assumption that the price of wood in line with the geometric Brownian motion established to determine the optimal rotation real options model. But in actual operation, there are many errors in real options method, i.e. geographical distribution of different species vary and are easy to produce errors. The demand of different tree species is not the same, and the error is easy to produce. Timber prices are not necessarily subject to geometric Brownian motion, are likely to follow other random process. Therefore, in the rotation process, using real option method inevitably brings a lot of trouble with errors.

Artificial trees planted by single tree ships this mode of operation taking into account their economic value obtained in different years the costs of operation were put into operation in the first year of planting and management and protection fee expressed as a C_0, the second year of operation includes cost of investment, management, and protection, expressed as C_1. Removing the management and protection fees rotation (year n) time needed to put into cutting costs $C_1 C_2$, assuming that all the wood available for sale amounted to F, while further consideration the course of business of the time value of money factor, under the assumption that monetary benchmark interest rate

r premise, according to the mathematical model solution further calculated annual income (a) which is the maximum a year, at this time is the best rotation period, and can be calculated at this time stand volume M.

An artificial timber forestation best rotations as follows: First, to fully understand the characteristics of artificial planting (usually a single species) based on its economic value and the corresponding mathematical model stand annual revenue value, the formula to (i) $A = (F - c_2)\dfrac{r}{(1+r)^n - 1} - c_1 - c_0 \dfrac{r(1+r)^n}{(1+r)^n - 1}$; secondly publicity (ii) basis for derivation n, n can be calculated what the maximum annual earnings value, that is the best rotation, then the result is $n = [\log_{(1+r)} \dfrac{c_0 r + r(F - c_2)}{c_0 r}]$, the maximum annual earnings a $A_{max} = c_0 r - \dfrac{c_0 r + r(F - c_2)}{r(F - c_2)}$; and finally, in accordance with established related expression trees DBH model ② $d_{t+1} = a_0 d_t^{b_0}$, crown diameter model (iii) $D = a_1 d^{b_1}$, density equation (iv) $N = \dfrac{10^4}{D^2}$, (v) $V = a_2 d^{b_2}$ plant timber volume factor of four, after substitution can be determined stand volume maximum expression for M

$M_{t+1} = N \cdot V = \dfrac{10^4}{m_1 d_t^{m_2}} \cdot m_3 m_4 d_t$, where in the index expression followed by $m_1 = a_0 a_1^2$, $m_2 = 2b_0 b_1$, $m_3 = a_0 a_2$, $m_4 = b_0 b_2$.

The calculation process is applied to the model by the formula:

(1) Trees DBH model: $\qquad\qquad d_{t+1} = a_0 d_t^{b_0}$ \hfill (6-43)

(2) Crown diameter model: $\qquad D = a_1 d^{b_1}$ \hfill (6-44)

(3) Density formula: $\qquad\qquad N = \dfrac{10^4}{D^2}$ \hfill (6-45)

(4) Individual timber volume: $\qquad V = a_2 d^{b_2}$ \hfill (6-46)

Inner Mongolia Forest Experimental results show that the poplar initial planting density 1163 plants per hectare for the nine-year-long program can get the best value for money when silviculture (Table 6-11).

Table 6-11　Inner Mongolia forest economic test

Years	D	Crown	Per hectare density	Profit (RMB ¥)
1	8.96	1.31348	5796	12634.79
2	9.81	1.51578	4352	21914.05
3	10.66	1.71808	3387	27444.89
4	11.51	1.92038	2711	30898.97
5	12.36	2.12268	2219	33056.08
6	13.21	2.32498	1849	34371.32
7	14.06	2.52728	1565	35119.02
8	14.91	2.72958	1342	35471.97

(continue)

Years	D	Crown	Per hectare density	Profit (RMB ¥)
9	15.76	2.93188	1163	35542.9
10	16.61	3.13418	1018	35407.56
11	17.46	3.33648	898	35118.2
12	18.31	3.53878	798	34711.67
13	19.16	3.74108	714	34214.49
14	20.01	3.94338	643	33646.06
15	20.86	4.14568	581	33020.82
16	21.71	4.34798	528	32349.68
17	22.56	4.55028	482	31640.95
18	23.41	4.75258	442	30901.08
19	24.26	4.95488	407	30135.11
20	25.11	5.15718	375	29347
21	25.96	5.35948	348	28539.94
22	26.81	5.56178	323	27716.48
23	27.66	5.76408	300	26878.69
24	28.51	5.96638	280	26028.26
25	29.36	6.16868	262	25166.59
26	30.21	6.37098	246	24294.84
27	31.06	6.57328	231	23413.98
28	31.91	6.77558	217	22524.83
29	32.76	6.97788	205	21628.07
30	33.61	7.18018	193	20724.31

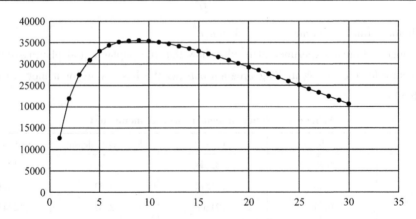

Figure 6-3 Inner Mongolia forest scatter plot economic test

Survey sites located in the artificial planting of Shandong Forest I-214 block shows that an area of one hectare in the survey contained 675 trees all of them are poplar trees. According to the survey, its canopy density reached to maximum or exceeded, and forest stand is saturated. At this point the total volume 247.35 m^3, the average individual volume 0.366 m^3, the average diameter of

21.3 cm, the average crown width of 11.2 m and the average tree height 10.29 is meters.

The poplar timber sales amount to approximately 1500 yuan/m^3; total investment of 5000 yuan of funds when planting in the survey area, the management fee is 1500 yuan/year, n years management fee of 1500 yuan and cutting costs 2000 yuan; Operating in the process of the monetary base rate of 6%.

Under the assumption that all the trees planted all timber sales, that is, the amount of sales (F) for 371025 yuan, it can be calculated: when about the eighth years, its annual revenue reached a maximum value of 44190.63 yuan.

6.4.2 Statistical calculation of maximum stock volume of over mature sub-compartment

Stand density and stand growth characteristics are the main controlling factors. It has important significance to study the influence law and to determine the appropriate density and the mixed degree, which is important for the cultivation and management of the forest. Survey results show that there are many problems in mature forest stands structure, such as stand structure is not complete, the structure of a single species composition. Stand density and mixed degrees is the basis for a reasonable stand spatial structure but also of the growth and development determinants of the size of the space the indicators significant impact Stands.

1) Mature forest density mingling optimal configuration management methods

To determine the optimal allocation method for the mixed degree of mature forest density, which provides a technical choice for the afforestation decision support or planning and design. For mature forest in the same position or the same grade status conditions index volume and mingling into something parabola. It is possible to obtain a mathematical model (1) $M = a_1 k^2 + b_1 k + c_1$, among them, a_1, b_1, c_1, It stands for the three parameters when mingling $k = -\dfrac{b_1}{2a_1}$. Stand volume maximum; at the same position or the same grade status index conditions, stand density and stand an average diameter of d to N diminishing relationship some regularity, it is possible to obtain a mathematical model (2) $d = a_2 b_2^N$ or $d = a_3 N^{b_1}$, wherein, a_2, b_2, a_3, b_3 stands for the four parameters, stand density, the smaller the average diameter. It can effectively understand when the number of stand density, the average diameter of the stand or stand volume reaches the maximum. Which point the mixed degree attains when the stand volume reached the maximum. In mature forest the same position or the same status index under the condition of stand volume and mingling into a parabolic relationship, the regression equation is established by data processing and in the same position or the same status index, the average diameter of the stand and the stand density into a certain regularity, through the data processing and the establishment of the regression equation.

2) Optimal stand density diameter distribution models

The natural state forests use natural thinning of trees and differentiation to adjust the density of the forest suits nature. Differentiation of sparse forest, natural forest growth and development is under the influence of certain environmental factors and their own elements of self-regulation of forest density phenomenon. But by natural thinning of stand density stand can only reach the maximum current density of space, rather than optimum density, because of environmental constraints and other conditions. The main measures of the adjustment of the diameter structure is thinning, thinning is to replace the natural artificial sparse, adjusting stand structure, lower density, change the stand growth environment by thinning. However in thinning we need to judge whether apply thinning, tending and how much.

Forestry workers in the long process of exploration found a balance uneven-aged forest tends to be expressed in a mean diameter distribution exponential equation:

$$N = c_1 e^{-c_2 d} \qquad (6\text{-}47)$$

Where in, N is number of trees per hectare density; d is the average diameter; e is the natural logarithm; c_1, c_2 are two constants.

After obtaining a piece of forest-related data, function fitting, get the function of D and N.

After receiving the above relationship, we can be shown forest stock volume of regional expression in the current unit:

$$M = V \cdot N \qquad (6\text{-}48)$$

Where V is the average diameter of a single tree volume, obtained according to a local volume table:

$$V = ad^b \qquad (6\text{-}49)$$

Density N can be obtained by a function after fitting:

$$N = c_1 e^{-c_2 d} \qquad (6\text{-}50)$$

So, $M = ad^b c_1 e^{-c_2 d}$ which is the unit stock volume mean diameter D, and M a membership function? In this set of functions through a simple derivation of the d, it can be obtained under the current forest environment and forest density equilibrium situation can be achieved with reference to the maximum amount M_{max} of accumulation units average diameter class d_r.

Example (Table 6-12):

Table 6-12 Nature reserve, fixed sample measurements

No.	Species	1999 inspection foot diameter/cm	2000 inspection foot diameter/cm	DBH increment	2000 volume
1	Fir	28.1	28.3	0.2	0.352
2	Fir	9.8	10.0	0.2	0.031
3	Fir	42.8	43.0	0.1	0.932
4	Fir	8.5	8.6	0.1	0.021
5	Fir	36.2	36.4	0.2	0.633
6	Spruce	52.0	52.3	0.3	1.478
7	Fir	25.3	25.4	0.1	0.273
8	Fir	15.8	16.1	0.3	0.093
9	Color tree	15.3	15.5	0.2	0.086

Here, we can think of one representative of each tree diameter class, and then calculates the volume of each tree according to local variable volume model $V = ad^b$.

By 3D angle gauge algorithms, and we can infer the value of existing observing unit density of this forest N, and unit accumulation amount M. After obtaining unit density, we can of exponential function fitting equation, where we take the angle gauge factor $F_g = 5$.

In the above data, we can find the 1999 unit stock volume $M_1 = 229.400$, 2000 Unit $M_2 = 235.090$ accumulation amount accumulated the amount of change in the value of $\Delta M = 5.690$ (Table 6-13).

Table 6-13 Stand density prediction of a nature reserve

$1/g(2000)$	Density N/ (plants/hm^2)	$V/g(2000)$	$M(2000)/$ (m^3/hm^2)	$1/g(1999)$	$V/g(1999)$	$M(1999)/$ (m^3/hm^2)
15.80	79	5.569	27.844	16.07	5.551	27.761
126.92	635	3.908	19.540	132.15	3.882	19.415
6.88	34	6.414	32.072	6.93	6.414	32.033
172.81	864	3.708	18.541	175.67	3.703	18.497
9.58	48	6.063	30.315	9.69	6.051	30.264
4.64	23	6.859	34.293	4.69	6.852	34.231
19.63	98	5.367	26.837	19.83	5.363	26.792
49.21	246	4.591	22.955	50.84	4.578	22.833
52.62	263	4.539	22.694	54.22	4.527	22.585

According to the judgment mature forest equation $\dfrac{M_2 - M_1}{M_1} \leqslant i$ (M_2 unit stock volume for the year, M_1 year accumulation unit volume) when less than i can be identified as mature forest, after a long investigation can understand the actual value of the constant i is about 0.015 change. We will take 0.015 for the i value. In the above data, $\dfrac{\Delta M}{M_1}$, then select the forest into the mature forest. Then in 2000, according to the amount of accumulation units, each unit corresponding to diameter classes calculated density N_i (Table 6-14).

Table 6-14 2000 each diameter class single tree DBH, volume, and density

No.	DBH/cm	Stock volume/m	Density/trees
1	28.3	0.352	667
2	10.0	0.031	7632
3	43.0	0.933	252
4	8.5	0.021	10951
5	36.4	0.633	371
6	52.3	1.478	159
7	25.4	0.273	859
8	16.1	0.093	2519
9	15.5	0.086	2725

The density and diameter classes conducted in matlab function fitting (Figure 6-4):

Function coefficient $c_1 = 55710$, $c_2 = 0.1932$.

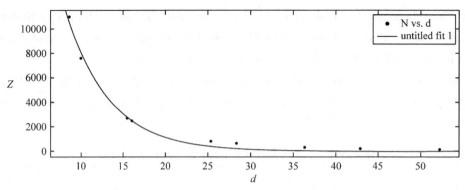

Figure 6-4 Density and diameter classes in matlab fitting results

Then in the formula $M = ad^b c_1 e^{-c_2 d}$ of x is evaluated, we get $d_r = \dfrac{b}{c_2}$, where we can see through the look-up table, this nature reserve a volume table of coefficients $b = 2.34$, and $c_2 = 0.1932$, so:

$d_r = 12.1$ cm

$M_{max} = 258.710$

Obtained after M_{max}, you can learn $N_{x_i} = \dfrac{258.710}{9 \cdot V_i}$, where after 258.710 m³ derivative obtained as a function of the theoretical maximum accumulation amount, V_i for the single tree diameter class accumulation amount. Optimum density in the following Table 6-15:

Table 6-15 Optimum density predictions

	King size class (38 or more)			Large diameter class (26~37.9)		
Reality density	57			127		
Optimum density	50			127		
	Tree No.	Reality density	Density adjustment	Tree No.	Reality density	Density adjustment
	6	23	19	5	48	45
	3	34	31	1	79	82
	The diameter class (14-25.9)			Trail grade (5~13.9)		
Reality density	607			1499		
Optimum density	746			2274		
	Tree No.	Reality density	Density adjustment	Tree No.	Reality density	Density adjustment
	7	98	105	2	635	934
	8	246	308	4	864	1340
	9	263	333			

Among them, the actual density of the whole forest is: 2290 (unit: plant/hectare), the optimal density: 3197 (unit: plant/hectare).

6.4.3 Statistical calculation of the optimum selection for the optimal thinning in sub-compartment

Establishment of the selective felling optimal model in a mature mixed uneven-aged forest reasonably determined in order to facilitate the growth of remaining trees. Stand optimization selective cutting of timber harvesting is reasonable to determine in order to obtain and maintain a non-spatial structure of wood while guiding ideal space structure; at the same time, selective cutting of trees achieve optimal stock according to market conditions. Since all aspects of both interdependent and mutually exclusive, may stand spatial structure, while meeting the requirements of each sub-optimal target is difficult, so multi-objective programming can get the best overall stand spatial structure.

To establish the optimal forest selective felling model, the first objective function is tending selective cutting to reduce the selective cutting costs. That is to meet the requirements of the formula:

In the formula, each diameter class n_i^0 is tending before selective cutting unit density, tending each diameter class n_i after selective cutting unit density. Z for the difference between the diameter classes of selective cutting and thinning, that is the number of trees manual processing integrated. Finally according to the goal of sustainable management of forest stands to set constraints, generally, there are two constraints:

(1) Selective cutting should not exceed the amount of annual growth, requirements that satisfy the formula:

$$\sum n_i \cdot \Delta U_i = \Delta M \tag{6-51}$$

(2) Selective cutting after tending to ensure accumulation amount unchanged that satisfies the equation:

$$\sum n_i \cdot V_i = M \tag{6-52}$$

(3) Tending selective cutting to ensure total no of tree unchanged, that satisfies the equation:

$$\sum n_i = N \tag{6-53}$$

According to the objective function, we can determine which type of tree species in the selective cutting stand as the most optimal logging when the constraint condition (1) and the constraint conditions (2) and (3) are satisfied simultaneously.

According to a nature reserve tending fixed sample data selection selective cutting of trees (Table 6-16):

Table 6-16 Nature reserve, fixed sample information tending selective cutting of trees

No.	Annual foot (2000)	Single trees involume(2000)	Growth	Unit density
1	28.3	0.352	0.0069	79
2	10.0	0.031	0.0014	635
3	43.0	0.933	0.0076	34
4	8.6	0.021	0.0004	864
5	36.4	0.633	0.0077	48

(continue)

No.	Annual foot (2000)	Single trees involume(2000)	Growth	Unit density
6	52.3	1.478	0.0198	23
7	25.4	0.273	0.0033	98
8	16.1	0.093	0.0035	246
9	15.5	0.086	0.0030	263

Linear programming equation based on the data listed in the table:

$$\min Z = \sum \left| n_i^0 - n_i - n_i \right|$$

$$\sum n_i \cdot \Delta U_i = 5.69$$

$$\sum n_i \cdot V_i = 258.71$$

$$\sum n_i = 2290$$

By Solving, we get:

$x_1 = 412$, $x_2 = 391$, $x_3 = 0$, $x_4 = 865$, $x_5 = 17$, $x_6 = 0$, $x_7 = 98$, $x_8 = 246$, $x_9 = 263$.

The positive sign = tending planting, minus sign = cutting. As can be seen from the table, in this forest thinning process, need to grow large diameter No. 1 trees 333. Cutting down small diameter No. 2 trees 244, large diameter No. 5 trees 31, especially large diameter No. 3 trees 34 and No. 6 trees 23 (Table 6-17).

Table 6-17 Nature reserve, a fixed sample density optimal tending felling

No.	Tending before selective cutting density/trees	Tending after selective cutting density/trees	The difference/trees
1	79	412	+333
2	635	391	-244
3	34	0	-34
4	864	864	0
5	48	17	-31
6	23	0	-23
7	98	98	0
8	246	246	0
9	263	263	0

6.5 Carbon sink analysis for Chinese forest vegetation

In recent years, the excessive combustion of fossil fuels and the intensification of carbon emissions have not only increased surface temperatures and aggravated natural disasters, but also reduced the nutrients zinc and iron in crops, which threatens human nutrition. Considering the important role of forests in the global carbon cycle, carbon storage and global climate change, research on methods for predicting the carbon sink potential of forest vegetation will provide a foundation for research on the mitigation of climate change by forests. The mean biomass density method (MBM) has historically been the primary method employed in global climate change research. In this method, the mean biomass density measured directly in the field is multiplied by the forest area to estimate forest carbon storage. The MBM is widely used to estimate forest biomass regionally, nationally and globally, but direct biomass density measurements tend to be greater than the average for a region. For this reason, many studies have implemented the biomass expansion factor (BEF), which defines forest biomass and forest volume as a fixed ratio and estimates the carbon storage in various regions via the mean ratio method (MRM). However, the BEF is a constant that varies with forest type, density and age. Fang proposed the continuous BEF method (CBM) to reflect the relationship between the volume and biomass of different forest types. and used this method to estimate the forest carbon storage in China. Stem volume per unit area, which reflects the influence of forest age, site class and stand density on forest biomass, is used in this method. Even without information on forest age and site class, forest biomass can be estimated according to area and forest volume data from the national forest inventory (NFI) In the past 20 years, Chinese scholars have conducted many studies on forest carbon storage using the CBM, and the estimation methods are well developed. However, there are still uncertainties in the prediction of future changes in forest vegetation carbon sinks; therefore, further improvements and innovations are needed. The study of dynamic vegetation change is the core of forest ecology and is of great significance for developing methods of predicting forest vegetation carbon sinks monitoring dynamic changes in forest vegetation over a large area and a long time period is crucial for accurate carbon accounting . Dynamic changes in forest vegetation are affected by environmental disturbances and subsequent regrowth processes to a large extent, which makes it difficult to predict forest vegetation carbon sinks. Therefore, it is necessary to study the relationships of dynamic changes in forest vegetation with geographical location, the meteorological climate, topography and soil conditions to improve the prediction efficiency and effects of forest carbon sinks. In this study, NFI-measured data collected from 7801 sample plots in 2003, 2008 and 2013 (tree species, diameter at breast height (DBH), tree volume, average age, latitude, longitude, altitude, average annual rainfall, average annual temperature, gradient, slope direction, slope position, and soil thickness) and forest ecosystem biomass data (longitude, latitude, altitude, average annual temperature, average annual rainfall, forest origin, species composition, average age, average DBH, average tree height, forest density, forest volume, arboreous layer biomass, undergrowth layer biomass, herbaceous layer biomass, undergrowth vegetation biomass, and dead plant biomass) were used to assess China's forest resources. The dynamic growth changes and carbon storage of arbor forests, economic forests and shrubbery forests from 2003 to 2050 were studied (Figure 6-5). The dynamic growth of 44 arbor forest species, 4 economic forest species (raw fruit material forest, raw food material forest, raw medical material forest, and raw chemical industry material forest) and the main shrubbery forest species were studied. The biomass conversion of 41 arbor forest species, 21 economic forest species and the main shrubbery forest species was studied.

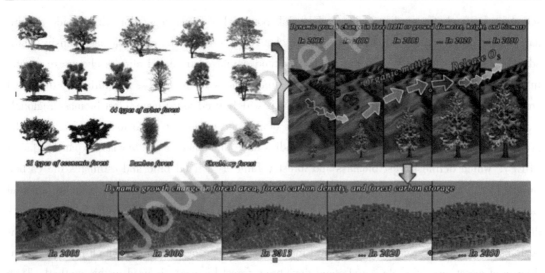

Figure 6-5 Dynamic growth, biomass conversion and carbon sequestration of forest vegetation in China.

6.5.1 Carbon sink measurement method for forest vegetation in China

Additional afforestation in 2020 and 2050 was predicted as follows. According to National Forest Management Plan data, compared with 2016 levels, the forest area will increase by 1.33×10^7 hm^2 by 2020 and by 4.17×10^7 hm^2 by 2050. In addition, based on the sixth, seventh and eighth NFIs, the proportions of total forest area accounted for by arbor forest, economic forest, shrubbery forest and bamboo forest will not change significantly. However, there will be a small increase in forest land and shrubbery forest, and the proportions of total forest land accounted for by arbor forest, economic forest and bamboo forest will generally remain stable. Assuming that there will not be any large-scale forest clearcutting or death in China over the next 30 years, the proportions of total forest area accounted for by forest land and shrubbery forest will maintain a small increasing trend over the next 30 years. In addition, the areas occupied by arbor forest, economic forest and bamboo forest will not change. Therefore, the total forest area in 2020 and 2050 could be allocated to each forest type according to the above proportions. In addition, the area ratio of tree species in the artificial forest could approximate the area ratio of tree species in the added arbor forests in the future. According to the area ratio of existing artificial forest species in the eighth NFI, the area of added arbor forests in 2020 and 2050 could be distributed to each tree species. The carbon storage of arbor forests in 2020 and 2050 was predicted as follows. First, based on the eighth NFI of arbor forest resources and natural environmental data and under the assumption that there will not be any large-scale natural disasters or extreme climate change in the future, Model (6-54) was used to predict the DBH growth of all kinds of trees in the country over a five-year period. Then, Model (6-56) was used to predict the volume growth of each stage in various plots, and the volume per unit area of the national sample plots was combined with the existing arbor forest area to calculate the volume of arbor forest at each stage. Notably, according to data from the sixth, seventh and eighth NFIs, arbor forests will be deforested and die annually. For example, in the sixth NFI, the average annual amount of clearcutting was 3.72 M m³, and the mortality was 0.72 M m³. Therefore, clearcutting was calculated according to the National Future Forest Management Plan, and mortality was calculated according to the proportion of forest resources. For example, from 2016 to 2020, the total amount of clearcutting is expected to be18.26 M m³. Between 2020 and 2050, the total amount of clearcutting is expected to be 109.55 M m³. In addition, the clearcutting and mortality at one stage must be deducted from the five-year growth forecast before the next stage can be predicted. Second, based on the assumption that new afforestation achieved by transplanting young trees (6~10 cm), Model (6-54) was used for DBH growth forecasts. Model (6-56) was used for forest volume growth forecasts, and the new afforestation area was used to calculate the forest volume per unit area in combination with the new afforestation arbor forest volume; clearcutting of arbor forest was proportional to new afforestation of arbor forest. Finally, the new forest volumes of arbor species in 2020 and 2050(relative to those in the eighth NFI) were calculated. The new biomass of arbor species in 2020 and 2050 was calculated using Model (6-57), and the carbon content of biomass ranged from 46.75% to 54.89% (Huang et al., 2007).

$$\ln\Delta Y_t^{(j)} + \Delta t = \ln A_j + 2 \cdot \ln Y_t^{(j)} - b_j \cdot Y_t^{(j)} + \lambda_L^{(i)} \cdot X_L^{(i)} + \cdots + \lambda_h^{(k)} \cdot X_h^{(k)} \qquad (6-54)$$

In the equation, j is the tree species; Y_t is the DBH (ground diameter), in mm; $\Delta(t+\Delta t)$ is the predicted growth after five years, in mm; A_j is the growth rate coefficient; b_j is the growth acceleration coefficient; L is the longitude influence coefficient; B is the latitude influence R is the rainfall influence coefficient. To eliminate the influence of the different dimensions of each factor, data normalization was first carried out:

(1) China is located in the northern half of the Eastern Hemisphere, in the southeast of the Eurasian continent, in the eastern and central part of Asia, and on the western coast of the Pacific Ocean. From north to south, the country stretches from Mohe (northern latitude: 53°33′) to James Shoal (northern latitude: 3°51′), spanning 49°42′(with a distance of 5500 km), and from west to east, it stretches from Pamirs Plateau (eastern longitude: 73°40′) to Fuyuan (eastern longitude: 135°3′), spanning 61°23′ (with a distance of normalization of latitude and longitude.

(2) Moving down the topography of China from west to east, the first step is the Tibetan Plateau (above 4500 m), the second step includes basins and plateaus (1000 m~2000 m), and the third step includes plains and hills (500 m). The first and second steps are divided by the Hengduan Mountains, Qilian Mountains and Kunlun Mountains, and the second and third steps are divided by Xuefeng Mountain, Wushan Mountain, the Taihang Mountains and the Greater Khingan Mountains. In this study, the highest and lowest elevations were used for the normalization of elevation data, as follows:

$$X_E = \frac{E - E_{\min}}{E_{\max} - E_{\min}} \qquad (6-55)$$

(3) The average annual temperatures in the Greater Khingan Mountains and Aksu Prefecture are the lowest (−3.8°) in the area, and those in Sanya and Huizhou are the highest (24.5℃). In this study, the maximum average annual temperature and minimum average annual temperature

were used for the normalization of temperature data, as follows:

$$\left(X_R = \frac{R - R_{\min}}{R_{\max} - R_{\min}}\right) \quad (6-56)$$

(4) Precipitation is abundant in southeastern China. Some areas (Taiwan, Zhejiang, Jiangxi, Fujian and Guangdong) have an average annual precipitation of approximately 2000 mm. Northwest China is relatively arid, and some areas (Qaidam Basin, Turpan Depression and Tarim Basin) have an average annual precipitation of less than 20 mm. In this study, the highest annual rainfall and the lowest annual rainfall were used for the normalization of rainfall data.

(5) The slope gradient was between 0° and 60°, and the slope direction was divided into 0°, 45°, 90°, ⋯, 345°. Slope position was divided into the upper slope position (1), middle slope position (0.625) and lower slope position (0). Slope and slope direction were normalized by $X_\alpha = \sin \alpha$ and respectively.

(6) The thickness of the soil layer is an important indicator of soil mineral element contents, development degree, soil fertility and soil erosion degree, which have an important impact on soil plant growth. According to the sample plot data used in this study, the thickness of the soil layer was between 30 and 300 cm.

$$M = \sum\nolimits_1^j c_j \cdot \overline{d}_j^{9j} \cdot \overline{H}_j^{fj} \cdot N \cdot k_j \quad (6-57)$$

$$\Delta M \approx M \cdot \left(g_j \cdot \frac{\Delta \overline{d}_j}{\overline{d}_j} + f_j \cdot \frac{\Delta \overline{H}_j}{\overline{H}_j}\right) \quad (6-58)$$

$$B = p_j M + q_j \quad (6-59)$$

$$\Delta B = p_j \Delta M \quad (6-60)$$

DBH measurements are often quick, convenient and accurate, while tree height measurements are time consuming and laborious. Therefore, in forest surveys, the tree heights of only some dominant species are measured, and the missing tree heights are often predicted by tree height models of different tree species (Sharma and Parton, 2007). M is forest volume, in m³/ha; ΔM is the change in forest volume, in m³/hm²; G_j, g_j, and f_i are the volume parameters of species j; b_j is the DBH conversion parameter of the tree species (Cheng et al., 2017); H_j is the average DBH of tree species i, in cm; H_j is the average height of tree species j, in m; N is the forest density of the sample plots, in trees/ha; k_j is the proportion of tree species i; Δd_j is the change in DBH of tree species j, in cm; ΔH_j is the height growth of tree species j, in cm; B is biomass, in Mg/hm²; ΔB is the change in forest volume, in Mg/hm²; B, and q is the biomass parameters. The carbon storage of economic forests and shrubbery forests in 2020 and 2050 was predicted as follows: First, based on economic forest and shrubbery forest resource data from the eighth NFI and site environmental data and under the assumption that there will be no large-scale natural disasters or extreme climate change in the future, Model (6-54) was used to predict the change in economic forest ground diameter in the first five years. Model (6-62) was used to predict the growth of shrubbery forests, and the biomass carbon content was calculated according to the 50% value. Model (6-63) was used to predict the biomass and carbon storage of economic forests and shrubbery forests at each stage. Second, under the assumption that the new economic forest and shrubbery forest areas would be similar to the existing economic forest and shrubbery forest areas in terms of planting proportions, the future change in ground diameter of economic forest was forecasted with Model (6-54). The future change in ground diameter of shrubbery forest was forecasted with Model (6-61), and Model (6-63) was used to predict new economic forest biomass and carbon storage in every phase.

$$\ln \Delta Y_t^{(j)} = \ln A_j + \frac{-b_j}{Y_t^{(j)}} + \lambda_L^{(i)} \cdot X^{(i)} + \ldots + \lambda^{(k)} \cdot h \qquad (6\text{-}61)$$

$$\Delta B = N \cdot p_j \cdot [(\bar{d}_j + \Delta \bar{d}_j)^2 - \bar{d}_j^2] \cdot \bar{H} \qquad (6\text{-}62)$$

$$B = N \cdot p_j \cdot \bar{d}_j^2 \cdot \bar{H} \qquad (6\text{-}63)$$

The factors shown in Model (6-62) are the same as the factors in Model (6-58); d_j is the average ground diameter, in mm; Δd_j is the change in ground diameter, in cm; H_j is the average height, in m; N is the forest density of the sample plots, in trees/ha; and P_j is the biomass parameter. The change in height of shrubbery forest was not obvious. The change in height of economic forest was not obvious because of artificial management and pruning and because economic forests need to be pruned to the same height to facilitate selection by managers. Therefore, the economic forest and shrubbery forest heights were calculated based on the average height of each forest species.

Model (6-54) and Model (6-62) are forest vegetation growth models; Model (6-54) is improved by the logistic equation (1838), and Model (6-62) is improved by the Mitscherlich equation (1919). Therefore, the growth rate of forest vegetation follows the sequence slow-vigorous-slow-stop. According to China State Forestry Administration(China-State-Forestry-Administration, 2014b), combined with NFI data, China's forests are still very young; therefore, the age of forest designated as mature(including near-mature forest, mature forest, and over-mature forest) is low (mostly between 20 and 60 years), and most of the mature forest is in the stage of vigorous growth. Group A includes *Pinus koraiensis*, *Picea asperata*, *Cypress*, *Tsuga chinensis*, etc. Group B includes *Larix gmelinii*, *Abies*, *Pinus sylvestris*, *Pinus densiflora*, *Pinus thunbergii*, etc. Group C includes *Pinus tabuliformis*, *Pinus massoniana*, *Pinus yunnanensis*, *Pinus kesiya*, *Pinus armandii*, *Pinus densata*, *Pinus densata*, etc. Group D includes *Populus tremula*, willow, *Eucalyptus*, *Sassafras tzumu*, *Paulownia*, *Casuarina equisetifolia*, *Pterocarya stenoptera*, *Acacia confuse*, etc. Group E includes birch, elm, *Schima superba*, *Liquidambar formosana*, soft broad-leaved forest, etc. Group F includes, *Quercus* spp., *Castanopsis fargesii*, *Cinnamomum camphora*, *Phoebe zhennan*, *Tilia*, sclerophyllous broad-leaved forest, etc. Group G includes cedar, *Cryptomeria fortune*, *Metasequoia*, etc. (Table 6-18).

The carbon storage of bamboo forests in 2020 and 2050 was predicted using the following steps First, based on bamboo forest resource data from the eighth NFI and the market demand forecast report of China's bamboo industry (Zhi-Yan-Research-Group, 2018), the changes in bamboo quantity and area were predicted. It was predicted that in 2020, there will be 123 M moso bamboos (4.86 M hm^2) and 802.3 M bamboos of other types (1.78 M hm^2). It was predicted that in

Table 6-18 Age survey criteria of dominant species

Tree species	Region	Origin	Young forest	Half-mature forest	Near-Mature forest	Mature forest	Over-mature forest
Group A	North	Natural	0~60	61~100	101~120	121~160	161~
	North	Artificial	0~40	41~60	61~80	81~120	121~
	South	Natural	0~40	41~60	61~80	81~120	121~
	South	Artificial	0~20	21~40	41~60	61~70	71~
Group B	North	Natural	0~40	41~80	81~100	101~140	141~
	North	Artificial	0~20	21~30	31~40	41~50	51~
	South	Natural	0~40	41~60	61~80	81~120	121~
	South	Artificial	0~20	21~30	31~40	41~50	51~
Group C	North	Natural	0~30	31~50	51~60	61~80	81~
	North	Artificial	0~20	21~30	31~40	41~60	61~
	South	Natural	0~20	21~30	61~80	41~60	61~
	South	Artificial	0~10	11~20	31~40	31~50	51~
Group D	North	Artificial	0~10	11~15	16~20	21~30	31~
	South	Artificial	0~5	6~10	11~15	16~25	26~
Group E	North	Natural	0~30	31~50	51~60	61~80	81~
	North	Artificial	0~20	21~30	31~40	41~60	61~
	South	Natural	0~20	21~40	41~50	41~70	71~
	South	Artificial	0~10	11~20	21~30	31~50	51~
Group F	All	Natural	0~40	41~60	61~80	81~120	121~
	All	Artificial	0~20	21~40	41~50	51~70	71~
Group G	South	Artificial	0~10	11~20	21~25	26~35	36~

2050, there will be 138.8 M moso bamboos (5.48 M hm^2) and 905.3 M bamboos (1.96 M hm^2) of other types and the biomass carbon storage of other bamboo types was calculated to be 47.86 t/hm^2 (Wang et al., 2001). Second, due to their rapid growth, high yields, good benefits, short life cycles, wide use and sustainable renewal, bamboos can be grown and felled in 3-5 years. For example, from 2003 to 2008, the carbon sink generated by the newly added area of bamboo forest was 1.77Tg C/years, and the carbon sink generated by the yield of bamboo forest was 4.03Tg C/year. From 2008 to 2013, the carbon sink generated by the newly added area of bamboo forest was 2.73Tg C/year, and the carbon sink generated by the yield of bamboo forest was 5.17Tg C/y. Therefore, the carbon sink generated by the newly added bamboo forest area was only a small part of the carbon sink of bamboo forests, and more of the carbon sink was related to bamboo yield. According to the market demand forecast report, the output of moso bamboo and other bamboo types is predicted to reach 16.1 M and 11.1 M plants/year, respectively, by 2020. The output of moso bamboo and other bamboo types is predicted to reach 2.11 and 15.1 M plants/year, respectively, by 2050. Model(6-63) was used to calculate the carbon storage of felled bamboo, and the biomass of other felled bamboo forest types was calculated to be 0.999 kg/plant (estimated to be 47.86 t/hm^2; the average density of other bamboo forest types was 47 970 plants/hm^2).

6.5.2 Measurement deviation analysis for Chinese forest vegetation carbon sinks

The methods for estimating the carbon storage/carbon sinks of existing forest vegetation are well developed, but suitable methods for predicting the carbon storage/carbon sinks of future forest vegetation are lacking (Xu et al., 2010). The NFI data included the impacts of forest improvement cutting, forest selective cutting, and forest replanting on carbon storage. The deviations generated in the prediction process were as follows:

(1) Using the eighth NFI data including influential factors (longitude, latitude, altitude temperature, rainfall, terrain, and soil conditions) in combination with Model (6-54) to predict arbor forest growth resulted in an RMSE (%) between 4.76% and 18.64%, in combination with Model (6-54) to predict economic forest growth resulted in an RMSE (%) between 7.68% and 14.59%, and in combination with Model (6-61) to predict shrubbery forest growth resulted in an RMSE (%) of 13.30%. Thus, process produced some deviation.

(2) Based on the increase in volume of the arbor forest, Model (6-60) was used to predict the biomass growth of arbor forest, and the RMSE (%) was between 7.66% and 37.64%. Based on the DBH, tree height and density of economic forest and shrubbery forest, Model (6-60) was used to predict the biomass growth of economic forest, and the RMSE (%) was 0.67% ~ 38.00%. When using Model (6-63) to predict shrubbery forest biomass growth, the RMSE (%) was 27.359%. Based on the DBH, height and density of moso bamboos, Model (6-63) was used to calculate the biomass of moso bamboos, and the RMSE (%) was 15.951%. The biomasses of other bamboos were calculated according to a biomass per unit area of 47.9M g/hm^2. Thus, the above prediction process produced some deviation.

(3) The carbon content of arbor forest, economic forest, shrubbery forest, and bamboo forest is often 50%. In this study, the arbor forest carbon content ranged from 46.75% to 54.89%; however, the carbon contents of economic forest and shrubbery forest were 50%. In contrast, the bamboo forest carbon content was 47%. Thus, the above measurement process produced some deviation. In addition, the prediction of bamboo forest carbon sinks involves some assumptions, which are mainly based on the newly increased area, newly increased quantity and bamboo forest output. Each of these factors is based on bamboo forest resource data and the market demand prediction report of China's bamboo industry. According to the proportions of moso bamboos and other bamboos, the total amount and area of newly added bamboo are allocated, and the prediction of moso bamboo and other bamboo yields will lead to some deviation in the calculation of bamboo carbon sinks.

(4) The prediction of newly added forest area was based on the National Forest Management Plan under the assumption that there will be no large-scale clearcutting or death of China's forests in the next 30 years. In addition, the proportion of total forest area accounted for by woodland,

shrubbery woodland and nursery land will continue to increase slightly with the trend of the past 15 years, while the proportion of arbor forest, economic forest and bamboo forest in the woodland will remain unchanged. The total forest area in 2020 and 2050 will be allocated to each forest type according to their current proportions. In addition, the proportion of newly added tree species in 2020 and 2050 will be allocated according to the proportions of existing plantation species. In the future, China's forests will be more scientifically managed and planned. To further improve forest quality, the proportions of forest types and tree species in the area will inevitably change. Therefore, the abovementioned distribution of newly added forest area is also one of the deviation results for the forecast sources

(5) Clearcutting was calculated according to the National Future Forest Management Plan, and mortality was calculated according to the proportion of forest resources. The amount of clearcutting was based on forest management planning. The planning scheme will be further improved, and the change in planning scheme will produce some deviations. The amount of forest mortality was not analysed in detail according to the proportion of vegetation but according to a certain proportion of the amount of forest mortality and the total amount of the forest at a large scale. Statistical analysis involving approximation and estimation will produce some deviations.

(6) Environmental changes in temperature, climate, soil and water will affect forest growth, biomass accumulation and other processes, which will indirectly affect the prediction results of studies such as this study. Whether the forestry development target given by the National Forest Management Plan (2016-2050) can be achieved and whether the market demand forecast report of China's bamboo industry is accurate in predicting the area, quantity and yield of bamboo forests will directly affect the forecast results of similar studies.

6.5.3 Filling the gap in research on carbon sinks of economic, shrubbery and bamboo forests

The major limitation of existing carbon sink research is that there have been no estimations of carbon storage or carbon storage changes in economic forests, bamboo forests, shrubbery forests or other forests. Estimating these aspects of carbon sinks will be important in future studies. In this study, the effects of longitude, latitude, altitude, temperature, rainfall, topography, soil conditions and other factors on the growth of economic forests and shrubbery forests were studied in detail. The growth of economic forests and shrubbery forests contributed greatly to the carbon sink capacity of forests, which had not been considered in previous studies. China's economic forest management started late, and after the implementation of key forestry ecological construction projects, such as returning farmland to forest, economic forest areas expanded rapidly, and the output increased significantly. China's economic forests are young and in a period of rapid growth, and its economic forests have significant carbon sink potential. Shrubbery forests play an important role in the conservation, restoration and reconstruction of desert ecosystems. The growth of shrubbery forests changed little, and the carbon sink potential of the existing shrubbery forests was low. However, in arid desert regions, the area of shrubbery forest expanded greatly, and the added shrubbery forest still had significant carbon sink potential. The potential carbon storage capacity of bamboo forest vegetation is small, and the growth of bamboo is fastest when the bamboo is young and slows with age. Therefore, considering the carbon sink from the perspective of the growth of bamboo forests has limitations. The carbon sink of bamboo forests should be measured based on bamboo yield, primarily because as China's bamboo yield continues to increase, the number of bamboo forests continues to increase. Considering the growth characteristics of bamboo forests in 3-5 y, the increases and decreases in bamboo forests will lead to a large carbon sink potential.

6.5.4 Carbon sink potential analysis for Chinese forest vegetation

Since China's reform and opening up, with the implementation of key forestry ecological projects, great achievements have been made in afforestation. For China, a large developing country, to develop its economy, afforestation is one of the most feasible and effective measures for absorbing carbon dioxide, offsetting the greenhouse gas emissions from some industries and reducing the international pressure on China to reduce emissions.

From 1995 to 2005, global forest vegetation absorbed 60 PgC ~ 87 PgC in the form of CO_2, which is approximately equal to 12% ~ 15% of the CO_2 emissions from fossil fuel combustion in this decade. From 1981 to 2000, 14.6% ~ 16.1% of the CO_2 emissions from China's fossil fuel combustion were absorbed by the carbon sink of forest vegetation, which was similar to the global average level. In September 2016, China formally became the 23rd party to complete the ratification of the Paris Agreement. China has implemented some of its commitments under the Paris Agreement three years ahead of schedule and will fully honour its commitments by 2020. In fact, as early as 2014, when the International Energy Agency stated that 'China has replaced the United States as the world's largest greenhouse gas emitter', the China National Information Center judged the current situation and determined future carbon emission trends. As shown in 6-19 of China's carbon emission analysis report, China's carbon emission intensity will continue to decline, and the total carbon emissions will show an annual negative growth trend after 2040. However, five years later, China pledged to peak its carbon emissions by 2030, and therefore, China's carbon emission analysis report overestimated China's emissions over the next 30 years. Assuming that China's total carbon emissions will follow the trend in Table 6-19, China is expected to emit 91.43 PgC of CO_2 from 2020 to 2050. According to the equation $(C) = (CO_2) \cdot 12/44$, carbon accounts for 24.93 PgC. Therefore, according to the predictions of this study, it is conservatively estimated that from 2020 to 2050, China's forest vegetation (arbor forests, economic forests, shrubbery forests and bamboo forests) will absorb 22.14% of the CO_2 emissions from fossil fuel combustion.

Table 6-19 Carbon emission projections in China

Year	Total carbon emissions/PgC	Average annual growth/%	Carbon emission intensity/g/RMB	Reduction in carbon emission intensity per decade/%
In 2020	2.54	2.26	38.4	40.4
In 2030	3.10	2.04	25.0	34.8
In 2040	3.24	0.42	16.8	32.8
In 2050	3.07	-0.52	11.9	29.4

References:

Brown S L, Schroeder P E, 1999. Spatial patterns of above ground production and mortality of woody biomass for eastern US forests [J]. Ecological Applications, 9(3): 968–980.

Brown S, Lugo A E, 1992. Aboveground biomass estimates for tropical moist forests of the Brazilian Amazon [J]. Interciencia. Caracas, 17(1): 8–18.

Chmura G L, Anisfeld S C, Cahoon D R, et al., 2003. Global carbon sequestration in tidal saline wetland soils [J]. Global Biogeochemical Cycles, 17: (4): 1111.

Coomes D A, Allen R B, 2007. Effects of size competition and altitude on tree growth [J]. Journal of Ecology, 95 (5): 1084–1097.

Cusack D F, Axsen J, Shwom R, et al., 2014. An interdisciplinary assessment of climate engineering strategies [J]. Frontiers in Ecology and the Environment, 12(5): 280–287.

Daubenmire R. 1954. Alpine timberlines in the Americas and their interpretation [J]. Butler University Botanical Studies, 11(8/17): 119–136.

Farrelly D J, Everard C D, Fagan C C, et al., 2013. Carbon sequestration and the role of biological carbon mitigation a review [J]. Renewable and Sustainable Energy Reviews, 21: 712–727.

Fang J, Wang G G, Liu G, et al., 1998. Forest biomass of China an estimate based on the biomass volume relationship [J]. Ecological Applications, 8(4): 1084–1091.

Friedlingstein P, Houghton R A, Marland G, et al., 2010. Update on CO_2 emissions [J]. Nature Geoscience, 3 (12): 811–812.

Goetz S J, Baccini A, Laporte N T, et al., 2009. Mapping and monitoring carbon stocks with satellite observations a comparison of methods [J]. Carbon Balance and Management, 4(1): 2.

Guo Z, Fang J, Pan Y D, et al., 2010. Inventory-based estimates of forest biomass carbon stocks in China: A comparison of three methods [J]. Forest Ecology and Management, 259(7): 1225–1231.

Hyvonen R, Agren G I, Linder S, et al., 2007. The likely impact of elevated CO_2 nitrogen deposition increased temperature and management on carbon sequestration in temperate and boreal forest ecosystems. A literature review [J]. New Phytologist, 173(3): 463–480.

Harmon M E, Ferrell W K, Franklin J F, 1990. Effects on carbon storage of conversion of old-growth forests to young forests [J]. Science, 247(4943): 699–702.

Holtmeier F, 2009. Mountain timberlines ecology patchiness and dynamics [M]. New York: Springer Science Business Media.

Ju W M, Chen J M, Harvey D, et al., 2007. Future carbon balance of China's forests under climate change and increasing CO_2 [J]. Journal of Environmental Management, 85(3): 538–562.

Körner C, 1995. Towards a better experimental basis for up-scaling plant responses to elevated CO_2 and climate warming [J]. Plant Cell Environment, 18(10): 1101–1110.

Lai L, Huang X, Yang H, et al., 2016. Carbon emissions from land-use change and management in China between 1990 and 2010 [J]. Science Advances, 2(11): e1601063.

Lal R, 2004. Agricultural activities and the global carbon cycle [J]. Nutrient Cycling in Agroecosystems 70(2): 103–116.

Lal R, 2007. Carbon sequestration [J]. Philosophical Transactions of the Royal Society B Biological Sciences, 363 (1492):815–830.

Li P, Zhu J, Hu H, et al., 2015. The relative contributions of forest growth and areal expansion to forest biomass carbon sinks in China [J]. Biogeosciences Discussions, 12(12): 9587–9612.

Myers SS, Zanobetti A, Kloog I, et al., 2014. Increasing CO_2 threatens human nutrition [J]. Nature, 510(7503):

139.

Qiu Z X, Feng Z K, Jiang J Z, et al., 2018a. Application of a continuous terrestrial photogrammetric measurement system for plot monitoring in the Beijing Songshan National Nature Reserve [J]. Remote Sensing, 10 (7): 1080.

Qiu Z X, Feng Z K, Wang M, et al., 2018b. Application of UAV photogrammetric system for monitoring ancient tree communities in Beijing [J]. Forests, 9 (12): 735.

CHAPTER 7 The Precise Wisdom of the Chinese Forest Management

In human society, science and technology have played an important leading role. Whether it is the agricultural revolution 10000 years ago or industrial revolution 200 years ago. The industrial revolution including modern power, electronics, computer-based modern industry revolution, and recently all the rage of the 3D printer, are no proof of science and technology first productive force of this axiom.

The basic, scientific, technical, and method ological problems of forest management science were considered from the *Qi Min Yao Shu* of the Northern Wei Dynasty. It was only one hundred years' history as a discipline, but also caused by the increasing demand for wood by the Industrial Revolution. As the leading discipline of forestry (also known as the core disciplines), she always and the times the most advanced science and technology, complex systems theory, information theory, cybernetics, nonlinear, large data, cloud computing, networking and so closely linked together. Moving around the world 3D printer, in essence, 3D laser scanning technology is used as the main object in space geometry, physical and biological characteristics of information extraction technology, advanced manufacturing technology, bio-bionic and other advanced technology systems integration. Usually, 3D printing technology has three levels, the first is homogeneous material geometric model print, for example, If your body scan with mud to print out the mud even; also physical print, according to different materials and electronic circuits, print submachine gun, Electronic equipment and equipment; the most high-end simulation for you to print out your body organs, such as the heart and liver, known as the biological 3D printing.

Can 3D printer print out regional or global forest ecological environment? Now it must be possible.

Looking back over more than 100 years of forest manager technology development process as summarized in Table 7-1.

Table 7-1 Development history of forest managers

Stage	Time period	Observation tools	Observation method	Observation accuracy	Business theory, mode and program
Simulation manager	1900—1950	Measuring ruler Compass	Measured tree height Measurement of DBH	5% 2%	1. Forest operating theory clear cutting program 2. Simple mathematical operations
Quantity forest manager	1950—2010	Aerial survey / remote sensing Ruler / feet Compass, GPS Angle gauge, 3S technology applications, Total station application	Measured tree height Measurement of breast diameter	5%~1% 2%~1%	Multi-dimensional forest management theory model Mathematical statistics computer modeling of forest management The business model
Wisdom forest manager	2010 & after	Generalized 3S technology, Cloud Computing Internet of things,	GNSS zoning RSRDOPS survey, Cloud Computing process Nonlinearity	According to the demand for design accuracy, such as precision measurement can be reached precision for diameter 1/200, tree height 1/500	LUCC woodland planning model Regional forest planning model Model of forest density selection for sustainable cutting Intelligent forest process management-platform

7.1 Forest manager technical problems and further perfect way

Forest manager is involved in the use of technology for forest management process, seeking solutions to problems in forest management including the planning & design, forest survey design, and business operation etc. This includes the investigation of the forestry production, monitoring and analysis the inventory of forest resources, organization and forest products. Forest management measures and finally compile the forest industry cases to guide the forestry production and management work.

In our country the forest managers have the following main tasks i. e. planning, investigation of forest resources, business planning and information resources management etc. Main forest management development in our country has been occurred at the following different stages first in the 1930's the preliminary research of forest management is conducted in China. In 1950's forest manager feels the need of social development in forest and development occurred due to this in the early days. After the foundation of our country, cutting and utilization of forest resources was in the forest management policy and this phase was the primary stage of our country's forest management construction system. In the 1980's, China's forest management system gradually developed to prosperity, recovery in timber production and simultaneously development in the ecological system occurred.

In recent years, with the development of market economy in our country the party central committee and the state council put forward the strategy of sustainable development and ecological civilization construction, to promote forest manager again to the new brilliance. Overall China's forestry development has obtained many achievements and ecological system gradually improved. The forest ecological system constantly plays an important role in the development of forestry industry system and also to enhance economic and social functions.

The ecological culture system rising prosperity these achievements made a significant contribution to the national economic and social development. With the rapid development of the modern economy, there are increasing contradictions between ecological environment problems. To meet the requirements of sustainable development strategy and the strategy of ecological civilization, innovation and development of forest many management technologies and methods were introduced.

7.1.1 Application of new technologies in forest managers in China

1) The number of forest research and construction aspects

(1) Not enough attention towards experiment. The establishment of forest tables like local volume table, standard volume table, ground diameter volume table, biomass of single tree on the table and some other forest table is not only needed in the field to collect information but also need to prepare the management system. They are very complex and rely on the support of a large amount of financial and technical availabilities, otherwise it is difficult to successfully complete them. Single entry volume table is simple to use but the measuring factor is the only diameter at breast height and is prone to error. Although the standard volume table can make difference in the same volume, the same diameter and tree height caused by the different reflect, but its use will increase the field investigation and statistical work, which also increase the cost accordingly. Chinese foresters are using older operating tables and now due to global change in forest timber species today with increasing complexity tables of forest management in urgent need of revision and correction of meticulous precision. But in recent years the Government gives enough attention to the construction of forest data, make the necessary investigations to revise them.

(2) Cutting and destructive experiments as the main body. Forest table required necessary information obtained from stem analysis. In the cutting process, if cut all of the design of the logging, coupled with the inevitable cut like felling damage wood and mechanical damage wood during the process of harvesting. This logging style neither considers the actual cutting operation feasibility nor failed to determine the number of production logging according to the design strength. So it is necessarily bound to exceed the design intensity, resulting in ultra intensity cut, so that the forest was greatly damaged. This traditional analysis method of wood forest table needed to be effectively improved should minimize the destruction by the more accurate forest table.

(3) All types of technical standard programs and improve space. Forestry data tables include the forest inventory and planning. Forest management is basically forestry scientific research based therefore constantly improvement in the table is needed to establish technical standards and programs. In early 80's Changkun formally introduced the standardization of forestry data table series of tentative. In early 90's Wu Fuzhen wrote a call ' The urgent task of present forestry is how to establish tables and perfect the system of Chinese with forestry tables on the basis of summarizing the table experience of domestic and international, combined with the requirement of our country's forest resources management, eventually to make series and standardization'. On September 21, 2012, Guizhou became the first province to establish the standard system of forestry tables in China. But China has a vast territory, forestry development in China fluctuates and due to old and the new system of conversions and other reasons, the standardization of forestry numerical tables is difficult. Forestry numerical tables present system technical specifications, test methods and criteria of unity, approval is not complete and does not seriously affect the table description investigation of forest resources.

2) Problems existing in investigation

(1) Tree height is difficult to precise automatic measuring. Tree height is one of the most important factors in the forest survey for evaluating site quality and tree growth. Tree height is also an important basis for forest not only reflecting land productivity but also used to determine standing wood volume and volume growth rates. Tree height is used in preparing volume table and is also one of the main factors that determine the amount of the stand, similarly is one of the forest inventory items. In traditional forestry tree height measurements, methods include visual inspection or by using the Blume-leiss Height Finder such as ultrasonic measurement instruments. But using these tools for measuring tree height site conditions should be better like the terrain should be relatively flattened so that it will become easy to used horizontal distance measured test results before they can meet the requirements. In addition, these height finding tools are greatly influenced by human body errors like shaking while using, so when measuring in the field its accuracy and efficiency will be affected and have significant limitations. Meanwhile due to measurement errors in measuring tree height directly affects tree growth estimates, unable to meet the ' precision forestry' and ' digital forestry' requirement. Therefore nowadays with the lack of forest resources, forest trees with high accurate, non-invasive measurement is particularly important and by using electronic theodolite, total station, electronic measure emerging high-precision measuring instruments such as laser guns becomes new development trend of modern forestry with precision measurement.

(2) Diameter measurement automation. Determination of diameter at breast height is one of the most basic factors which considered plant the most intuitive, DBH is measured at the height of 1.3 meters, traditionally DBH measured manually by using standard measurement tools such as measuring tape, caliper etc. Diameter at breast height (DBH) and biomass represented tend to be affected by light, temperature, precipitation, latitude and altitude and the combined effects of many environmental factors. Diameter at breast height measurement has been used for a long time; its accuracy is low and unable to achieve real-time monitoring. In recent years with technology devel-

opment increasingly more workers make efforts to explore simple shortcut of the DBH measurement method, such as trees measurement ring, trees measuring diameter feet (DBH measurement instrument), using parallel laser beam indirect measurement DBH method. The measurement accuracy compared with the previous method of marking, paint work measure has improved tremendously, but still cannot completely achieve the DBH envisaged a class survey in automatic measurement.

(3) Low degree of automation with sample wood reduction determination. In the continuous forest inventory work sample area and sample tree reduction is a key part of external investigation work. Sample area and sample tree is to be reset so that the measured sample area and sampletrees show consistency with the previous samples. Samples accurate reduction of wood in order to ensure continuous forest inventory data continuity, accuracy, and comparability, which reflect the growth and decline of forest resources in a scientific way and changes. To reset the complexity of traditional staff often through a local forest station or township and village cadres to contact and let it guide for people on the way. Due to complicated topography with high mountains and steep slope many plots were located in remote and inaccessible places causing great inconvenience to reset. Because of restricted technical conditions sample reduction rates are generally lower in the past influencing the accuracy of the survey. Thereby increasing sample forest reduction technology has become the key to reset the fixed sample survey questions; Low degree of automation with sample wood reduction determination become an urgent problem to be solved in the investigation of forest resources.

3) Question of the second investigation

(1) Visual based. Forest resources second investigation is based on the state-owned forest, nature reserve, forest park forest business units or county administrative, it is the forest resources investigation which is used to meet forest business program, general design, the need of forestry division and planning design. It is also an important basis for guiding and regulating scientific forest management to set regional economy development planning and forestry development planning, implementation of compensation for ecological benefits of forest and forest resources asset management.

The second type is mainly through visual inspection the measurement obtains the relevant information. Through visual inspection one can account for a large proportion such as various types of forest timber growing stock and volume all types of woodland area. When the measured information from visual information and sampling results exceeded the allowable deviation, a large number of visual information survey result will become inaccurate that makes duplication and complexity in the survey.

(2) RS and UAV application difficulties. With the development of remote sensing technology and as well as the popularity of computer technology the applications of remote sensing in forestry will alsoincrease. Although it has made great achievements in various fields but as compared with developed countries forestry remote sensing application in China has not yet attained the stage of full functional development. Forestry users are still using relatively low remote sensing data. Due to the limitations of spatial resolution of remotely sensed data and high prices of high precision re-

mote sensing data, so to get the suitable data is difficult. Application of remote sensing data tends to have a lot of restrictions, faced a dilemma in the survey. UAV applications are expensive and difficult to deal with scale.

(3) Growth measurement is not accurate. Forest growth is a second class investigation of forest resources must provide one of the most important data. At present, there are many methods to calculate the growth of forest and the results of different calculation methods may be quite different. Due to lack of precise measurement data year after year with no advanced stand number density method for determination of the estimation, forest management causing serious consequences, especially in the case of excessive felling, high forest biomass estimation results that will lead to the destruction and depletion of forest resources. Thus ensuring accurate determination of forest is an important aspect of such investigations.

4) Three types of survey questions

(1) Experience-oriented afforestation design, lack scientific program. Afforestation of survey design namely in the investigation on the basis of natural and economic conditions, scheduled according to the afforestation the of land suitable for forest, tree species, and various silvicultural techniques and their implementation. The design stage of afforestation technology is mostly based on local experience of afforestation. Although combining the principles of afforestation, tree science, forest ecology and suitable tree planting is a good combination but many problems will occurs like the afforestation tree species selection, afforestation density, tree species composition, site preparation methods, afforestation methods and tending measures other than that the amount of seedlings needed to be the amount of labor costs and actuarial there are many problems. While some workers tried 3S technology in the planting design but due to the limitation of the understanding of this new technology, the practice of scientific afforestation could not be completed with very good design.

(2) Cutting design information support and supervision. Compared with other survey design, forest design requires greater use of 3S technology to ensure accurate design. Designed cutting to cutting survey data as the basis and cutting stand standard survey, cutting conditions investigated, update investigation etc. The investigation process required relatively long time and complex content.

(3) Species information research. The principle of suitable land and tree for afforestation should adhere to current research on afforestation techniques. But the dry type of tree species are unable to meet the needs of multiple species from different parts, even with the trees in accordance with local environmental indicators but the type is single. Monoculture plantation also known as 'pure forest', it is relative to the 'mixed forest'. Monoculture plantation has its own advantages the general form of monoculture forest is only single-storied forest and have an ecological relationship which is relatively simple. But monoculture plantation has the disadvantage that single species and single plantation can easily lead to serious plant diseases and insect pests, even resulting in a serious decline, a deeper level of the disease can also cause a severe decline in biodiversity. The natural forest vegetation of the area is complex and diverse; species diversity can provide more

habitats, rich water conservation function of the forest. Single tree species and also few tree species in large plantations not only caused the decline in biological diversity can also lead to the deterioration of the ecological environment in the corresponding, the function of these forests cannot play their role better and stability is always under threat.

5) Forest management issues

(1) Insufficient attention to forest management. Forest management is science and on the otherhand, it is also a business. Forest plays an important role in ecological, economic and social benefits. Generally, economic and social development require legal forestry regulations policy, forest resource status and beside these social, economic and natural conditions. Forest business management subject prepared aimed for the betterment of forest resources, forest protection and by implementing medium and long-term planning and using measures of planning and design. Many subjects on the preparation of forest management plans are not enough and the important role often appeared crude and self-management of the program was unable to improve management level of the forest management plan. Forest management did not pay attention to control mixture ratio, age class (diameter) structures, especially the number density of key issues, qualitative analysis, quantitative research, practical management lack precision guidance, supervision, and inspection.

(2) High quality of the preparation of management plans, business relationscoordination in program management. Many local forest management plan prepared for the timber production is centered on the influence of traditional business models. While there is the lack of the understanding and analysis of the actual situation of the business units and the social economic environment and changes in the market the plan pertinence is not strong, the focus is not prominent and cannot contact specific analysis of actual problems. The required capital and efficiency evaluation is too idealistic, resulting in some planning projects goals cannot achieve.

Many unclear divisions of duties between the various functions of the forestry department, lack of accountability and coordination between departments. These problems can seriously affect forest management plan. Some places lack the vision and global awareness of the problem in the long run. Without prescribed procedures revising the program deployment causing management plan implemented in practice.

(3) Planning and design of forest mere formality, mainly qualitative, the lack of a quantitative optimization in practice, difficult to implement. No parties are realistic because of scientific management of forest benefits and effects.

6) Common problems

(1) Theory of lag or lead. Three type of investigation theory or the establishment of forest management plan, especially the basic theory of forest management there are some corresponding theories of lead or lag. Theory ahead, the existing conditions are not enough cannot be achieved. Theory of lag cannot keep up with the pace and can't meet the current needs of forest survey and management. Forest management workers in our country always have theoretical questions concern the most real lack of attention on the benefits of forest management techniques and do not have professional practice, deep, high-tech support for a long time. China has been using the number of

mature forests, natural maturity as the main technical indicators to determine the forest logging quota, the preparation of forest management programs, the development of forest operations procedures, guide the practice of forest management. Due to the strong implementation of a natural forestry national policy that bans the commercial logging policy, many types of research has not focused on forest production products.

(2) Equipment of forest management techniques is relatively backward and lack of capital investment. Financial investment is the common problem of forest class I, class II, and class III. For decades due to lack of funds for these classes of forest resource investigation fieldwork techniques and equipment have been without a major change. The density measuring instruments have problems like largely diameter tape, angle gauges, and topographic maps. In recent years with the development of remote sensing & GIS and GPS technology brought vitality to the forest but in a broader sense rather than a smaller, but it cannot really provide much guidance for sub-compartment. Development of corresponding information systems is a matter of concern as sufficient funds are required to continue to update equipment improved technology. The domestic production of MINI integrated measure instruments promotion is very difficult.

(3) Systems are not compatible, poor co-ordination, integration. 3S Technology and its application in forestry have injected new vitality into the development of forestry but inthe survey design and actual operation process. Different staff in practice application of 3S technology are slightly different such as access to outside information, data processing methods, updates to the data are inconsistent resulting in forest resources inventory and the stock map information is not consistent. The forest resource investigation and management system based on GIS are not the same. Although the each system can give full play to their own advantages, but poor compatibility between system information integration cannot be achieved, inconvenient for forest resource management.

7.1.2 Counter-measures

1) Observation scheme for modernization

Modern observation scheme can save a lot of manpower, time and thus reducing the cost. For a long time, artificial forest investigation of observation scheme is stronger, needs further investigation in the forest and the future observations shall gradually results into more modern solutions. Since the 1980's the 3S technology becomes more popular and developed, high integration of computer hardware, the reliable technical guarantee for the modernization of the forest observation program. Using 3S technology including UAV camera system on forest survey the traditional observation that exist in the form of complicated calculation, summarizing and drawing work has been eliminated. Now forest survey is divided and can be complete as macro investigation, medium investigation and micro investigation of the forest resources dynamic display.

2) Electronic observation instrument

In the process of measurement, forest manager shall gradually increase the use of electronic instruments. Now widely used electronic observation instruments including MINI tree measuring instrument, measuring tree electricity by the Total Station and CCD measuring instrument, with 3S technology makes the forest manager precise and modern in instrument learning.

(1) Mini UAV aerial system. In forestplantations, the use of the mini UAVs having an ordinary or digital camera for quick small tree measurement information is accessed by processing the Ariel photographs within an hour as compared to the conventional survey which takes about 2 weeks to complete the survey.

(2) CCD standing tree meter. Set electronic theodolite precision electronic angle gauge, stumpage volume precision measurement and tree height automatic measuring in one, without cutting down, no parse tree research forest management tables.

(3) The MINI tree super station meter. Handheld electronic total station having electronic three-dimensional angle gauge(with slope correction), electronic compass, which can measure tree height, average high measurement, density, diameter and height measurement basic measurement magnetic (distance, azimuth angle and tilt angle). GPS positioning is used for the small class demarcation, crown density measurement etc.

(4) Mobile phone/tablet records with the cloud computing system. First, second or third transfer audio or manual type recording, online or offline cloud computing or real-time results and the forest manager after the formation of an integrated platform for internal and external integration platform.

3) Data processing information

Due to modernization in the process of taking an observation, the amount of labor is greatly reduced and also the quality of the observation is improved. This solves the problems existing in the traditional process of observation. Observation in the process of modernization is mainly embodied in the planning and design field observation (data acquisition), drawing (building) as well as in the industry acceptance evaluation phase. In the observation process the establishment of fixed angle point, DGPS program of precision calibration plots (points). The invention of electronic stereo angle gauge, achieve a measurement of basal area and number of trees, density, stand average height, volume and biomass of precision measurement. The use of forest resources survey cloud forest resources data after processing and drawing platform of forest information cloud computing data acquisition and processing system realize the observation process modernization.

4) The forest and global change

In the 21st century globalization is speeding up the forest change mainly the research content and research direction for forest managers and scholars should be in focus with globalization. In order to adapt to the historic policy change from forestry production to ecological construction, taking the sustainable development of forestry and strengthening the basic strategy of ecological civilization construction, we should actively carry out the relevant international conventions, international treaties, try to meet the needs of the global forest resources assessment, and constantly improve China's forest resources monitoring system. So that China's forest resources monitoring system to become a national monitoring system, local-level monitoring system and the ecological environment special monitoring system as a whole integrated monitoring system; as much as possible the use of land resources on the 3rd satellite (stereo mapping satellite) data, analysis and study of ground survey data, using modern means, such as GIS, GPS forest vegetation Change and global change, to provide accurate and effective data for the government.

The changing world and development of the external environment is both the opportunity and challenge for forest managers. Forest management pattern alsoneeds development with the passage of time. Make full use of the modern knowledge theory of innovation and technological innovation. Forest manager constantly has to create new ideas, new technologies, new methods for the modern forest management theory to inject new vigor and vitality, realize the sustainable development of forest management. The theory and practice of modern forest observation mode are an indispensable work in forest managers' activities and it will become the basic way for forest managers to develop in the future and it should be put on the agenda as soon as possible. Forest management is an external subject of forestry work and is an important mean of modern forestry development. Scientific forest management system will promote forestry development enormously. According to the current situation of forestry in China the author found problems of forest management and constantly to correct, improve and to promote the sustainable management of forest system to contribute the sustainable development strategy in our country.

7.2 System innovation project of the forest survey

1) Scheme of measuring fine tree information (Figure 7-1)

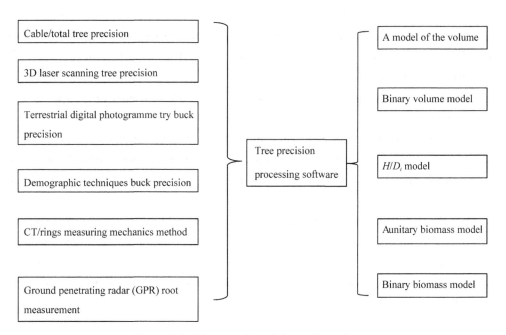

Figure 7-1 Tree precision information solution

2) Forest type of survey information programme (Figure 7-2)

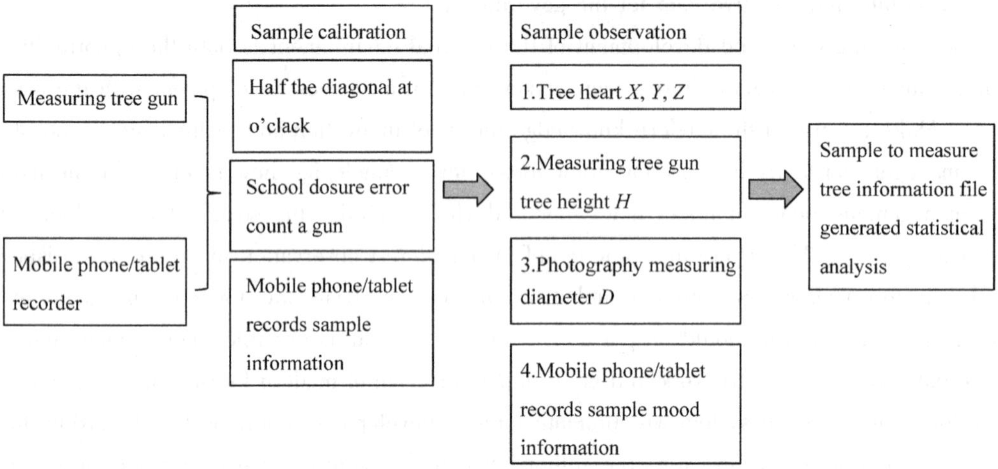

Figure 7-2 Forestry type of survey information programe

3) Second-class information technology program

Forest farm level can be divided into three programs:

(1) Each sub-compartment a plane angle point;

(2) ZY3 Stereo image pairs + 200 3D Angle gauge points;

(3) Unmanned aerial vehicle (UAV) 3D photography + 50 points 3D Angle gauge. The second class software platform includes automatic electronic zoning (forest area → forest compartment → small compartment) and area adjustment. Soil and site conditions information database, terrain information (slope, slope, slope) calculation and storage, forest Vegetation information database and statistics, mapping, tabulation and so on.

(4) Three types of survey design informatization plan. Cutting design survey can be divided into two programs

(i) Combine D_i, h_i of standing tree information and the varieties of trees (x, y, z, D, h) According to the needs of tree species in the market, through the linear programming theory we select the optimal tree diameter and then make the waste rate to least and by doing this the yield reaches to the highest point.

(ii) Get the percentage of sub-compartment tree species with the 3D angle, N, $\overline{D}$ and radial order distribution, through the observation of wood sample like point no 1 to solve the problem In the design of afforestation, we mainly combine the expert knowledge of artificial afforestation, computer afforestation expert knowledge, research and afforestation design platform from site investigation, other than that tree species and mixed ratio spacing design and design effect expression.

7.3 To the wisdom of forest management

Cloud Computing is another great changing after the big change in the mainframe computer to client server (B/S structure) in the 1980s.

Cloud Computing is a product of Distributed Computing, Parallel Computing, Utility Computing, Network Storage Technologies, Virtualization, Load Balance and the development of traditional computer and network technology.

At the early stage of the service is dominated by IaaS(Infrastructure-as-a-Service), Users can access the service from the perfect computer infrastructure through Internet. Through IaaS, forestry workers cannot have the necessary server equipment, cloud server rental operators through cloud servers, greatly reducing the cost and maintenance of the server and network security of human resources.

With the continuous development of cloud services, to meet the needs of the market, the PaaS have been produced (Platform-as-a-Service). PaaS is actually referring to take the software development platform as a service submitted to the user by SaaS mode. Therefore, PaaS is also an application of SaaS model. However, the emergence of SaaS can accelerate the development of PaaS, especially to speed up the development speed of SaaS application.

SaaS: the software is the service. It is amode of providing software through the internet, the user does not need to buy software, but to rent a provider of web-based software, to meet their own needs(Figure 7-3).

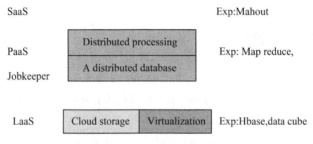

Figure 7-3 Two kinds of service model

From the emergence of these two models we have entered into a large data era, the internet's most important capital data. In today's forestry software development, we through the construction of a unified platform to provide users with data processing computing. In the wireless network environment this providing forest workers to record data, data analysis tools, so that forestry workers through mobile phones, notebook computers and other portable devices, in the field at the same time. On the other hand, data mining and analysis using large data technologies (eg, Hadoop, NoSQL, etc.) by means of massive data distribution storage technology and massive data management technology by forestry workers in the field work, further promote the forestry field scientific research.

There is no greater change in technology than the changes that internet technology brings to human life, learning, economic recreation and culture. Similarly, the internet is also a profound impact on forest management, although still in the testing phase but show a seductive future. We can make bold predictions about the state of modern technology affecting forest management:

1) The forest growth monitoring

The forest growth model is based on the destructive test analysis of wood ring on. Global change and newly cultivated tree varieties require systematic observation of today's forest growth, from a technical point of view, this aspect of technology development in two levels.

Break through rings non-destructive measurement techniques, are based on the mechanics (HengLi drilling) extraction speed signal (rings out of soft wood, speed), there is more work need to thorough research into the practical and computed tomography (CT) technology, main is portable and the observer, the harm of human body, but also limited its development, on the other hand, is based on the electric / whole station / 3D laser scanning / 3D photography-oriented fine measuring technology. To find the diameter of standing tree, tree height information extraction. Application of ground penetrating radar, lossless, accurate, automatic and efficient determination of tree root distribution.

Using modern mathematical methods, such as the artificial neural network and genetic algorithm, make full use of the precise data of standing trees.

It is important to add new instruments, new technologies and new methods in the first, second and third types of surveys, and to produce efficient and economical technical methods.

2) The forest information transmission

Communication in the forest area is still a technical and economic problem. Mobile communications 4G based on mobile phones / Tablet PC-based cloud computing We chat platform still needed to develop.

3) Forest planning and design

In the future, forest planning and design should be done in three aspects:

(1) Establish the LUCC foundation, 3S technology as a means to ecological environment and energy conservation as the goal of forest land spatial planning and regional coordination and coordination of regional economic and social development.

(2) On the basis of optimizing the design of forest land, forest managers should take up the task of forest design, coordinate the design of forest species with regional environment, economy and society, and fully embody the spirit of humanism.

(3) The detailed design of the forest management plan is mainly to control the proportion of small-scale mixing, age distribution, diameter distribution, especially the best density design. Through the density design, a certain period of time to select a small classes of wood cutting the most, the largest growth the best environmental effects.

4) Forest management operations

Main technology of the current forest management work in our country is under the guidance of GNSS, GIS and design scheme of mechanical. How proper optimization scheduling, in the case of selective felling, want automation to solve the problem are as follows:

(1) Whether the design to the cutting of trees is consistent? Y/N

(2) The design of D/H is consistent? Y/N

(3) Accurate observation is consistent? Y/N

(4) Logging? Y/N

(5) Logging quantity completed? Y/N

Finally the observation – the design – the work, in the field and outside the field.

5) Forest building materials design

China has firmly implemented the ban policy on natural forests. As a big country of timber demand and consumption, we still need to study the problem of timber cutting and processing. Such as the limits of the plantation cutting, we need to seriously study the logging survey design, output rate statistics and related technologies, methods and systems. The classic survey method is to estimate the size of each wooden rule, experience and estimate the model building materials, a larger workload, can improve efficiency? D, H, x, y, z and other information, but also the application of MINI station measuring the upper diameter of the functional observation statistic stand counting wood large, medium and small diameter building material number. Average diameter H, mean diameter at breast height D, diameter at breast height D_i, travel rate, cost rate and so on through the forest one-dimensional and two-unit volume library, HD model and D-hi model. Perhaps only 1%~2% of the workload of each wooden measuring foot, but the overall accuracy is quite, it is a worthy of promotion of technical methods.

6) Forest environmental effect and its response

Forest has ecological, economic and social benefits in industrial society and information society. How to exaggerate the ecological benefits of forests, the effect and positive response can not be over emphasized. We have built the environment observation network in Beijing forest through the establishment of seven data observation set up center and two network data processing center. Seven data observation centers were deployed out of which Matsuyama 1, Jiufeng 4, Beijing Forestry University 1, Beijing University of Technology 1. Five species of *Mt.* Songshan and Jiufeng were selected as the field ecological stations, which were selected from five types of coniferous forest, broad-leaved forest, mixed coniferous and broad-leaved mixed forest, shrub-grass mixed forest and grass land. Temperature, humidity, PM value, NO content, O-content, SO_2 content in the undergrowth or grassland environment with m <H<2m. Beijing Forestry University and Beijing University of Technology, two stations as a collection of urban data and forest data comparison. 2 data processing centers were set up in the Beijing Forestry University and Beijing University of Technology, to collect the data collected and processed. And the topographic data, tree growth data (tree height, diameter at breast height) of Jiufeng and Matsuyama were analyzed. Field observation station and data processing center between the network through real-time data transmission. Through long-term data collection and processing found the following scientific laws:

(1) Use the atmospheric composition parameters, such as NO, SO_2, and PM2. 5 in different forest types (coniferous forest, broad-leaved forest, mixed forest) content are the same, so as to explore the rule and its related degree.

(2) Use the measured parameters of atmospheric composition data to explore the growth of vegetation (mainly refers to the tree height, diameter at breast height data of several years of growth) whether there is a correlation with the atmospheric environment and how is the extent.

(3) To explore whether there is a correlation between different atmospheric environment and vegetation biomass, carbon sink and other parameters, and how is the extent.

The results of observation can be applied in:

(1) Using measured temperature and wind speed data to take appropriate measures of cold resistance, lodging resistance of young forest.

(2) Detect camera real-time data transmission by infrared, forecast of a forest fire, forest fire will bring losses to a minimum.

(3) Using the N, P, K content in soil, to evaluate the soil fertility, so as to facilitate according to the 'suitable to the tree' principle, select the appropriate tree species for planting.

(4) Using the image data of the placement of the camera, the discovery of wild animals and the scope of activities to detect, so as to guide the protection of wildlife.

(5) Using the measured data of O_2 and O^- in the forest, to develop the tourism industry, and promote the development of forestry.

Planning Through long-term observation to explore the relationship between the forest and the surrounding environment, further evaluation of forest ecological value, measured by the data in order to improve the production efficiency, and ensure the maximum benefit of Forestry forest sustainable management to provide a theoretical basis for reasonable.

7) Polygon method of sample plot measurement

The calculating formula of stand volume:

$$M = \frac{1}{4}\pi f_\theta \sum_{i=1}^{n} P_i D_i^2 (H_i + 3)/S \tag{7-1}$$

$$P_i = \frac{\alpha_i}{2\pi} \tag{7-2}$$

$$\alpha_i = \pi - \frac{1}{2}\pi \times \text{sgn}(\Delta y) - \arctan\left(\frac{\Delta x}{\Delta y}\right) \tag{7-3}$$

$$\begin{cases} \Delta X = X_{i+1} - X_i \\ \Delta Y = X_{i+1} - X_i \end{cases} \tag{7-4}$$

$$S = \frac{1}{2}\sum_{i=1}^{n} | X_i(Y_{i+1} - Y_{i-1}) | \tag{7-5}$$

Through the statistical analysis, it is concluded that the best shape of the polygon is eight shape (Figure 7-4).

Average per hectare forest stand high calculation:

$$\overline{H} = \frac{\sum_{i=1}^{n} P_i H_i}{\sum_{i=1}^{n} P_i} \tag{7-6}$$

A hectare of stand density calculation:

$$N = \left(\sum_{i=1}^{n} P_i/S\right) \times 10^4 \tag{7-7}$$

The sample size distribution:

$$N_i = \frac{P_i}{\sum_{i=1}^{n} P_i} N \tag{7-8}$$

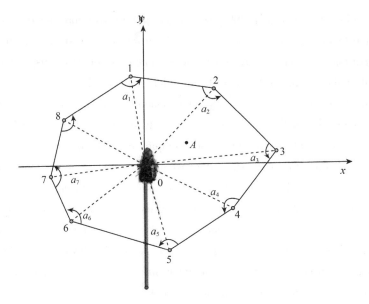

Figure 7-4 Schematic diagram of polygon plot

$$\sum_{i=1}^{n} N_i = N \qquad (7\text{-}9)$$

7.4 Prospect: forest management technology development

7.4.1 Forest management planning overview

How to make the forest management science to become general and understandably more popular, the forestry engineering technology which in one profession in the industry generally important, technology and methods still need to attach great importance to China's forest management industry.

Forest management is not a pure theory or concept nor it is a kind of management science or economics method. It is a comprehensive, complex and systematic science integrating the forest ecology, resource economics, environmental management, modern surveying and mapping technology, modern information technology and sustainable development theory. Any pure theory, partial technology, language management, economic thinking can not guarantee the efficient and healthy development of forest management system from the system.

All along the domestic and foreign scholars have devoted themselvesto the research of theories, models, techniques, equipment and operation methods of forest management and continuously promote the technical level of forest management. Throughout the forest management disciplines development, i.e. theory, methods, goals are complementary to each other with the scientific and technological, economic and social development of forest management based on the classic theory of produce. Undoubtedly in the middle of the 18th century in Germany and other countries, many theories were developed regarding forestry management. Today we criticise these theories but this did not affect as these originates more than 200 years before and are still core theory in forest management, with operation technology method such as wheel cutting method, staging method, volume bisection method, area bisection method, age level, the economic law of stand, inspection check method, the generalized method of forest, has long been widely concerned.

Many theoretical models of contemporary forest management theories have some expression differences but the core idea is the forest as a is not only the production of wood but a combine economic, ecological and social benefits in various complex ecosystem, mainly representative theories such as return to nature forest theory and constant continued thinking, multi-benefit leading theory, ecosystem management theory, the theory of classification management and so on, the idea of the relationship on the whole known as the forest ecosystem management. Unfortunately, this theory does not have much to do with the component and can be operated, for everyone recognized the specific technical operation mode and implementation method. Obviously, as the world recognized social, economic and environmental development strategy, the theory of sustainable development is the ultimate goal of forest ecosystem management.

Obviously, the sustainable forest management is a modern management goal of forest managers. There are two main theories of forest management: the theory of forest sustainable utilization and the theory of forest ecosystem management the former is the primary theory, the later is the advanced theory of evolution.

If the theory of sustainable utilization of forests is the primary management theory of forest management the core idea is static, homogeneous, continuous and shallow, simple understanding forest management and forest ecosystem management is dynamic, heterogeneous and discontinuous, non-superficial, dimension, complex, considering the ecological, economic, social and environmental effects of expression of forest management. Obviously, forest ecosystem management theory at least in applied mathematics expression did not specifically recognize model and model still need to do more practice investigation, mathematical modeling and optimal solution and operation methods in-depth study.

7.4.2 Prospect of forest survey technology

Forest survey is in urgent need to improve and develop the technology methods and equipment system. The contents of forest survey mainly include three aspects first one is the forest area, the second one is the dynamic growth of the forest and the third one is the evaluation of forest environmental effects. The technical system established for forest survey is within the limits of the national norms, standards, technology, methods, equipment, security and other aspects of a scientific and technological system. Based on the global ecological environment changes and the state government the public concern on the ecological status of the forest, the global, international, political and government issues led this investigation to a hitherto unknown height.

At present, our country's 1st class, 2nd classes, 3rd kinds of investigation is a perfect country and place and the spot enterprise's investigation system. Main problems are obvious such as a class of even-aged forest proofs extra protection accident or accredited only for individual tree height, a dynamic 3D complex problem is simplified to plane problem and inconvenient height in the actual operation, progress and inaccurate. A second class survey in good conscience is not comprehensive, not deep, going through the motions, by experience, particularly the lack of dynamics of the annual growth rate, between 1st, 2nd and 3rd kind of investigation does not cooperate, each other did not complement each other, sequential control system and so on. One by one look at the following.

1) Forest survey system should have to be an optimized classification

Develop in the direction of the network to fully participate in the scrutinizing of CABR, namely the forest survey system from the current one, two, three types of optimization and integration to new Ⅰ, Ⅱ, Ⅲ grade. Grade Ⅰ (forest model), grade Ⅱ (forest) and grade Ⅲ (forest census). Wheregrade Ⅰ am mainly based on the non-cutting down (or a small amount of logging) based on regional and national forest table (volume, area, biomass, carbon storage, etc.) development. The general should be every 10 years for a time the State Forestry Bureau organization and coordination of the Provincial Forestry Bureau. The state-owned forest enterprises and county (city) forestry bureau forest table update and retold in work and to the non-destructive observation mathematical modeling and information management. Grade Ⅱ scrutinizing development to UAV aerial monolithic photography, the accurate measurement of the forest area by (MIN1 tree measuring station finder) like size, sampling, sample years of precision measurement, wood seized every ruler for body measurement system, corresponding to the main business the second continuous sample observation all three kinds of investigation. For the second class survey, finished by grade Ⅲ survey development mainly to low resolution RS such as TM and ZY-3 annually, determination of forest area in order to establish a national sample and forestry enterprises of continuous observation samples of annual fine measurement results to determine the annual forest area and accumulation data and dynamic growth.

At present, the government is interest on the annual growth rate and the public is extremely concerned about this, scholars believe that can save half of the fixed sample plots and also get the same accuracy of CFI results. From the theory and the cost of inputs is worthy of recognition and support, but considering the forestry situation, the conditions, and the future development give up half of the observed years of fixed sample plots is very regrettable. Therefore it is still to maintain the status, re-innovation and then enhance the accumulation of different conditions of the data.

2) Integration of observation equipment and electronic digital automatic observation process

Compared with the geological, surveying and mapping, petroleum, mining and other fields, forest surveyequipment, and technical methods are at least 10 years behind.

We should generalize to total station instrument/electric, close single photogrammetry, 3D laser scanning is a dominant tree lossless refinement measuring method with the mechanic's principle of the ultrasonic CT/nondestructive observation of tree rings, vegetation root system of radar and other series of precise measuring techniques. To promote the collection MEMS angular measurement, laser rangefinder and CCD photography, GNSS positioning in various handheld superstation forest location, measuring the diameter, height integration of wood seized every ruler. In 3D angle gauge is dominant with five tree stand of observation angle scale, size ratio, mixed ratio, high density / pitch / stand / biomass / accumulation of reserves such as Rmax / technology, algorithm, and method system. Smartphone / tablet to investigate the dominant measurement recorded on the integration of 3S technology application is the development of the main body of internal and external integration of forest management platform construction problems.

3) Forest survey technology program system

System design is shown in the 2-1 diagram. The survey programs may consider viewing every five years a retest of the country even the proofs by the State Forestry Administration into level II examination. According to the concept of standard quantitatively by new technology methods, instruments, and means of information needed to enhance the level of technology in order to play in the overall control of the forest survey operations. As for the first level forest model research and development, once every 10 years, in order to have the annual data accurate, organized by the provincial forestry department, in the fixed plots in the province each year by 20% of the sample to the uniform implementation of five tree plots in the central observation tree for the target tree, get the annual dynamic change and value control. There are three types of enterprises according to the current forest survey II detailed sample observation method.

7.4.3 Forest precision management outlook

Objectively speaking, we are not lacking good forest management theories, models, and programs for several decades but the regret is more of our stay in the slogan stage at the surface poor implementation and supervision is not strict. Many of our business practice are qualitative and quantitative means simple, most implemented in the sub-compartment, such as selective cutting cannot accurately implement what to do in a tree without much experiment. The point here is to talk about the forest management problem and the future of how precise and accurate by afforestation some thoughts and ideas work camp.

1) Forest land planning model

Objective function: ecological, social and economic benefits

Constraints: Coverage constraint; Economic condition constraints; Constraints on localities and land types; Tree species, cultivars and origin constraintsForest planning / grid model

Small class precision management model

Small class accurate business model

Question 1 How to choose cutting? Selective cutting which trees? Which path? Selectively cut which tree? How much?

Question 2 How to do afforestation? Mixed ratio? Age difference?

2) About the problem of optimal selective cutting mathematical deduction

(1) A stand with 3D anglegauge and F_g coefficient measured in the $i = 1, 2$ tree, $\cdots, M$, observation angle point, $j = 1, 2, \cdots, n$ tree species count of the wood of the first K ($k = 1, 2, \cdots$, Tree height, diameter at breast height i. e. D_{ijk}, H_{ijk}, L. Known the number of test type J-tree species is f_j.

(2) The interval of large, middle and small diameter of the forest stand was $[d_1, d_2][d_2, d_3][d_3, d_4]$.

(3) Calculate the t time ($T = 1, 2, \cdots$, Observation results of D diameter of T tree species in the first j tree species:

$H_{tjd} = 1/m \sum H_{jd}$ $N_{tjd} = 1/m \sum N_{jd}$ $V_{tjd} = 1/m \sum V_{jd}$

$H_{jd} = 1/l \sum H_d$ $N_{jd} = 1/l \sum N_d$ $V_{jd} = f_j N_{jd} g_{1.3j}(H_{jd}+3)$

(4) Stand prediction model

$H_j d_{t+1} = a_1 d_{jt}^{b1t}$ $n_j d_{t+1} = a_2 d_{jt}^{b2t}$ $d_j d_{t+1} = a_3 d_{jt}^{b_3(t+1)}$

(5) The best-known stand volume ratio of

$V_{\text{large}} : V_{\text{medium}} : V_{\text{small}} = m_1 : m_2 : m_3$, such as ( )

(6) Calculation of $V_{\text{large}} : V_{\text{medium}} : V_{\text{small}}$, select the imbalance of selective cutting, selective cutting V_{large} proportion of parts and so on the maximum benefit selection and selection of tree species and ratio.

(7) The precise cutting wood according to the principle of competition each choice, cutting volume.

(8) Repeat (1)-(6), next year's selective cutting.

(9) Maintain $m_1 : m_2 : m_3$ relationship and achieve sustainable management.

3) Basic problem

(1) At certain proportion n, d, H relationship with the growth and time.

(2) From 7 to infer the best $V_1 : V_2 : V_3 = ?$

(3) From 2 to infer the optimal choice cutting diameter order.

(4) From 3 to infer the optimal choice of cutting tree species and target selective cutting.

Figure 7-5 forest survey technology method equipment integration plan make a comprehensive view of the current global forestry science and technology development, combined with China's national conditions and forest conditions. It is not hard to think that China's forest management and information technology has become forestry primary core problem but in past it has been neglected. Now we have paid much attention to the problem, we through modern biotechnology in forest tree genetic improvement and breeding do many researches, has obtained satisfactory results; we also developed the world's largest artificial forest system and it is a great achievement; the reform and open policy and the market economy for forest products processing using open terrible day alone thick. Looking to the future, modern information technology, and forest management will be effectively combined to usher in the fall of China's forest management technology.

7.4.4 Wildfire prevention and control plan

This plan comprehensively uses modern mathematical, physical and ecological research data to solve the current hot spot problem of forest, mainly including three aspects: the design of forest fire prevention and control network scheme, fire risk forecast, fire extinguishing scheme and post-disaster loss evaluation, using space satellite remote sensing (RS) technology, unmanned aerial vehicle digital photography technology (UAVDPS), satellite navigation and positioning (GNSS)

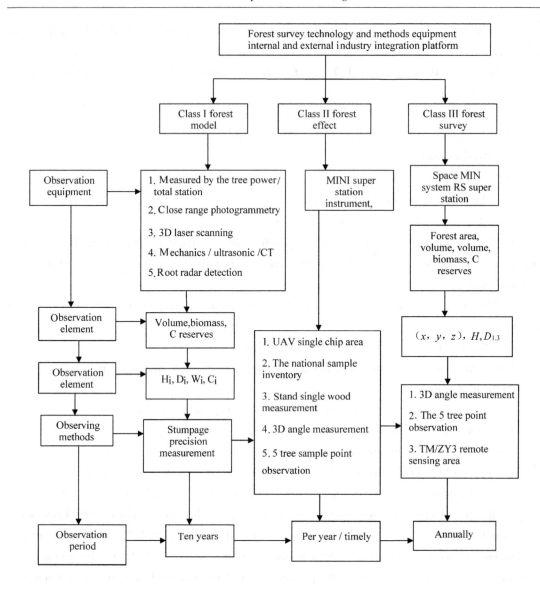

Figure 7-5 Forest survey technology method

technology, geographic information system (GIS) technology and computer information technology (big data, artificial intelligence and cloud computing).

7.4.4.1 Plan design of forest fire prevention and control network

A perfect forest fire prevention and control network shall be established in the forest region in a sky-ground integrated manner. Satellites and unmanned aerial vehicles to carry out monitoring was used in sky, and an efficient and comprehensive monitoring network with watchtowers, forest rangers and local residents was used in ground to make effective information processing and key decisions on the prevention, control and extermination of forest fires, thus forming a complete and effective forest fire emergency command and monitoring system.

1) Firerisk zoning

According to the particularity of forest region, forest fire prevention zoning is carried out by dividing forest zone first and then determining fire prevention zone. The main application data include vector layers such as Resource-3 satellite images, sub-compartment data, forest facies maps, DSM (Digital Elevation Model), administrative division maps, roads and water systems, etc. Combined with the consistent classification standard of forest zoning system, forest fire prevention zoning is implemented with sub-compartment as the final zoning unit. According to the natural terrain and terrain characteristics, vegetation characteristics, combined with administrative divisions at the same time, the formation of terrain conditions, soil conditions, vegetation types homogeneous single, continuous fire compartment, to achieve forest fire prevention zoning. The zoning principles are as follows:

(1) For sub-compartments in forested areas, forest fire prevention zones shall be defined at a distance of 500 meters from forests, unproven forested lands and forest edge meadows, scrub lands, sparse forested lands and wasteland. However, forest areas with bare farmland or less than 1500 meters in width should be expanded into forest fire prevention zones to facilitate management and ensure safety.

(2) For urban areas and some township centers in mountainous areas, except for the locations of villages (towns) with large population and large farmland concentration and large villages, all other areas can be designated as forest fire prevention zones within the scope of forest type nature reserves, with the exception of the forest fire prevention zones that can be determined within 500 meters from the edge of the forest. It is mainly aimed at forest land zoning around the urban area, and no specific forest fire prevention zoning is carried out for the central urban area.

(3) For plain areas, the forest fire prevention zone shall be determined with a distance of 300 meters from forests, forest edge meadows, shrubby sparse forest lands and wasteland, and the areas with bare farmland less than 1000 meters in length or width in the above combustible vegetation zone shall be included in the forest fire prevention zone.

2) Firepoint detection

(1) Satellite remote sensing monitoring. EOS-MODIS, of the 36 channels, there are 7, 20-25 channels with forest fire detection capability, which obviously reflects the heat on the ground. MODIS is used to monitor forest fires on the basis that the ignition point is hotter than the surrounding fire point, and the judgment is based on the relationship between the intensity of thermal radiation and temperature and wavelength.

NOAA meteorological satellite has five channels, of which the third channel is unique to the satellite and is sensitive to temperature. NOAA-AVHRR can provide medium resolution remote sensing images covering the whole world, with a wide wavelength range, and is suitable for forest fire monitoring in large areas. Its monitoring principle is basically the same as MODIS, but the accuracy of monitoring results is somewhat different from MODIS.

(2) UAV monitoring. It has the advantages of low operation cost and high flexibility in executing tasks. In order to obtain high-resolution spatial data as the application goal, through the inte-

grated application of 3S technology in the overall system, the real-time earth observation capability and the rapid processing capability of spatial data are realized simultaneously. The unmanned aerial vehicle is used to monitor the occurrence and spread of forest fires, thus improving the accuracy and real-time operability of observation.

(3) Watchtower monitoring. Forest fire monitoring occupies the most important position in the whole forest fire prevention information system. By integrating various forest fire monitoring technologies and using the forest fire observation tower to monitor the whole city in real-time, it is more economical than aerial monitoring and realizes 360 real-time monitoring at the same time. The watchtower distribution plan aims at the maximum visibility and the minimum blind area.

(i) Reasonable site selection. It is very important to scientifically and reasonably choose the construction site of the observation platform, and the maximum practical efficiency of the observation platform should be given priority. Therefore, the observation platform should be built in a place where a large area of key forest areas can be observed. There are no peaks or other obstacles that hinder the observation, and the visibility conditions are good.

(ii) Determine the altitude. The structure of the observatory should be suitable for the ground. A steel watchtower should be built where there are over-mature forests on the top of the mountain. The watchtower room must be higher than the highest crown of the surrounding trees, and the higher part must not be less than 2 m.

(iii) Determine the number. Take the whole forest region as a unit, make overall arrangements, and form a regional network. Due to the limitation of human vision and telescope magnification, the maximum observation radius of the general observation platform is not more than 20 km, and the maximum observation range is not more than 120000 hm^2. Within this range, due to terrain restrictions, there must be certain blind areas. In the larger Dalin area with an area of more than 100000 hm^2, more than two observation platforms need to be established to form an observation network. In a watch net, any three adjacent watchtowers should not be built in a straight line. The connecting lines of the three tables shall form a triangle, and the largest internal angle of the triangle shall not be greater than 120.

On the basis of understanding the basic technical data of the observation platform, the visibility distribution system of the observation platform is established by using the main principle of visibility analysis in the geographic information system, combining the topographic characteristics of the forest in the studied area (slope, aspect, vegetation information, etc.), RS remote sensing images of the studied area (SPOT, TM and other satellite image data), and combining the characteristics of the number of local forest fire fighting staff and work efficiency. In order to obtain the most accurate information about forest fires in the shortest possible time, a reasonable number and height of lookouts should be established within the scope of visibility. At the same time, in the blind area of lookouts, the scope of manual work and the number of workers should be determined according to the efficiency of manual inspection.

7.4.4.2 Forest fire forecast

1) Data processing

When establishing the forest fire prediction model, in addition to the fire point data, a certain amount of random points should be established as non-fire points to participate in the fitting. Fire points and random points together form sample points, and ArcGIS is used to establish random points. When establishing random points, it is necessary to ensure that the random points fall on the forest land. Therefore, based on the 2015 national land use data, the extracted forest land range is used to create random points within this range. Because the random points need to be matched with the daily meteorological data, after the random points are created, the random points need to be assigned with dates in excel, and the creation of the random points should follow the double randomness in time and space.

In ArcGIS, the Tyson polygon method is used to match weather stations and sample points (including fire points and random points). A Tyson polygon is a subdivision of a spatial plane. It is a continuous polygon composed of a group of perpendicular bisector connecting two adjacent point segments. It is characterized in that each polygon contains and only contains one sample point. Any position in the polygon is closest to the sample point of the polygon and far from the sample point in the adjacent polygon. Tyson polygons are established by taking meteorological stations as sample points, and the fire points and random points falling in each polygon are corresponding to meteorological station sample points by using spatial connection tools in ArcGIS. After determining the meteorological stations corresponding to fire points and random points, match the meteorological data in excel according to the meteorological stations and time corresponding to sample points.

2) Forest fire forecast model based on logistic regression

Logistic regression analysis (LR) belongs to a kind of generalized linear model. Its dependent variable values are discontinuous and can be classified into two or more categories. Its independent variable can be continuous variable or classified variable. Logistic regression is commonly used in disease diagnosis, economic risk prediction and other fields, such as analysis of risk factors leading to disease, prediction of the probability of disease occurrence, etc. The forest fire prediction model is similar to the disease diagnosis model. When forecasting the occurrence of forest fire, the result is a two-category variable, namely 'occurring' and 'not occurring'. Taking the influence factors related to the occurrence of forest fire as independent variables, the contribution value of meteorological, vegetation, social and other fire risk factors related to forest fire to the occurrence of forest fire can be studied, and the forest fire probability under various conditions can be calculated.

Let the probability of occurrence of forest fire be p, the probability of non-occurrence of forest fire be $(1-p)$, and the logistic expression of occurrence probability of forest fire and fire danger factor be:

$$\ln\left(\frac{p}{1-p}\right) = \beta_0 + \beta_1 X_1 + \beta_2 X_2 + \cdots + \beta_n X_n \tag{7-10}$$

3) Forest fire prediction model based on stochastic forest algorithm

Random Forest(RF) algorithm is a statistical learning theory, which can solve both regression and classification problems. The dependent variable of the forest fire prediction model established in this study is whether forest fire occurs or not, with occurrence of 1 and no occurrence of 0. Therefore, using random forest to predict forest fire is to solve a two-class problem with random forest.

Random forest uses bootstrap resampling method to extract multiple samples from the original samples, and then models each sample with decision tree. When dealing with regression problems, the average value of each decision tree output is taken as the final coefficient, and when dealing with classification problems, the final prediction category is determined by voting. Its advantage is that it can handle a large number of input variables, can evaluate the importance of variables, and generally does not appear over-fitting

N is used to represent the forest fire data and m is used to represent the number of independent variables. The random forest algorithm uses bootstrap resampling method to randomly extract n-tree sample sets with N capacity from N sample data. Based on this, n-tree classification regression trees are created, and mtry independent variables (mtry$\leqslant$m) are randomly extracted at each node of each classification regression tree. Select the variable with the strongest classification ability to branch, and each tree does not need any pruning, allowing it to grow freely to the maximum. Finally, n-tree decision outputs are obtained. In the classification problem, the mode of n-tree decision outputs is selected as the final prediction result of the random forest algorithm.

Random forest algorithm can also calculate the importance score of each variable to evaluate the importance of each fire risk factor in the model. Finally, the specific dependency relationship between dependent variables and independent variables can be analyzed through partial dependence plots.

7.4.4.3 Fire fighting plan and post-disaster loss evaluation

1) Determine thespread trend of forest fire

(1) Data acquisition. In the research area, in coordination with on-site ignition tests and planned burning activities, unmanned aerial vehicles or manned machines are used to load digital photogrammetry systems to record, photograph, process and map the time sequence of forest fire spreading process in the fire scene, and extract fire scene information. The RTK/RTD GPS high-precision positioning system is adopted to measure the shape, area and perimeter of the burned area and investigate the resources, providing theoretical basis for high-precision modeling and post-disaster assessment(Figure 7-6). Collect historical forest fire data, multi-level resolution remote sensing images and corresponding scale topographic maps and other basic data of the study area, extract basic information of forest combustible, tree species, site factors and other basic information, and collect meteorological data and socio-economic data of the local and surrounding areas.

In view of the uncertainty of forest fire data in the past, R, MATLAB and other software are used to study the algorithm suitable for forest fire spread prediction. Gross errors in historical forest fire data are eliminated, interpolation of missing data, selection of attribute data, processing of in-

 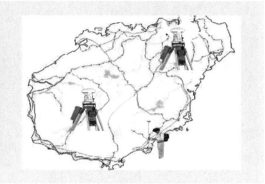

Figure 7-6　Measurement of the shape, area and perimeter of the burned area

stance data and preprocessing of data reduction are carried out. Through a large number of algorithm comparisons, research on algorithms suitable for forest fire knowledge discovery can not only discover hidden and objective rules of forest fire occurrence, spread and fighting, provide scientific basis for forest fire management and fighting decision-making, but also determine the main factors that affect forest fire spatio-temporal model, and prepare for the effective establishment and modification of spatio-temporal model. Through the combination of neural network algorithm and genetic algorithm, the characteristics of multi-parameter and fast convergence of neural network and global optimization of genetic algorithm are fully utilized to realize complementation. Fuel and terrain data are extracted through GIS, and a forest fire space-time spread model based on remote sensing pixel probability integration method is established.

(2) Forest fire spread model

(i) Improvement of Rothermel forest fire spread model. The basic idea of Rothermel model is that the spreading process of forest fire is actually a process in which unburned combustible materials are continuously ignited in front of the flame. The flame area transfers heat to the unburned matter in front by means of radiation, convection and conduction. When the unburned combustible absorbs heat and heats up to the ignition point, these combustible materials are ignited and the flame spreads to the area. Since the Rothermel forest fire spread model requires certain assumptions, the entire fire scene cannot meet the use conditions of the Rothermel model. In this study, a large range of fire scene areas are divided into several grid units, which meet the conditions of the Rothermel model. The concept of 'octree' is introduced, and then the overall fire spread can be dynamically calculated from the overall consideration of the forest fire spread situation, and the three-dimensional visual expression of the forest fire process can be carried out through GIS.

(ii) The space-time spread model of forest fire based on neural network. According to the environmental factors of the fire scene and the parameters of the forest fire spread model, it is determined that the input nodes are 6 factors, namely wind speed (V), wind direction (α_1), fuel accumulation per unit area (M), slope (δ), aspect (α_2) and air dry humidity (H), which are expressed by vectors as follows:

$$S = [V, \alpha_1, M, \delta, \alpha_2, H] \qquad (7\text{-}11)$$

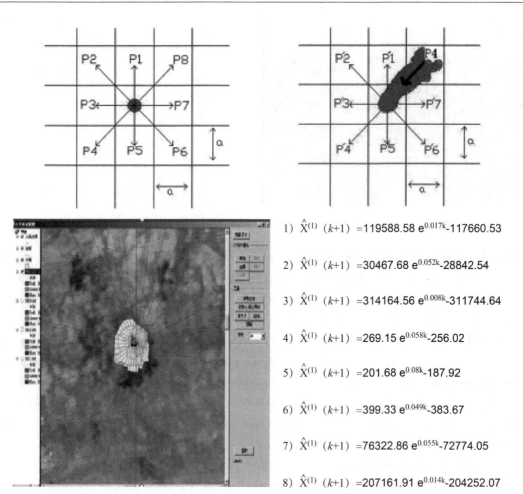

1) $\hat{X}^{(1)}(k+1) = 119588.58\ e^{0.017k} - 117660.53$

2) $\hat{X}^{(1)}(k+1) = 30467.68\ e^{0.052k} - 28842.54$

3) $\hat{X}^{(1)}(k+1) = 314164.56\ e^{0.008k} - 311744.64$

4) $\hat{X}^{(1)}(k+1) = 269.15\ e^{0.058k} - 256.02$

5) $\hat{X}^{(1)}(k+1) = 201.68\ e^{0.08k} - 187.92$

6) $\hat{X}^{(1)}(k+1) = 399.33\ e^{0.049k} - 383.67$

7) $\hat{X}^{(1)}(k+1) = 76322.86\ e^{0.055k} - 72774.05$

8) $\hat{X}^{(1)}(k+1) = 207161.91\ e^{0.014k} - 204252.07$

Figure 7-7 The dynamic model of fire spread and its three-dimensional visualization

There are 8 output nodes, which indicate the forest fire spreading speed from the wind speed direction and the fire spreading speeds $V_1, V_2, \ldots, V_8$ in 8 directions at 45 intervals in the clockwise direction, thus constituting the forest fire plane octree spreading model. The number of hidden layer nodes is determined by the number of input and output layer nodes, i. e. the average value of the number of input and output layer nodes is taken, so the number of hidden layer nodes is 7. In this way, a three-layer neural network-based model training and forest fire spread simulation network structure is established as shown in the figure 7-8.

2) Isolation belt design

The dividing principle of isolation belt: (i) perpendicular to the main wind direction. At the forefront. (ii) Set up the first fire isolation zone perpendicular to the main wind direction, with the largest protection area and the best effect. (iii) The setting position is downward on the ridge or upward in the valley. These places have the slowest fire development and less vegetation, and are easy to control. (iv) Setting density shall be determined in combination with actual terrain. Generally cannot exceed 5 kilometers. (v) The width is 40–60 m, which is different from grassland,

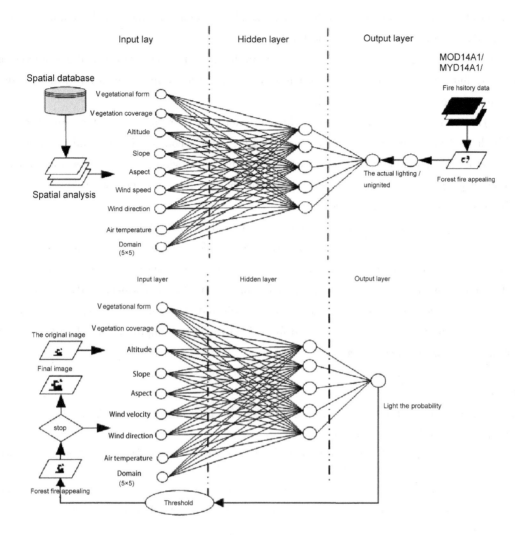

Figure 7-8 The space-time spread model of forest fire based on neural network

trees and shrubs.

3) Fire prevention tree species design

In order to save human and financial resources for fire prevention, biological fire prevention belts should be built at the borders of fields and roads entering mountains that are prone to fire, i. e. passive fire prevention should be changed into active fire prevention, and the stand structure should be optimized and improved to increase forest accumulation. Among various tree species, common fire prevention plants include: *Ginkgo biloba*, *Schima superba*, *Viburnum odoratissimum*, *Pittosporum tobira*, *Nerium indicum*, etc.

4) Post-disaster loss evaluation

(1) Assessment of the area affected. In coordination with the forest fire ignition test and planned fire removal, RTK/RTD GPS high-precision positioning system is adopted to measure the shape, area and perimeter of the corresponding fire site and investigate the resources, and a calcu-

lation model for the fire area of the forest fire is established so as to quickly and accurately determine the damaged area and estimate the fire loss.

(2) Post-disaster state evaluation. According to the area of affected forests and the number of casualties, forest fires can be divided into four states: general forest fires, larger forest fires, major forest fires and especially major forest fires.

(3) Economic loss evaluation. The evaluation criteria for economic losses after forest fires are roughly divided into four parts: forest resources losses, direct economic losses, indirect economic losses and forest environmental resources losses. For different regions, different tree species have different evaluation criteria. At present, the establishment of relatively uniform evaluation criteria has become the focus of most studies.

(4) Ecological restoration and reconstruction. According to the forest species, tree species, forest age and damage degree, the affected forests are classified and treated. Measures such as reforestation, replanting and replanting, artificial promotion of sprouting and closing hillsides to facilitate afforestation are respectively adopted. Tree species structure is reasonably adjusted in afforestation, tree species with strong stress resistance are selected, forest structure is optimized, mixed forests are actively built, forest stability and forest biodiversity are improved, and forest ability to resist natural disasters is enhanced.

References

Coops N C, Tompalski P, Nijland W, et al. , 2016. A forest structure habitat index based on airborne Laser scanning data [J]. Ecological Indicators, 70 (SI): 644.

Danskin S, Bettinger P, Jordan T, 2009. Multipath mitigation under forest canopies a choke ring antenna solution [J]. Forest Science, 55(2): 109-116.

DeConto T, Olofsson K, Gorgens E B, et al. , 2017. Performance of stem denoising and stem modelling algorithms on single tree point clouds from terrestrial laser scanning [J]. Computers and Electronics in Agriculture(143): 165-176.

Diamantopoulou M J, Milios E, 2010. Modelling total volume of dominant pine trees in reforestations via multivariate analysis and artificial neural network models [J]. Biosystems Engineering, 105(3): 306-315.

Donager J J, Sankey T T, Sankey J B, et al. , 2018. Examining forest structure with terrestrial lidar suggestions and novel techniques based on comparisons between scanners and forest treatment [J]. Earth and Space Science, 5 (11): 753-776.

Dong H, Zhang L J, Li F R, 2015. A three-step proportional weighting system of nonlinear biomass equations [J]. Forest Science, 61(1): 35-45.

Dulamsuren C, Hauck M, Bader M, et al. , 2009. Water relations and photosynthetic performance in Larix sibirica growing in the forest-steppe ecotone of northern Mongolia [J]. Tree Physiology, 29(1): 99-110.

Ebermayer E, 1876. Die gesammte Lehre der Waldstreu mit Rücksicht auf die chemische Statik des Waldbaues. Unter Zugrundlegung der in den Königl. Staatsforsten Bayerns angestellten Untersuchungen [M]. Berlin: Springer.

Erdody T L, Moskal L M, 2010. Fusion of LiDAR and imagery for estimating forest canopy fuels [J]. Remote Sensing of Environment, 114(4): 725-737.

Fang J Y, Piao S L, Tang Z Y, et al. , 2001. Interannual variability in net primary production and precipitation [J].

Science, 293(5536): 1723.

Fei L, Yan L, Chen C, et al., 2017. OSSIM: An object-based multiview stereo algorithm using SSIM index matching cost [J]. IEEE Transactions on Geoscience and Remote Sensing, 55(12): 6937-6949.

Ferraz Filho A C, Soares Scolforo J R, Ferreira M Z, et al., 2011. Dominant height projection model with the addition of environmental variables [J]. Cerne, 17 (3): 427-433.

Forsman M, Borlin N, Olofsson K, et al., 2018. Bias of cylinder diameter estimation from ground-based laser scanners with different beam widths: A simulation study [J]. Isprs Journal of Photogrammetry & Remote Sensing, 135: 84-92.